Ecological Restoration in International Environmental Law

W0112978

Human activities are depleting ecosystems at an unprecedented rate. In spite of nature conservation efforts worldwide, many ecosystems including those critical for human well-being have been damaged or destroyed. States and citizens need a new vision of how humans can reconnect with the natural environment. With its focus on the long-term holistic recovery of ecosystems, ecological restoration has received increasing attention in the past decade from both scientists and policy-makers.

Research on the implications of ecological restoration for the law and law for ecological restoration has been largely overlooked. This is the first published book to examine comprehensively the relationship between international environmental law (IEL) and ecological restoration. While IEL has developed significantly as a discipline over the past four decades, this book enquires whether IEL can now assist states in making a strategic transition from not just protecting and maintaining the natural environment but also actively restoring it. Arguing that states have international duties to restore, this book offers reflections on the philosophical context of ecological restoration and the legal content of a duty to restore from an international law, European Union law and national law perspective. The book concludes with a discussion of several contemporary themes of interest to both lawyers and ecologists including the role of private actors, protected areas and climate change in ecological restoration.

Anastasia Telesetsky is Associate Professor at the University of Idaho, US.

An Cliquet is Associate Professor at the University of Ghent, Belgium.

Afshin Akhtar-Khavari is Associate Professor at Griffith University, Australia.

Routledge Research in International Environmental Law

Available titles in this series:

Law and Practice on Public Participation in Environmental Matters
The Nigerian Example in Transnational Comparative Perspective
Uzuazo Etemire

Climate Change and Human Rights
An International and Comparative Law Perspective
Ottavio Quirico and Mouloud Boumghar

Human Rights Approaches to Climate Change
Challenges and Opportunities
Sumudu Atapattu

Responsibilities and Liabilities for Commercial Activity in the Arctic
The Example of Greenland
Edited by Vibe Ulfbeck, Anders Møllmann and Bent Ole Gram Mortensen

Stratospheric Ozone Damage and Legal Liability
US Public Policy and Tort Litigation to Protect the Ozone Layer
Lisa Elges

Ecological Restoration in International Environmental Law
Anastasia Telesetsky, An Cliquet and Afshin Akhtar-Khavari

Sustainable Development Principles in the Decisions of International Courts and Tribunals
1992–2012
Marie-Claire Cordonier Segger and H.E. Judge C.G. Weeramantry

Forthcoming titles in this series:

Natural Resources Law, Investment and Sustainability
Shawkat Alam, Jahid Hossain Bhuiyan and Jona Razzaque

Environmental Law and the Ecosystem Approach
Maintaining ecological integrity through consistency in law
Froukje Maria Platjouw

Ecological Integrity and Global Governance
Science, Ethics and the Law
Laura Westra

Ecological Restoration in International Environmental Law

Anastasia Telesetsky, An Cliquet and Afshin Akhtar-Khavari

Routledge
Taylor & Francis Group

LONDON AND NEW YORK

First published 2017 by Routledge

2 Park Square, Milton Park, Abingdon, Oxfordshire OX14 4RN
52 Vanderbilt Avenue, New York, NY 10017

Routledge is an imprint of the Taylor & Francis Group, an informa business

First issued in paperback 2018

British Library Cataloguing in Publication Data
A catalogue record for this book is available from the British Library

Library of Congress Cataloging in Publication Data
Names: Telesetsky, Anastasia, author. | Cliquet, An, author. | Akhtar-Khavari,
Afshin, author.
Title: Ecological restoration in international environmental law / Anastasia
Telesetsky, An Cliquet, and Afshin Akhtar-Khavari.
Description: Abingdon, Oxon [UK] ; New York : Routledge, 2017. | Series:
Routledge research in international environmental law | Includes bibliographical
references and index.
Identifiers: LCCN 2016029641
Subjects: LCSH: Restoration ecology--Law and legislation. | Nature
conservation--Law and legislation. | Biodiversity conservation--Law and
legislation. | Environmental law, International
Classification: LCC K3478 .T43 2017 | DDC 344.04/6--dc23
LC record available at https://lccn.loc.gov/2016029641

ISBN: 978-1-138-79683-6 (hbk)
ISBN: 978-0-367-19344-7 (pbk)

Typeset in Galliard
by Taylor & Francis Books

Contents

Tables

Acknowledgement

This project started with a conversation in Baltimore at the 10th Annual Colloquium of the IUCN Academy of Environmental Law in 2012, when we realised that we were each working on the topic of restoration. At this conference we were each intrigued by the lack of interest from environmental lawyers in the subject of restoration and this sparked our interests to work on a book that would begin to fill this gap. Since then we have each become more passionate about the topic of restoration in environmental law by sharing information with each other and debating the subject matter over many Skype conversations across three continents – managing the time zones across Idaho, Brisbane and Ghent is no easy feat. Routledge was kind to quickly offer us a contract to publish our book and we met again in Manchester in 2015 for the meeting of the International Society for Ecological Restoration (SER) where we fine-tuned the book project. At this conference we also took the opportunity to organise and present our respective papers at the first law panel at an SER conference. We were convinced after our participation at this conference that lawyers had a great deal of work to do in supporting and enabling ecologists to think about the implications and applications of their work.

The field of restoration ecology is heavily engaged with defining what restoration is and this subject also consumed our attention a great deal. We did not always agree that the definition of ecological restoration supported through the work of the SER was inclusive enough of a range of practices, but we were always in agreement that the historical trajectory of an ecosystem had to be respected and supported for its own sake. International law and legal instruments generally, and paradoxically, do not seem as concerned with definitions, but then this is more reflective of the lack of interest that the law has had in the subject matter of restoration.

Whenever we talked to lawyers about our project we seemed to face the same reaction, which is that environmental law has already had a long history of engaging with the subject of restoration. These reactions made this book even more important in that we wanted to illustrate that restoration is not just about remediation or rehabilitation but that it is an idea and an approach in terms of how we think about the recovery of the natural world in a broad sense. As such, restoration is not just a response when harm is done to an environment but, rather, we think that by taking away the capacity of nature to recover on its own we are also doing harm to the environment – and this is a different way to think about the natural environment than just through the lens of protection and conservation which environmental law has been built on.

No book project can be completed without having to thank a range of colleagues and friends. We want to initially thank Routledge and because this book took some time to mature we have dealt with a range of people, including Mark Sapwell and Olivia Manley. Additionally, we also want to thank our colleagues who in some way or another contributed to ideas and materials that are now part of this book, including Douglas Fisher, Benjamin Richardson, Louise Kotze, Lee Godden, Blake Hudson, Don Anton, Rak Kim, Kris Decleer and Hendrik Schoukens. We would also like to thank Kris Decleer for providing us with pictures for the cover of this book. We want to thank all of the anonymous restoration practitioners around the world tirelessly doing their bit to revitalise this world. Lastly, but also significantly, we need to thank members of our family who supported and tolerated the late-night and early-morning Skype conversations and mishaps: Matthew Church, Kris and Lore Decleer and Nikki, Lachlan, Zoe and Elise A.-Khavari.

Finally, given our interest in this subject we couldn't avoid publishing our views as we went along and we want to acknowledge that the following publications are in some way a part of this book:

- A. Akhtar-Khavari and A. Telesetsky, 'From protection to restoration: a challenge for environmental governance' in D. Fisher (ed), *Research Handbook on Fundamental Concepts of Environmental Law* (Edward Elgar 2016);
- A. Cliquet, K. Decleer and H. Schoukens, 'Restoring nature in the EU: the only way is up?' in C.-H. Born, A. Cliquet, H. Schoukens, D. Misonne and G. Van Hoorick (eds), *The Habitats Directive in its EU Environmental Law Context: European Nature's Best Hope?* (Routledge 2015);
- A. Akhtar-Khavari, 'Ecosystem services, fear and the subjects of environmental human rights' in A. Grear and L. Kotze (eds), *Research Handbook on Human Rights* (Edward Elgar 2015) 508–530.

The law is stated as of May 2016.

Anastasia, An and Afshin

1   Introduction

E.O. Wilson raised the 'rallying cry for advocates of ecological restoration'[1] in 1992 by writing that, 'The next century, will, I believe, be the era of restoration in ecology.'[2] Increasingly, ecologists are identifying a variety of ways to intervene in reducing or stopping environmental harm or ecosystem degradation. Globally states and private parties invest in large landscape restoration projects ranging from the reforestation of Rwanda to the reconstruction of depleted wetlands in the Gulf of Mexico. Ecologists, hydrologists and biologists have subscribed to the 'era of restoration' and have formed professional organisations such as the Society for Ecological Restoration with several thousand members that publish scientific journals such as Restoration Ecology to help the field continue to advance. More recently, social scientists and humanities scholars have contributed to restoration research with books on the economics of natural resource capital and the ethics of restoration.

However, before continuing on this cheery note it must be said that in darkness even a speck of light can be seen and admired. Let us explain why we have to suddenly cast a shadow on this positive start to this book. The amount of global ecological restoration work needed is daunting. At the end of May in 2016 the United Nations Environment Programme (UNEP) presented a report at a ministerial policy review session, which it titled Healthy Environment, Healthy People. In this report UNEP identified that:

> 15 out of 24 categories of ecosystem services are in decline … and four of the nine planetary boundaries (climate change, loss of biosphere integrity, land-system change, altered biogeochemical cycles (phosphorous and nitrogen)) have been crossed. Approximately 15,000 species (or 21 per cent) of global medicinal plant species are now endangered as a result of overharvesting and habit loss.[3]

1 N. Barrett, 'The promise and peril of ecological restoration: why ritual can make a difference' (2011) 32(2) *American Journal of Theology and Philosophy* 139–155.
2 E. Wilson, *The Diversity of Life* (Harvard University Press 1992) 340.
3 United Nations Environment Programme, *Healthy Environment, Healthy People. Thematic Report for the Ministerial policy review session of the Second session of the United*

The UNEP report observes that the effects of deteriorating environmental conditions on the health of people represent 23 per cent of the total human deaths globally.[4] Despite its strengths, the report focused on the consequence of degradation for human health and did not cover, for instance, ecosystem losses. Species are disappearing at unprecedented rates and possibly faster than natural systems can adapt to these losses.[5] The loss of what might otherwise be considered 'redundant species' has uncertain impacts because species loss may have influences on future buffering capacity for specific ecosystems against environmental changes.[6] With these kinds of statistics and facts describing more a norm than an exception to the rule, humanity's search for solutions and new ways of thinking about the human-nature relationship is, as one would expect, occupying much scholarly time.[7]

Ecological restoration is clearly making its mark as an intervention that aims to be holistic and capable of restoring self-sustaining ecosystems containing the characteristics of 'past or least-disturbed landscapes'.[8] The Society for Ecological Restoration has defined ecological restoration as 'an intentional activity that initiates or accelerates the recovery of an ecosystem with respect to its health, integrity and sustainability'.[9] Restoration is also a global proposition. In researching this book, the three authors residing in Australia, Belgium and the United States became aware of the Herculean efforts of hundreds and thousands of individuals reshaping landscapes to recapture lost or disappearing ecological values. Although the idea of ecological restoration is further discussed in Chapter 2, a few examples of ecological restoration initiatives will illustrate the potential and breadth of approaches that exist for restoring the natural environment. These examples illustrate the diversity of efforts that qualify as ecological restoration depending on the restoration goals for the project, restoration methodology and the types of actors involved.

Nations Environment Assembly of the United Nations Environment Programme, Nairobi, 23–27 May 2016 (2016) p. 26 (footnotes omitted).

4 Ibid., 24.

5 L. Burkle et al., 'Plant-pollinator interactions over 120 years: loss of species, co-occurrence and function' (2013) 339 (6127) *Science* 1611–1625 (describing the loss of 50 per cent of bee species using historic data sets and concluding that both the quantity and quality of pollination services are continuing to decline).

6 F. Chapin et al., 'Consequences of changing biodiversity' (2000) 405 *Nature* 234–242 (noting that 'We are in the midst of one of the largest experiments in the history of the Earth').

7 For an example of two broad studies that contextualise their work in restoration see: N. Dudley, *Authenticity in Nature* (Earthscan 2011); and W. Jordan, *The Sunflower Forest: Ecological Restoration and the New Communion with Nature* (University of California Press 2012).

8 M. Palmer and J.B. Ruhl, 'Aligning restoration science and the law to sustain ecological infrastructure for the future' (2015) 13(9) *Frontiers in Ecology and the Environment* 512–519.

9 Society for Ecological Restoration International Science & Policy Working Group, *The SER International Primer on Ecological Restoration* (Society for Ecological Restoration International 2004), <www.ser.org>.

The pictures on the cover of this book, for instance, poignantly illustrate the potential of restoration, even in densely populated regions such as Belgium, with its highly fragmented and degraded natural heritage. They show the ecological restoration of a species of rich fen meadow and marsh relic. The project is situated in the lowland wetland nature reserve 'Vallei van de Zuidleie' (Flanders, Belgium). In the 1960s the area had been filled in with sludge from the dredging of the world-famous canals of Bruges, which were highly polluted in those days, and an impediment for tourism development. The wetland was also used as a dump site for domestic waste. In the period between 1992 and 2005 restoration work, financed by the Flemish government, removed the sludge and domestic waste down to the original peat soil surface and resumed the traditional mowing management without fertilisation. The management of the area is in the hands of 'Natuurpunt', a nature conservation non-governmental organisation (NGO), where both professionals and volunteers cooperate to manage the area. Shortly after the restoration work was finished, the compacted peat layer started to grow again. Highly endangered communities of plants characteristic of low-productive base-rich fen meadow and transition mire were able to recolonise the area including thousands of dazzling wild orchids. Some of the emblematic flora and fauna of the restored area is depicted on the cover of this book. Despite the positive results so far, restoration of hydrological conditions in the surrounding landscape, which have been highly degraded by intensive agriculture, will be essential to complete all restoration efforts and ensure a sustainable future for the area.[10]

In the United States, ecological restoration efforts have been ongoing for decades. For example, efforts started in 1934 at the University of Wisconsin Arboretum in the United States to re-establish an 'original Wisconsin' landscape that might reflect some of the landscape qualities that existed before European settlement. When the tall-grass prairie restoration project was initiated at the Arboretum, one of the early advisers was asked how long it would take to achieve restoration; the answer was 'about a thousand years'.[11] While subsequent scholars think a millennium might be optimistic, scientists and volunteers continue to invest their talents and time into this legacy project.[12] At the heart of the Arboretum's restoration, the 73-acre Curtis Prairie, named for one of the earliest champions of ecological restoration research, is the product of thousands of visionaries, experiments and hours of hard work including recruits from the Civilian Conservation Corps. A former working farm and pastureland, today the Prairie is home to 230 native species including delicate flowering gentians and orchids. The ongoing ecological restoration success of the Curtis Prairie has built momentum and the Arboretum is

---

10 K. Decleer, 'Vallei van de Zuidleie: Leiemeersen (Oostkamp)' in K. Decleer (ed), *Ecological Restoration in Flanders (Belgium)* (published on the occasion of the 6th European Conference on Ecological Restoration, 8–12 September 2008, Ghent, Belgium, INBO 2008) 48–49.

11 T. Blewett and G. Cottam, 'History of the University of Wisconsin Arboretum Prairies' (1984) 72 *Transactions of the Wisconsin Academy of Sciences Arts and Letters* 130–144, 143.

12 Ibid.

also investing restoration efforts in the 47-acre Greene Prairie, 14-acre Wingra Oak Savanna remnant and the 53-acre Southwest Grady Oak Savanna. Additional projects that seek guidance from Wisconsin's past to create a future for Wisconsin where residents are connected with ecologically healthy landscapes are also under way for wetlands and deciduous forests. While these projects have achieved milestones, they also continue to face ongoing challenges including incursions of urban storm-water run-off and invasive species. Even so, the overall trend for restoration on sites such as the University Arboretum is positive. In a piece celebrating the 75th anniversary of Curtis Prairie, Wegener et al. queried whether it was accurate to state that Curtis Prairie is 'restored'. They concluded that while 'restoration is rarely finished ... there is much to celebrate'.[13] These sentiments capture the spirit of many working in the US ecological restoration efforts.

The breadth and range of restoration activities in Australia is remarkable given the size of the population in that country. An example is the approach taken to the Murray-Darling Basin (MDB) in Australia which is its largest river system with a total length of 1.061 million km^2.[14] The complex MDB waterways move through several Australian states making them jurisdictionally complex and very difficult to manage across political boundaries. The MDB is a critical resource for many competing stakeholders. For example, the Basin's rivers have 16 wetlands in them listed under the Ramsar Convention on Wetlands of International Importance Especially as Waterfowl Habitat (1971),[15] making the ecological function of the river one of international significance.[16] The river also sustains a very large number of communities along its length, including over 30 indigenous Australian nations. The MDB is significant for Australia's economy. In 2005–6 a report by the Australian Bureau of Statistics indicated that 66 per cent of the water consumed in Australia for agriculture was in the MDB.[17] All of these competing interests have led to an over-allocation of water with implications for the economy, livelihoods of communities and ecology (e.g. increased salinity) of the river and its riparian areas. In 2007, Australia adopted the Water Act (WA) to manage complex and often competing interests.[18] While the WA's primary purpose is managing the water resources of the MDB, the WA is also designed to restore the 'ecological values and ecosystem services of the Murray-Darling Basin'.[19] Given the size of this venture its success has yet to be properly realised. Buying water back from farmers – through a federal

13 M. Wegener, P. Zedler, B. Herrick and J. Zedler, 'Curtis Prairie: 75-year-old restoration research site' (2008) *Arboretum Leaflets*, Leaflet 16, 3.

14 J. Pittock and D. Connell, 'Australia demonstrates the planet's future: water and climate in the Murray-Darling Basin' (2010) 26(4) *International Journal of Water Resources Development* 561–578.

15 996 *UNTS* 245.

16 R. Kingsford et al., 'A Ramsar Wetland in crisis – the Coorong, Lower Lakes and Murray Mouth' (2011) 62 *Australia Marine Freshwater Research* 255–265.

17 Australian Bureau of Statistics, *Water and the Murray-Darling Basin – A Statistical Profile* (2005–05), <www.abs.gov.au/ausstats/abs@.nsf/mf/4610.0.55.007>.

18 Water Act 2007 (Cth) s18C.

19 Water Act 2007 (Cth) s3(d).

scheme – to sustain the long-term restoration of the river system and its riparian ecology is an important intervention and commitment to a very large and complex series of connected ecosystems.

While the operational details vary greatly in each of these examples, the general objectives for each of these projects is the same. So while on the one hand the Belgian and the American civil society managed projects offer discrete projects that reflect ongoing experimentation and the Australian government-led project covers a much more dispersed space for restoration, all of the projects are attempting to revive ecological values that have become compromised by human activities. While the approaches vary from physical removals of soils to annual burnings of grasslands to market transactions to ensure that water remains in the river system, all of the projects offer a means for reconnecting a variety of communities from indigenous groups to university students back to a functioning landscape. Negotiations among long-term and short-term human needs are at the heart of each of these projects as restoration proponents seek to explain the necessity of using heavy machinery in a place that appears natural, burning prairie grass in a community where fire is historically feared and re-allocating water from agricultural industry back to the river itself.

While each of the examples from Australia, Belgium and the United States should be regarded as a success story for ecological restoration, there is still much to be done to advance the 'era of restoration' heralded by E.O. Wilson. Some of the work will be technical work including developing best practices to manage invasive species, reversing trends of anthropogenic harm and designing large landscape projects that will remain viable over time given other ecological trends.

Other work will require changes in governance. This book argues that less global attention has been focused on developing appropriate governance for ecological restoration interventions through, for example, developing the law. As this book intends to demonstrate, while specific restoration outcomes will differ across the globe, there are key governance lessons that underwrite the success of ecological efforts that should be incorporated into restoration governance efforts. For example, ecological restoration efforts do not reside exclusively in the domain of science. While wetland biologists and hydrologists have the needed expertise to ensure success, there are also numerous cultural dimensions to restoration projects that require applying knowledge from social sciences and from communities. Law is one tool, albeit an often imperfect tool, in organising a governance structure for ecological restoration efforts.

Connecting ecological restoration and law

In theory, ecological restoration does not need law or lawyers. Ecologists such as the botanists and field scientists working at the University of Wisconsin Arboretum have made great progress in returning the vitality of the tall-grass prairies with no legal obligation to do so. Restoration efforts through projects such as the Curtis Prairie could continue on an ad hoc basis with only the occasional involvement of a policy decision-maker. After all, it could be argued that projects ultimately depend on the quality of restoration leadership rather than the quality of law.

While leadership definitely matters to the success of a restoration project, this book argues that it is too late given the extent of human impacts on the ecosystems now to simply pursue this ad hoc approach to restoration planning with each project being developed in isolation. There needs to be geographical connections made across projects to ensure that individual restoration efforts go beyond gardening initiatives that look aesthetically pleasing but fail to protect critical ecological values. Aldo Leopold sagely wrote that 'If the biota, in the course of aeons, has built something we like but do not understand, then who but a fool would discard seemingly useless parts? To keep every cog and wheel is the first precaution of intelligent tinkering.'[20] In practice, law should be a form of 'intelligent tinkering'.

Law offers two specific advantages in the context of ecological restoration. First, it offers a driver for action to get parties to begin to 'tinker'. States 'tinker' actively when they understand that they have obligations to engage in certain behaviour. The reason for mainstreaming environmental law and particularly international environmental law into restoration efforts is the potential of law to support the effective management of the recovery of the natural world. Environmental law can help coordinate, systematise and regulate conduct to both avoid and correct harm and damage.

This idea of environmental law as a 'driver' for active engagement in the environment is a recent development in the evolution of law. Environmental law began as a social tool to prevent certain activities such as seal-hunting or pollution dumping. It has only relatively recently evolved to operate as a common tool of protection with key areas set aside for restricted uses, such as wetlands of international significance. The transition of environmental law to creating rules of engagement with the environment will hopefully usher in a new era of norms, institutions and structures that are reflexive enough to accommodate changes and opportunities. This is not just simply a reference to more environmental law through principles rather than rules. It is more about how environmental law has to create opportunities and open up spaces for individuals and communities to more deeply interact and engage with the natural environment to support its recovery and restoration. It will not be sufficient to simply focus on preventing new sources of pollution or managing existing conservation and preservation areas. Instead, environmental law will need to evolve as an increasingly proactive field focused on building new relations and connections to reverse damage to environments.

Second, law offers 'intelligent' means for reconnecting humans with a larger natural community. Law is typically a conservative tool that operates largely through legislative accretion and avulsion. In the case of ecological restoration, what this means is that law may offer an 'intelligent' progressive tool for gradually reviving lost or disappearing ecological values that evolves in step with other

20 A. Leopold, 'Conservation' in L. Leopold (ed), *Round River* (Oxford University Press 1966) 146–147.

human norms and institutions. The power of law as a tool for influencing behaviour is embodied in its ability to assist society in changing directions by articulating changing social norms. For example, for centuries in many Western democracies, women and minorities were by law second-class citizens. When socio-political relations between men and women and between races began to change, laws began to change to reflect these newly understood relationships. The same trend appears to be emerging in the field of ecological restoration as international and national decision-makers recognise that existing preservation and conservation efforts alone will not be adequate in reviving human connections with the larger natural community. While these preservation and conservation efforts must continue to be embodied in international and national law, there is room for additional 'intelligent' means to support the revival of damaged ecosystem services or the renewal of cultural connections with important places.

Part of any 'intelligent' approach requires a sensitivity to understanding when to act and how fast to act. Given the condition of some key natural places, action should be taken before further ecological values are lost. Yet, decision-makers must be prepared to exercise precaution before embarking on a restoration project to avoid a situation where action taken in haste results in future complications. For example, while various herbicides and pesticides might be applied to immediately manage difficult invasive species, the long-term effects of this approach may counsel against this. Existing environmental law mechanisms such as impact assessment and other planning tools may provide necessary checks and balance mechanisms to ensure that resolving one ecological issue doesn't create future problems that may be even more difficult to resolve. One elder from the White Mountain Apache Reservation in the New Mexico, United States counsels against rushing into restoration efforts and instead advises, 'Go slowly; listen to the land and it will tell you what to do.'[21]

It is important, however, to note that restoration is also a departure from existing environmental law that inadvertently supports hands-off passive restoration. For example, conservation, protection or preservation efforts embodied in some environmental laws assume that nature can and will recover by itself if certain kinds of human activities cease.[22] Other environmental laws premised on sustainability infer that the natural world itself is independently capable of replenishing, renewing or maintaining viable and healthy ecosystems.[23] For many conservation and sustainable development laws, there is little recognition that as ecological

21 J. Long, A. Tecle and B. Burnette, 'Cultural foundations for ecological restoration on the White Mountain Apache Reservation' (2003) 8(1) *Conservation Ecology* 4, <http://www.consecol.org/vol8/iss1/art4/>.

22 See, e.g., national and sub-national nature preserves laws including marine protected areas that focus on reducing particular human activity such as fishing or forest product extraction.

23 Hall has argued that one of the interesting assumptions behind George Perkins Marsh's views on restoration is that 'nature by itself could not adequately repair damaged cause by humans': M. Hall, *Earth Repair: A Transatlantic History of Environmental Restoration* (University of Virginia Press 2005) 7.

conditions have changed those potentially complex human interventions in natural systems may become increasingly inevitable if we hope to speed up the recovery of some ecological systems.

To the extent that ecological restoration law offers the possibility of a structure for 'intelligent tinkering', there is room for a new practice of law to pragmatically address some of the recurring problems for international and national environmental law including the ongoing fragmentation among discrete legal frameworks such as the Biodiversity Convention. Focused on holistic outcomes, ecological restoration thinking may provide a broader focus for states to help integrate existing environmental commitments and respond in more timely fashion when environmental interventions may be necessary. One of the key contributions that law may provide is to offer a critical examination of sometimes divergent ecological restoration practice across the world and help to identify best practices that may be implemented effectively across jurisdictional boundaries. This book recognises that 'intelligent tinkering' may not be enough. When Leopold was writing, states were not facing the reality of global climate impacts with its complex drivers. Globally, we may need 'extremely intelligent tinkering' that requires a new way of thinking about how laws and legal institutions can adapt to known changes such as climate change and respond to long-term uncertainties. While it is not possible to predict the future impact of linking ecological restoration and law, linking the two practices may provide the necessary long-term catalyst for undertaking large, costly and complex projects to renew human connections with the larger natural community.

Ecological restoration in the context of an evolving proactive environmental law offers a seed of hope. Where climate change reflects a classic complex 'wicked problem' where every proposed solution creates potentially unexpected social challenges, perhaps ecological restoration can become a 'wicked solution' where the addition of new functionality and complexity back into ecosystems offers unexpected possibilities and opportunities for adaptation. If it is possible for restoration to be a 'wicked solution', law will be a critical factor in shaping the implementation of viable but effective ecological restoration strategies as law mediates across a variety of human relations underlying ecological restoration efforts.

Structure and purpose of this book

While there is great enthusiasm around the promotion of restoration activities where conservation activities have failed to protect environmental values, to date, there have been no books written about the relationship between the concept and practice of ecological restoration and law. This book, examining the intersection between ecological restoration practices and the law, intends to fill this gap in knowledge as well as to expose environmental and international legal scholars to an emerging area of law. This topic will become of increasing importance in the coming decades as states undertake restoration activities in order to resuscitate damaged ecosystem services and to adapt to climate change.

Based on the best available ecological science, this book starts with the premise that effective restoration work should restore basic ecosystem function to an area, should be self-sustaining, should be conducted at the largest available spatial level and should be designed to be long-lasting. Many activities that are currently classified as restoration work do not meet these criteria, in part, because there has been no legal framework requiring restoration activities to be ecologically effective.

This book does not pretend to offer discrete solutions to the complex questions that define the field of ecological restoration and law. These solutions will be negotiated over time between policy-makers and ecologists. Meanwhile, there are many questions that not just scientists and lawyers need to consider but also society at large. If we are going to engage in restoration, which appears to be the current trend of environmental law, then we have a plethora of issues that nations, sub-nations and communities will need to answer. Below is a list of questions that nations and communities are currently grappling with in developing restoration policies and projects:

- Why should we restore?

  - Are we restoring primarily to recover ecological functioning?
  - Are we restoring in order to enhance human welfare by recovering ecosystem services?
  - Are we restoring because we believe that we can change the impacts of the Anthropocene?
  - Are we restoring to recalibrate human relationships with the larger natural community?
  - Are we restoring as means of creating more robust novel ecosystems capable of surviving future environmental threats?
  - Are we restoring because of concerns with social and ecological justice?
  - Are we restoring because we owe something to future generations?
  - Are we restoring because we believe in rewilding?

- What should we restore?

  - Which ecosystems and which species are we going to restore?
  - Do we focus our investments on those species that provide humans with the greatest benefit?
  - Do we decide that some ecosystems are beyond the possibility of restoration?

- Where should we restore?

  - Given limited resources, do we focus only on protected areas?
  - Do we have to include areas outside of protected areas in order to promote connectivity?
  - Do we invest restoration efforts primarily on terrestrial resources?

- Do we restore global commons?
- Should we prioritize restoration within urban areas to build more human resource capacity for restoration?

- How are we going to restore?

  - What baselines or reference system will we use?
  - Will the creation of novel ecosystems qualify as legal restoration?
  - Should we have quality standards (e.g. favourable conservation status) or harmonised codes to guide restoration effort?

- Who decides what, where, and how we restore?

  - Should it be the state who decides what happens and where?
  - What role should indigenous communities and peoples have in restoration activities?
  - How should communities and cultures be involved in decision-making and actual restoration efforts?
  - Should funders for projects be allowed to determine how and where restoration work is done?

This book is divided into three parts. The first part sets the scene and discusses the importance of and justification for the idea of restoration for environment law. The second part describes the legal approaches taken to restoration by environmental law in international, European and some national legal systems. Ecological restoration as an idea is still maturing in international law. This second part also has chapters discussing the approaches to restoration by European and some national legal systems to enable a more complete discussion of the nature of the evolution of restoration in international law. The third part discusses and examines some themes that have particular relevance and significance for understanding and implementing restoration strategies in relation to the law. Given that international law doesn't contain a single instrument dealing with restoration, this last part allows this book to contribute to discussions in particular areas of concern to or importance for environmental law.

The first part, which contains Chapters 2 and 3, sets the scene for the discussion of the role and potential of international law in developing an approach to restoration. Both chapters aim to justify restoration as an important idea and practice. Restoration generally requires that human beings intentionally engage in the recovery of nature. However, what we do, where we do it and how we do it can vary from one country to another and not just because environments are different, but also because culture, ideas and economic development and other social factors can influence the choices that are made in restoring nature. Chapter 2 explores these ideas by defining restoration as compared to other approaches that deal with the intentional process of assisting with the recovery of nature. In particular ecological restoration is compared with rewilding and novel ecosystems because they are also proposed as holistic approaches to recovery. Chapter 3 examines several reasons for taking restoration seriously in the context of international environmental law. In particular the chapter discusses why ecological restoration is important for enhancing the survivability of the human population in the Anthropocene epoch; providing ecosystem services for human well-being; and broadening and deepening the environmental justice agendas to support social justice for current and future generations of human beings.

The second part of this book is designed to address the question of how restoration is currently embodied in existing law. While humans have been engaging in various activities that may be characterized as restoration for centuries, this book focuses on contemporary ecological restoration efforts. Starting with a discussion of international policy instruments including the Stockholm Declaration and the Rio Declaration plus general principles of environmental law, Chapter 4 offers a context for the more formal commitments to state-based restoration activities embodied in treaties and national law. What becomes clear from early international environmental negotiations is that even though states never defined restoration, the concept of restoration has become an essential part of a continuum of environmental conservation efforts. This chapter will also advance some observations about how restoration efforts might be informed by general environmental principles such as the polluter pays principle and the precautionary principle.

Using treaty texts and subsequent state practice in the form of Conference of Parties' recommendations and decisions, Chapters 5 and 6 trace the emergence of ecological restoration as a state obligation. A key question for both chapters is whether states understand the duty to restore as an obligation of conduct or an obligation of result. Chapter 5 covers explicit and implicit species and habitat restoration obligations in a broad spectrum of treaties including the Ramsar Convention on Wetlands and the Convention on the Law of the Sea. Chapter 6 describes international obligations to restore under the Convention on Biological Diversity and how these obligations are reflected in the non-binding but influential Aichi Biodiversity Targets. This chapter concludes with a discussion of a new direction for international cooperation for restoration with government-facilitated multi-stakeholder efforts to restore large landscapes such as the Global Partnership on Forest Landscape Restoration.

Wrapping up the second part of the book, Chapters 7 and 8 offer multiple examples of how international obligations to restore have been domesticated. The measure of success for many international environmental treaties is the ability of a given state to take general obligations and implement them in national laws or policies. Domestication of international law is essential for states that do not regard international law as self-executing. Chapter 7 focuses on the regional governance experience of the European Union (EU) in harmonising basic legal standards that promote restoration among the EU Member States. EU law not only implements international law, but is also a unique body of regional law with very specific implementation mechanisms both at the EU and Member State level. The chapter focuses specifically on the restoration obligations in the Birds and Habitats Directives that are the EU's core biodiversity legislation. These Directives, as well as their interpretation by the European Court of Justice, are probably some of the most advanced regional transboundary legal instruments on nature protection in the world, protecting the world's largest ecological network.[24] Recent case law by the European Court of Justice has shed some more light on the restoration obligations of the Member States. The chapter also elaborates on the EU Biodiversity Strategy, which contributes to the achievement of the Aichi Biodiversity Targets at the EU level by setting specific restoration targets for the EU. Chapter 8 looks at several very different domestic legal approaches to restoration efforts in the United States, China, Brazil, South Africa and Ecuador that at least partially fulfil the state's international duty to restore. The Ecuadorian example of domestication is particularly interesting as ecological restoration is recognised as a constitutional duty for the state and citizens of Ecuador.

The third part of the book offers a thematic approach to understanding how law and ongoing restoration efforts interact. Chapter 9 recognises that a great deal of restoration work is not done by the state but in partnership with the state or individually by landowners or communities. This chapter provides a brief overview of issues that involving restoration efforts by private actors including state-mandated restoration, payment for ecosystem service programmes and civil society guidance on restoration. The chapter concludes by asking whether there needs to be some shared normative or possible legal best-practice standards developed in order to ensure that restoration activities by both state and non-state actors are ecologically meaningful.

Chapter 10 focuses on ecological restoration within protected areas as the designation and management of protected areas are one of the key strategies to protect biodiversity. It explores specific duties and targets in international and EU law for parties to restore in protected areas. The chapter also describes the obligations to create connectivity between protected sites to address ongoing ecological fragmentation. The chapter finds that even though international and regional law requires parties to undertake some degree of connectivity measures, implementation on the ground is often missing. The chapter concludes by examining compensation requirements that may include restoration as a legal response to the loss of protected areas. Although these obligations may offer a financial and practical push for restoration measures to be implemented, compensatory restoration from an ecological point of view should only be implemented as a last resort action.

Chapter 11 explores the crucial issue of what role restoration can or should play in light of climate change trajectories as a mitigation and adaptation tool. While there is the potential for the United Nations facilitated REDD+ programme to enhance the quantity of resources that will be directed towards restoration, this chapter queries how states will appropriately safeguard vulnerable biodiversity to ensure that restoration efforts incorporate not just short-term mitigation gains but

---

24 See extensively on EU nature legislation: C.-H. Born, A. Cliquet, H. Schoukens, D. Misonne and G. Van Hoorick (eds), *The Habitats Directive in its EU Environmental Law Context: European Nature's Best Hope?* (Routledge 2015).

also long-term ecological objectives. The chapter further looks into the challenges for restoring nature while faced with the additional pressure from climate change, in order to determine if international biodiversity law is ready to cope with these pressures. The book concludes with a short reflection on future directions for ecological restoration and law.

Part I

Setting the scene for international law and ecological restoration

2 What is ecological restoration?

Introduction

This is a book about restoration, but what is restoration? It depends on who you ask. Is restoration simply a process that describes a variety of different relationships between humans and nature aimed at supporting the recovery of the natural environment? Are all restoration-oriented activities equally valuable simply because they assist with the recovery of vegetation and plants? For instance, would the removal of invasive species from a particular area be sufficient to constitute restoration if it was to assist an ecosystem to be healthier even if the underlying causes of ecosystem degradation are not addressed? Is restoration a process for thinking about what we do to nature or should its definition focus on the outcomes that need to be achieved?

If restoration is at times more than 'preservation' or 'conservation' then what is it?[1] Etymologically, the word restore comes from the Old French 'restorer' which means 'to give back' or to 'build up again' or 'repair'. The *Oxford English Dictionary* defines 'restoration' as the 'act of restoring to a former state or position ... or to an unimpaired or perfect condition'. Implicit in the *Oxford English Dictionary* definition is a return to a past where something or someplace was whole. While these definitions may be too ambitious in light of the inability to bring back an ecosystem to a 'former state or position' because of global irreversible environmental changes, the term restoration is often used generically to refer to the recovery of ecosystems or biotic features of the natural world. In some respects, these definitions raise as many questions as they answers.

Restoration is a term that has come to have a variety of meanings depending on the discipline. For the purposes of this book, ecological restoration is understood as a holistic effort to assist ecosystems to continue to develop along healthy historical trajectories.

The definition of restoration is important when an ecosystem needs to be managed through a recovery process. Defining restoration enables states and restoration practitioners to agree on a selection of shared goals that citizens hope to achieve in assisting an ecosystem to recover (e.g. increasing the availability of

---

1 Restoration can contribute to conservation strategies but it is also more than just about the maintenance of ecosystems, which is why we characterise the term differently. For more discussion on this see below.

ecosystem services, reducing or reversing biodiversity loss). Without agreeing on what restoration encompasses, they cannot set appropriate goals against which outcomes can be measured and states will not promote restoration as a worthwhile initiative to take in support of the natural environment.[2]

In the legal discipline, the term restoration is rarely defined but is frequently used under the assumption that everyone understands what it stands for. As ecological restoration becomes more prevalent, definitions become increasingly important to improve social and political efforts to create a mainstream legal framework for future ecological interventions and facilitate dispute settlement that may arise over restoration obligations. When practitioners and politicians use the word 'restore', the public may be receiving an incorrect message on the feasibility of reviving the complex ecosystems we have lost due to human impacts. The use of the term 'restore' may provide false hopes that a system can 'recover' to some past state of health and integrity. In many instances, unfortunately, it is difficult or even impossible to bring back past environmental conditions and the field of 'restoration' law and policy needs to be clearer about what it is striving to achieve to ensure conscientious and effective ecological decision-making.

This chapter explores the meaning of ecological restoration in order to argue for a normative practice of environmental law that extends beyond the protection and maintenance of the natural world and encompasses recovery processes. It aims to define and contextualise the concept of ecological restoration within a range of other approaches within restoration ecology that focus on the holistic recovery of the natural world and its ecosystems. Initially, the chapter describes the meaning of ecological restoration and focuses in particular on the way in which historical information about an ecosystem is used. It then describes two other holistic approaches to restoration, which conceptually approach the subject from different starting points. The chapter concludes that the definition of restoration can make a difference in terms of the goals that are set for restoration projects including the recovery of an ecosystem to some point along its historical trajectory.

The history of the idea of restoration

The idea that we can restore or help with the recovery of nature is not new. Going as far back as the middle of the nineteenth century the popular and very influential conservationist George Perkins Marsh published *Man and Nature*[3] that vigorously promoted the idea of restoration.[4] In the opening sentence to the preface of his

---

2 J. Bullock et al., 'Restoration of ecosystem services and biodiversity: conflicts and opportunities' (2011) 26(10) *Trends in Ecology and Evolution* 541–549.

3 G. Marsh, *Man and Nature* (Scribner 1864).

4 For an example of a work discussing Marsh and his influence on restoration see, M. Hall, 'Restoring the countryside: George Perkins Marsh and the Italian land ethic (1861–1882)' (1998) 4(1) *Environment and History* 91–103; and M. Hall, *Earth Repair: A Transatlantic History of Environmental Restoration* (University of Virginia Press 2005) 7 ("[E]nvironmental restoration is an old pursuit, much older than a few decades.")

book, first published in 1864, he wrote that 'The object of the present volume is: … to suggest the possibility and the importance of the restoration of disturbed harmonies and the material improvement of waste and exhausted regions.'[5] The discussion and inclusion of restoration in his book was not accidental. While living in Italy in 1861, Marsh recognised the significance of restoration for the approaches of the Italians versus the Americans in managing the natural world.[6] During Marsh's time in Italy he drew from concepts developed by Italian scholars to better understand the importance of human agency in either repairing or improving on the existing capacities and functionalities of the natural environment.[7]

The modern approach to restoration has many historical antecedents.[8] For instance, Frederick Olmsted back in 1878 used native vegetation, when commissioned by the City of Boston, to improve the ecological health of the Back Bay area.[9] Scholar William Jordan has commented about how 'active attempts to recreate whole, ecologically accurate examples of historic landscapes or ecosystems – were undertaken in the United States as early as the 1920s'.[10] This is not to suggest that individuals did not engage with restoration ideas before the 1920s and, more importantly, in other parts of the world. In fact, restoration practices to improve soil fertility for agriculture have been around for a long time, even before the industrial revolution. Indigenous peoples around the world have also been engaging in forms of restoration activities to sustain their use of ecosystem services.[11] These earlier approaches can be criticised for not being holistic even though they represent early efforts to actively recover parts or features of the natural world.

Despite the age-old usage of restoration strategies to help nature recover, it is only in the last two-and-a-half decades that we have seen exponential increases in the number of peer-reviewed articles discussing restoration ecology.[12] The field of restoration ecology is very broad and is concerned with a significant and diverse range of issues that relate to restoration. Within this field, concerns for ecosystems and 'ecological restoration' have emerged more recently. Some credit the work of Bradshaw in 1983 with the beginnings of discussions within restoration ecology

---

5 G. Marsh, *Man and Nature* (Scribner 1864) iii.
6 See Hall, 'Restoring the countryside' 91–103; and Hall, *Earth Repair*.
7 An example is the concept of 'bonifica' which according to Hall means 'improvement' and informed Marsh's view on the possibilities and potential of restoration: Hall, 'Restoring the countryside' 96–97.
8 D. Egan, 'Historic initiatives in ecological restoration' (1990) 8(2) *Restoration and Management Notes* 83–90.
9 Ibid., 83.
10 W. Jordan, *The Sunflower Forest: Ecological Restoration and the New Communion with Nature* (University of California Press 2003) 2.
11 See S. Stevens (ed), *Conservation through Cultural Survival: Indigenous Peoples and Protected Areas* (Island Press 1997).
12 See T. Young, D. Petersen and J. Clary, 'The ecology of restoration: historical links, emerging issues and unexplored realms' (2005) 8 *Ecology Letters* 662–673.

with what he called at that time the 'reconstruction of ecosystems'.[13] In that work he emphasised the active role that human beings needed to play in reconstructing ecosystems. Since the late 1980s, the international Society for Ecological Restoration (SER) has become a central player in promoting 'ecological restoration' as a field within the broader scientific discipline of restoration ecology.[14]

Restoration as central and additional idea for to how we think about the maintenance and protection of the natural world

The recovery of nature is of concern to many different academic fields of research, whether it be ecologists, conservationists, engineers or social scientists. Although scientific disciplines have a tendency to break down life systems into smaller components to study them, they have yet to regularly succeed in understanding and re-creating holistic structures and ecosystems.[15] Ecology scholars like Liberty Hyde Baily and Aldo Leopold both recognised that the value of specific parts and features in an ecosystem were important but that their real significance came from how they were integrated as parts of a system.[16] The interconnectedness of parts within nature leaves ecosystems potentially vulnerable if some part or function is harmed or damaged. This is why recovery of ecosystem functions and structures is of increasing concern to scientists – they need and want to be able to pre-empt the possibility that the failure of one part may have detrimental effects on the whole, whether that be of a habitat of significance, an ecosystem or an Earth system.

While the work of a range of disciplines are linked through words like conservation, preservation and the protection of nature, the fields of restoration ecology and ecological restoration mark a sizeable departure from these other efforts, which can also assist with the recovery of nature.[17] Although concepts like preservation, conservation and protection can all be used to explain the need for some restoration-oriented activities, there are limitations in defining restoration activities using these terms.[18] For example, while preservation can include saving pockets of redwoods from the logger's axe, it can also extend to the removal of invasive

13 See A. Bradshaw, 'The reconstruction of ecosystems' (1983) 20 *Journal of Applied Ecology* 1–17, which was the published version of his speech to the British Ecological Society in 1982.

14 J. van Andel and J. Aronson, 'Getting started', in J. van Andel and J. Aronson (eds), *Restoration Ecology: The New Frontier* (2nd Edn, Wiley-Blackwell 2012) 4.

15 An example is the Earth system scientists' failure to achieve much, if anything, in their attempt to recreate life systems as part of the Biosphere 2 project.

16 See Z. Jack, *Liberty Hyde Bailey. Essential Agrarian and Environmental Writings* (Cornell University Press 2008); and A. Leopold, *A Sand County Almanac and Sketches Here and There* (Oxford University Press 1949).

17 See A. C. Leopold, 'Living with the land ethic' (2004) 54(2) *BioScience* 149–154.

18 J. Passamore, *Man's Responsibility for Nature* (Duckworth 1974) 101 (preservation is 'the attempt to maintain in their present condition such areas of the Earth's surface as do not yet bear the obvious marks of man's handiwork and protect from the risk of extinction those species of living beings which man has not yet destroyed').

species from wetlands to enable an ecosystem to maintain its own indigenous species and ecosystem integrity. The legal concept of 'preservation' as a freezing of a particular landscape in perpetuity, however, is now recognised as antiquated. Ecosystems are dynamic and require some capacity to change in order to adapt to a variety of threats. The failure of a 'preservationist' strategy for environmental protection becomes even more evident when reflecting on how climate change and invasive species are influencing the ecosystems and ecologies of both pro-tected and non-protected habitats. Given the pervasiveness of human impacts in the world, it has become more difficult to 'preserve' nature. At best, we can maintain critical aspects of nature.

A better long-term protection strategy as compared with preservation has been conservation, which can been defined as the 'saving of natural resources for later consumption',[19] but can also be enlarged to protect non-use values such as exis-tence values. Implicit in the definition of conservation is the concept that there is something of value to save which has some inherent ecological integrity. 'Con-servation' activities may include both public and private lands,[20] but today serious conservation efforts are generally associated with delineating protected areas.[21] Some may disagree and suggest that environmental law is all about conservation, defined broadly as concerned with the maintenance of the natural environment rather than its protection. The belief in the irreplaceability of the natural environment makes conservation (and also preservation) significant for the way that environ-mental law has developed and continues to develop. Scholars have criticised the conservation ideology. Jordan, for instance, has written about how the language of conservationists suggests that the 'influence of human beings on natural land-scapes is invariably negative and destructive: though we may take from such a landscape, we can never give anything back to it'.[22] Wilson has similar problems with conservation but has phrased his arguments in terms of the problem that conservation has with being forward looking. He has argued instead that approa-ches to conservation lack direction or goals, which means that it is often

---

19 Passamore, *Man's Responsibility for Nature* 73.
20 See, e.g., Endangered Species Act, 16 USC Sec 1531 (1973), Sec. 3. 'Conservation' is the 'use of all methods and procedures which are necessary to bring any endangered species or threatened species to the point at which the measures provided pursuant to this Act are no longer necessary. Such methods and procedures include, but are not limited to, all activities associated with scientific resources management such as research, census, law enforcement, habitat acquisition and maintenance, propagation, live trap-ping, and transplantation, and, in the extraordinary case where population pressures within a given ecosystem cannot be otherwise relieved, may include regulated taking.'
21 See N. Dudley (ed), *IUCN, Guidelines for Applying Protected Area Management Cate-gories* (IUCN, 2008); where at 7 it defined a protected area as 'a clearly defined geo-graphical space, recognized, dedicated and managed, through legal or other effective means, to achieve the long term conservation of nature with associated ecosystem services and cultural values'.
22 W. Jordan, *The Sunflower Forest: Ecological Restoration and the New Communion with Nature* (University of California Press 2003) 2.

reactionary.[23] Young is more explicit about the variations between conservation and restoration and sees a difference in the mindset of the two fields. The conservation mindset is one of more or less permanent loss. The restoration mindset is one of recovery after temporary loss. Put in a more simplified way, Young sees restoration ecologists as tending to be optimistic, whereas conservation biologists are more pessimistic.[24]

One unacknowledged but perhaps very important assumption behind conservation is that, when left alone, the natural world can restore itself. This presents conservationists with the responsibility for maintaining the natural world, but this is less realistic in current circumstances where – given the arguments about breaches in planetary boundaries by Earth system scientists – this is becoming more and more difficult to achieve.[25] There is no substitute for preserving a good-quality habitat and its maintenance and management is a priority. However, in many parts of the world, there is a lack of unaltered habitats, or the remaining habitats cannot sustain biota and need to be improved or expanded. Thus, restoration is an integral part what conservation can achieve.

As will be explained further in this book, restoration can effectively supplement conservation strategies but also departs from conservation strategies. The next section defines restoration to support the wider argument that environmental law has to move beyond protection and maintenance to more rigorously addressing restoration goals and objectives.

Ecological restoration

The term restoration is often used generically to refer to the recovery of an ecosystem or the biotic features of the natural world. The approach to recovery can be broken down into various stages depending on the kind of degradation that requires a response. Within restoration ecology Andel and Aronson have categorised four kinds of restoration activities including:

(1) **near-natural restoration**, aiming at almost non-assisted natural recovery,
(2) **ecological restoration**, that is, the return to some historic **reference system**, representing pre-disturbance conditions, be it natural or seminatural,
(3) **ecological rehabilitation**, that is, the improvement of **ecosystem functions** without necessarily a return to predisturbance conditions, and (4) **reclamation**, that is, conversion of heavily degraded land such as post-mining areas to a productive condition.[26]

23 E. Wilson, *Half-Earth: Our Planet's Fight for Life* (W.W. Norton & Company 2016).
24 T. Young, 'Restoration ecology and conservation biology' (2000) 92 *Biological Conservation* 73–83.
25 See F. Biermann et al., 'Down to Earth: contextualizing the Anthropocene epoch' (2015) *Global Environmental Change*, <http://dx.doi.org/10/1016/j.gloenvcha/2015.11.004>.
26 van Andel and Aronson, 'Getting started' 7.

Lawyers have yet to think more critically about the meaning of restoration for their profession. They are not accustomed to applying the concept of ecological restoration but are more familiar with terms such as rehabilitation, remediation, reclamation or mitigation because they appear in legal instruments or court documents. The word remediation for instance is a legal response expected from someone who has caused damage to a site but the obligation on the individual is simply to return an area back to the condition before the harm was caused to it. There is usually very little consideration for reviving previously existing ecological conditions in an area unless the enforcing authority is driven to, and has the capacity to articulate the exact nature of the harm that is being remediated. In situations where a site has been transformed by human activities, such as mine sites, there is little that can be done to achieve full recovery in terms of taking a habitat back to what it was like before the degradation. It is in these situations that rehabilitation may in fact be the only option available, although ecological restorationists would suggest that the timeframe for recovery in these circumstances would have to be longer to account for the severity of the damage to the soil in particular.

As a result of a lack of a shared understanding of what an obligation to restore entails, the use of the term 'restoration' within law and policy often fails to precisely describe what legal practitioners or ecologists hope to achieve in their work. This lack of a shared understanding may be due in part to the lack of engagement by lawyers and policy-makers with basic ideas and concepts from restoration ecology. Legal definitions that form the basis of obligations are often sharpened and refined through litigation and conflict resolution mechanisms, which is why the term restoration, to the extent it is ever defined, is conceived of in domestic legal systems as a remedy for a breach of an obligation.

The lack of clarity among lawyers about how to define restoration is not that different, however, to the issues that ecologists also have with articulating the content of a rich concept like restoration. As a long-standing member of the International Union for the Conservation of Nature (IUCN) and an observer organisation to the Biodiversity Convention (CBD) and the Ramsar Science and Technical Review Panel, the SER has been influential in defining ecological restoration for international policy discussions. The SER has been important in terms of defining and promoting the field of ecological restoration. It produces occasional briefing notes, policy position statements and technical documents such as the SER International Primer on Ecological Restoration,[27] and the guidelines on ecological restoration.[28] The SER International Primer on Ecological Restoration contains key concepts and fundamental principles upon which ecological restoration is based. It includes the most widely-cited definition of ecological restoration.[29]

27 Society for Ecological Restoration International Science & Policy Working Group, *The SER International Primer on Ecological Restoration* (Society for Ecological Restoration International 2004), <www.ser.org>.
28 A. Clewell, J. Rieger and J. Munro, *Guidelines for Developing and Managing Ecological Restoration Projects* (2nd Edn, Society for Ecological Restoration International 2005).
29 <http://www.ser.org/page/SERDocuments>.

In its 2004 Primer on Ecological Restoration (SER Primer), SER defined restoration as the 'process of assisting the recovery of an ecosystem that has been degraded, damaged, or destroyed'.[30] In this definition the terms 'assist' and 'recovery' have significance and are general enough to support a variety of activities aimed at ecosystems regaining their health, integrity, or ecological functions.[31] The SER Primer further elaborates that: 'Ecological restoration is an intentional activity that initiates or accelerates the recovery of an ecosystem with respect to its health, integrity and sustainability ... Restoration attempts to return an ecosystem to its historic trajectory.'[32]

The SER Primer makes it clear that 'ecological restoration' or 'ecosystem restoration' is not the same as 'restoration ecology'. Ecological restoration is an interdisciplinary practice of doing restoration work, which should include incorporating the experiences, political ideals and values held by people and their communities. For the SER, 'restoration' must be an intentional activity focused on ecological structures such as species composition and functions like nutrient cycling. In contrast, restoration ecology focuses on the science of restoration, which includes trial and error approaches to improving the health and integrity of ecosystems. Restoration ecologists have defined their work as the discipline of scientific inquiry that helps 'to move a damaged system to an ecological state that is within some acceptable limits relative to a less disturbed system'.[33] Restoration ecologists strictly define ecological restoration as 'an attempt to return a system to some historical state'.[34]

The SER is careful in its Primer to distinguish 'restoration' from other ecological interventions. The Primer indicates that 'restoration' is not the same as rehabilitation, reclamation, mitigation, ecological creation/fabrication or ecological engineering. Each of these interventions can contribute to restoration goals but these interventions are qualitatively different. For example rehabilitation is defined as having the goals of 'the reparation of ecosystem processes, productivity and services, whereas the goals of restoration also include the re-establishment of the pre-existing biotic integrity in terms of species composition and community structure'.[35] Reclamation refers to superficial efforts to return limited functions to a landscape including 'stabilization of the terrain, assurance of public safety,

30 Society for Ecological Restoration International Science & Policy Working Group, *The SER International Primer on Ecological Restoration* (Society for Ecological Restoration International 2004) Section 2.
31 S. Allison, *Ecological Restoration and Environmental Change: Renewing Damaged Ecosystems* (Routledge 2012) 5.
32 Society for Ecological Restoration International Science & Policy Working Group, *The SER International Primer on Ecological Restoration* (Society for Ecological Restoration International 2004) Section 1, p. 1.
33 M. Palmer, D. Falkand and J. Zedler, 'Ecological theory and restoration ecology' in D. Falk, M. Palmer and J. Zedler (eds), *Foundations of Restoration Ecology* (Island Press 2006) 1.
34 Ibid.
35 Society for Ecological Restoration International Science & Policy Working Group, *The SER International Primer on Ecological Restoration* (Society for Ecological Restoration International 2004) Section 10.

aesthetic improvement, and usually a return of the land to what, within the regional context, is considered to be a useful purpose'.[36] Mitigation is 'an action that is intended to compensate for environmental damage'.[37] Creation is the construction of a new ecosystem that would not have historically been at a location. This is not considered 'restoration' because it involves too much human intervention in terms of introducing processes that are independent of the ecological life-cycles of a given area. Finally, ecological engineering is the 'manipulation of natural materials, living organisms and the physical-chemical environment to achieve specific human goals and solve technical problems'.[38]

These distinctions among possible interventions are generally lost upon the law, particularly because, as will become apparent in later chapters, the law rarely regulates what qualifies as appropriate restoration. The law typically doesn't promote a particular approach to implementing ecological interventions. The difference between ecological restoration as compared with other ecological recovery efforts such as rehabilitation is that ecological restoration focuses on actions that strive to return an ecosystem to a pre-existing or historical ecological trajectory referred to as a 'historical trajectory'. Increasingly, the term 'reference systems', or 'reference points', is being used to more precisely designate the condition or historical trajectory that an ecosystem has to be returned to.[39] These ideas are discussed next, as they are central to ecological restoration. Following this discussion, two ways of interpreting the SER approach will be discussed to highlight how notions of historical trajectory or reference points are being applied in two important national initiatives in Australia and Brazil.

Historical trajectory and reference points

The idea in the SER Primer that ecological restoration has to 'return an ecosystem to its historic trajectory' is not an uncontroversial assumption. In fact, ecologists have noted that it is 'generally accepted that a return to past ecosystems, indeed a return to the past in general is, strictly speaking, not possible; history cannot be repeated'.[40] The idea of an ecological trajectory has been defined by the SER as describing the 'developmental pathway of an ecosystem through time ... The trajectory embraces all ecological parameters. Any given trajectory is not narrow and

---

36 Ibid., Section 10.
37 Ibid.; but see the difference between mitigation and compensation in Chapter 10, section on compensation.
38 Society for Ecological Restoration International Science & Policy Working Group, *The SER International Primer on Ecological Restoration* (Society for Ecological Restoration International, 2004) Section 10.
39 On the idea of reference systems see: L. Balaguer, A. Escudero, J. Martin-Dugue, I. Mola and J. Aronson, 'The historical reference in restoration ecology: re-defining a cornerstone concept' (2014) 176 *Biological Conservation* 12–20; and B. Pruitt, S. Miller, C. Theiling and C. Fischenich, 'The use of reference ecosystems as a basis for assessing restoration benefits', <http://citeseerx.ist.psu.edu/viewdoc/download?doi= 10.1.1.310.2493&rep=rep1&type=pdf>.
40 van Andel and Aronson, 'Getting Started' 7.

specific. Instead, a trajectory embraces a broad yet confined range of potential ecological expressions through time'.[41]

The term 'trajectory' refers to an holistic approach to the work that has to be done in ecological restoration. It is the idea that the state or reference point used for restoration needs to account for a range of factors and influences on the habitat being restored. The importance of this is also captured in another definition developed by the European Community in 2011 in its Biodiversity Strategy Impact Assessment. It defined restoration as: 'The return of an ecosystem to its original community structure, natural complement of species, and natural functions.'[42] The need to restore to an ecological trajectory is therefore central to the SER definition. The challenge arises, however, when restoration practitioners have to decide on the historical point of reference to use for the purposes of restarting an ecosystem on its historical trajectory.

A reference system or site for an ecosystem is different to the current state that an ecosystem is in. This is because the point of restoration is often to take an ecosystem past its degraded state, which is assumed to be its current condition. A reference system will not necessarily be the 'natural state' of the ecosystem because the natural state of an ecosystem may be different to what the ecosystem was like before it was degraded in some way by anthropogenic influences.[43] This means that a pre-degradation state and a natural state could both be historical reference points for restoring an ecosystem. A pre-degradation reference point includes human beings actively involved in the trajectory of the ecosystem whereas a natural state reference point does not always presume that. The trajectory that an ecosystem is returned to is therefore a choice amongst a range of potential reference points or systems. The choice in reference points is not always acknowledged in international policy-making. The Intergovernmental Science-Policy Platform on Biodiversity and Ecosystem Services (IPBES), for instance, in its second plenary meeting in 2015 adopted a definition of restoration that only refers to the 'recovery of an ecosystem from a degraded state'.[44] The historical reference point in this definition is the pre-degraded state of the ecosystem, but there may be multiple incidents of degradation. This type of definition offers a level of flexibility for decision-makers that might end up ignoring important aspects of the natural state of an ecosystem.

41 Society for Ecological Restoration International Science & Policy Working Group, *The SER International Primer on Ecological Restoration* (Society for Ecological Restoration International 2004) Section 4.

42 European Commission Biodiversity Strategy Impact Assessment 2011, <http://ec.europa.eu/environment/nature/biodiversity/comm2006/pdf/2020/1_EN_impact_assesment_part1_v4.pdf>.

43 On the idea of the natural state see, T. Newbold et al., 'Global effects of land use on local terrestrial biodiversity' (2015) 520 *Nature* 45–50.

44 Plenary of the Intergovernmental Science-Policy Platform on Biodiversity and Ecosystem Services, 3/7, 12–17 January, p. 1, <http://esa.org/ipbes/wp-content/uploads/2015/02/landdegscoping.pdf>.

What is common for both the SER and IPBES definitions of restoration is that they choose reference points for restoration that are to be historically determined. According to the SER Primer, restoration could be carried out by reference to a range of historical trajectories. For IPBES returning an ecosystem back to its pre-degradation state requires some agreement on the difficult question of what kind of degradation is relevant and significant. A pre-degradation state could be situated 10, 20 or 100 years in the past and someone must make a judgment call regarding what historical reference point will matter for restoration activities.[45] As compared to the SER Primer's definition, the IPBES definition is simpler to implement because it only requires the identification of a pre-degradation state rather than a historical trajectory. Neither definition is explicit, however, about the requirements to achieve 'ecological integrity'.[46]

The historical trajectory of an ecosystem is not easy to identify because one has to ask which indigenous version of the ecosystem we are interested in and what are likely to be the features of that system that will enable it to become self-perpetuating. For instance, if we were restoring an Australian landscape would we aim to restore it to its trajectory from pre-European settlement days? How would we adequately evaluate what it was like at that time to then be able to replicate it? Would it even be realistic to presume that the ecosystem under consideration has not been subject to myriad natural pressures to change either subtly or dramatically? The alternative approach we could take would be to restore the same ecosystem back to its pre-degradation stage even if the area of concern had simply been used for agriculture. Rackham has suggested that the best way to account for the historical trajectory of an ecosystem is to identify the losses that accompany the degradation. Important losses include the loss of historical vegetation and wildlife, but also the loss of meaning from the degradation.[47] The challenge with this view, however, is that restoration has to be about ecological recovery and by focusing on loss we tend to complicate the decision-making process by focusing on what humans value in the ecosystem.

An important challenge with returning an ecosystem to its historical trajectory or its supposed natural state is that ecosystems are dynamic and respond to both internal and external influences. Returning nature to some untouched or wild state before human beings had influenced it is almost impossible. Current temperatures, pollution levels and soil conditions for instance may prevent achieving either a natural state or a pre-degradation state. Identifying a natural baseline or reference point may be difficult to include in projects because of a great deal of uncertainty regarding ecological histories.[48]

---

45 Ibid.
46 Allison sees 'ecological integrity' as a key element in ecological restoration alongside having to achieve the recovery of the historical trajectory of an ecosystem (Allison, *Ecological Restoration and Environmental Change* 13).
47 O. Rackham, *The History of the Countryside* (Phoenix 1986).
48 B. Normander et al., *State of Biodiversity in the Nordic Countries* (Norde Press 2008) 25.

There are several arguments supporting the return of an ecosystem to its historical trajectory – as compared to creating a new trajectory which is discussed further below in the section on novel ecosystems. The first is that restoration managers do not have to make choices about what to leave in and out of the management of the restoration process. There is no room to creatively interpret the potential or capacity of an ecosystem to survive as long as the goal of the project is to take an ecosystem back to some natural state baseline.[49] One of the challenges with selecting what to restore in an ecosystem is the possibility of ignoring the influences of 'minor' species or 'inconspicuous processes'.[50] Gross has argued along these lines that restoration has to help with the recovery of the ecosystem along pathways that are not controlled by human beings.[51] Second, a return to a historical trajectory may avoid the homogenisation of landscapes currently caused by increasing climate change and invasive species.[52] Using historical trajectories may foster recovery efforts that are not always driven by what is immediately beneficial for human beings. Finally, historical authenticity, which is implicit in the concept of historical trajectories, helps restorationists avoid criticisms that they are replacing natural value with technologically and economically driven possibilities and priorities.[53]

Outcome-driven ecological restoration

Given the challenges in identifying a historically accurate reference point when defining the kind of restoration activities to be undertaken in relation to an ecosystem, SER has focused ecological restoration on returning an ecosystem to a historical trajectory. More recent efforts by institutions to define restoration have explored the complexities around identifying the historical trajectory of an ecosystem in terms of a reference point for restorationists. In attempts to define restoration more clearly or precisely, institutions in Australia and Brazil have focused on the outcomes that have to be achieved for an ecosystem to be considered to have returned to its historical trajectory. This is important because they have defined the reference points for restoration projects around important outcomes rather than an open-ended process whereby decision-makers have to decide on what is the natural state for an ecosystem. The approach taken to ecological restoration in Brazil and Australia are briefly discussed to highlight how they have grappled with practically implementing the concepts of 'historical trajectory' and reference systems used for achieving that ideal.

49 Jordan, *The Sunflower Forest* 23–24.
50 Ibid., 23.
51 M. Gross, 'Beyond expertise: ecological science and the making of socially robust restoration strategies' (2006) 14 *Journal for Nature Conservation* 172–179.
52 See Allison, *Ecological Restoration and Environmental Change* 42; P. Kareiva, 'Domesticated nature: shaping landscapes and ecosystems for human welfare' (2007) 316 *Science* 1866–1869; M. McKinney and J. Lockwood, 'Biotic homogenization: a few winners replacing many losers in the next mass extinction' (1999) 14 *Trends in Ecology and Evolution* 450–453.
53 M. Hunter, 'Benchmarks for managing ecosystems: are human activities natural?' (1996) 10 *Conservation Biology* 695–697.

Ecological criteria for success in Brazilian restoration projects

In Brazil, as is the case with a lot of other countries around the world, ensuring that restoration projects achieve ecologically sound outcomes has been difficult. Given their ambitious restoration goals enacted in the Law of Native Vegetation Protection and Restoration, Federal Law 12.651/2012, Brazil has been pushed to work harder at gaining consensus around indicators that can measure success.[54] It has been estimated that in accordance with Brazilian laws around 21 million hectares of land has to be restored nationally.[55] To ensure that the restoration goals are met, Resolution SMA 32 was passed in 2014 by the Sao Paolo provincial government enabling restoration within São Paolo to be measured against three indicators, namely:

- 'Ground coverage with native vegetation (percentage)';
- 'Density of native plants spontaneously regenerating (number of individuals per hectare; height H >50 cm and circumference at breast height)'; and
- 'Number of spontaneously regenerating native plant species (number of species with H > 50 cm and CBH < 15 cm)'.[56]

These indicators enable regulators to measure the success of the restoration initiatives in the state of São Paulo. The emphasis on native vegetation and plants is driven by the outcome that the government aims to achieve, which is ecosystems that can continue to function without interventions. The use of native species presumes that they are more likely to achieve self-perpetuating restoration for the state of São Paulo. Although it does not specify that restoration has to be historically authentic, the resolution focuses on outcomes that can set the ecosystem back on its historical trajectory. It does this by comparing projects based on pre-defined restoration objectives rather than creating a cumbersome process where each project is measured against a particular past or a specific kind of reference site or pre-degraded state or condition.

Australia and the principles for securing local indigenous ecosystems

In 2016 the Australasian branch of SER adopted National Standards for the Practice of Ecological Restoration in Australia (the 'Standards') that aim to guide approaches to restoration.[57] Central to the Standards is the recognition that 'local indigenous ecosystems' have intrinsic value that needs to be restored. The focus

54 See R. Chaves, G. Durigan, P. Brancalion and J. Aronson, 'On the need of legal frameworks for assessing restoration projects success' (2015) 23(6) *Restoration Ecology* 754–759.
55 Ibid., 755; referring to the work of B. Soares-Filho, R. Rajao, M. Macedo, A. Carneiro, W. Costa, H. Rodrigues and A. Alencar, 'Cracking Brazil's forest code' (2014) 344 *Science* 363–364.
56 These are translations from Chaves et al., 'On the need of legal frameworks for assessing restoration projects success', 756.
57 See <www.seraustralasia.com/standards/introduction.html>.

on the 'outcomes' of restoration projects is critical to these Standards and this is partly because they recognise that ecological restoration does more in terms of the recovery of ecosystems than what is ordinarily captured by 'rehabilitation' type activities.

The Standards adopt a number of principles that are meant to guide the activities of restoration practitioners. Principle 1 in particular indicates that historical information about an ecosystem being restored must be taken into account. It requires that ecological restoration practices should be based on the identification of local indigenous reference ecosystems. A description of Principle 1 highlights a range of evidence that can assist a practitioner in restoring along a historical trajectory:

> A fundamental principle of ecological restoration is the identification of an appropriate **reference ecosystem** to guide project targets and provide a basis for monitoring and assessing outcomes. The reference ecosystem can be an actual site (**reference site**) or a conceptual model synthesised from numerous reference sites, field indicators and historical and predictive records. The reference ecosystem can include local indigenous species (plants, animals and other biota) as well as species from neighbouring localities that have recently naturally migrated e.g. due to a changing climate (see definition of 'local indigenous ecosystem' in glossary). Where local evidence is lacking, regional information can help inform identification of likely local indigenous ecosystems. Identifying a reference ecosystem involves analysis of the **composition** (species), **structure** (complexity and configuration) and **function** (processes and dynamics) of the ecosystem to be restored on the site. The model should also include descriptions of successional states that may be characteristic of the ecosystem's decline or recovery.[58]

These Standards allow the outcomes of ecological restoration to be judged against a range of specific criteria that focus attention on the ecosystem itself. The focus on the connections between the composition, structure and function of the ecosystem deepens the reliance on historical and indigenous reference points for restoration practitioners. Significantly this principle also recognises the challenge of historically focused recovery because an ecosystem is not static. Restoration practitioners must define their objectives based on what the Standards call 'successional states'. This may require a certain amount of speculation.

The Brazilian and Australian example highlight how reference points for restoration projects need to be defined to assist an ecosystem to continue on its historical trajectory. What makes the SER definition of restoration significant is not just that an organisation that has global support from a range of academic and practitioner communities has adopted it, but that it creates both normative conditions and evaluative standards against which the success and outcomes of

58 <www.seraustralasia.com/standards/principle1.html> (emphasis in original).

projects can be assessed. Setting a historical benchmark for a recovery project enables restoration practitioners to communicate more clearly about what needs to be done to achieve restoration goals. However, restoring to some historical or indigenous state can have its challenges in that achieving a degree of authenticity may trigger a range of political, economic and social issues that may influence how restoration activities are performed.

Conclusion: ecological restoration and the SER

The SER definition of restoration operates from the assumption that historically authentic ecosystems – before they were degraded by some kind of anthropogenic influences – contain features that future generations will benefit from even though the current generation cannot accurately identify the key features. While this may be true, some scholars have pointed out the limits of ecological restoration and call for setting realistic expectations with regards to the results of ecological restoration.[59] Restoration may succeed with some taxa, but fail with others. Ecological restoration as currently practiced may not be holistic with the focus of a large number of ecological restoration projects appearing to be only on plants.[60] This raises the question of whether restoration is effective for achieving the recovery objectives proposed by the SER. Because the historical state of an ecosystem can be difficult to identify and can set a difficult standard to achieve, new approaches to defining ecological restoration – such as those in Brazil where components and outcomes of the process are identified – can appear attractive and easy to engage with at the grassroots level. In the next section of this chapter other approaches to ecological recovery that are discussed extensively in the literature on restoration are briefly outlined.

Alternative approaches to restoration

One recurrent theme in discussions about the meaning and purpose of restoration comes from conflicting views between those who view the world from the perspective of ecosystems made of biotic and abiotic matter and those who recognise the value in nature only through the perspective of human needs. This debate is not simple to resolve, because of the conflicting preferences and values that divide opinions.

As a result of this division of opinion over the purpose of restoration, there are also varying approaches to what we understand as appropriate interventions to renew or repair an ecosystem. These approaches may be more pragmatically driven as opposed to theoretically sound. Some alternative ways of thinking about restoration which may influence how we define the process include:

---

59 E.g. J. Ehrenfeld, 'Defining the limits of restoration: the need for realistic goals' (2000) 8(1) *Restoration Ecology* 2–9.
60 For instance, Allison, *Ecological Restoration and Environmental Change* 144.

1   rewilding a habitat or a protected area by focusing on keystone species;
2   human interventions in rebuilding the integrity or health of an eco-
    system without worrying about returning an ecosystem to its historical
    trajectory;
3   modifying or inventing nature or an ecosystem to give it the capacity to do
    something that will suit a particular purpose, whether that is for the benefit of
    nature or human beings.

These other restoration-type activities focus attention on the recovery of an
ecosystem and some of its functions, but do so for a variety of reasons and goals,
such as making a landscape more beautiful, or ensuring productivity to secure
food production. The purpose and role of restoration is discussed in Chapter 3.
In this section, the creation of 'novel ecosystems' and also 'rewilding' are
described and discussed as two prominent conceptual and pragmatic approaches
that differ from the SER-defined approach to ecological restoration. They are
both potentially classifiable as nature-based but because the human interventions
in these processes are targeted they can also be deployed for certain political and
pragmatic purposes.

As discussed above, the SER is not oblivious to alternative ways of dealing
with ecological recovery. A sustainable forestry initiative or revegetation project
that reduces soil erosion may result in a 'novel ecosystem' by replacing historical features of an ecosystem with new and sometimes more ecologically
productive natural features.[61] Rewilding, describing a process whereby a keystone species is introduced into an environment, may build complexity back
into an ecosystem where conservation efforts have been adversely affected.[62]
The introduced species can support the recovery of native animals, plants and
vegetation. While both these recovery efforts support the active involvement of
human beings in the recovery of ecosystems they fall short of some of the
theoretical and policy-driven ambitions inherent in the SER Primer's definition
of restoration to connect humanity back to its historical relationships with
nature.

Novel ecosystems

More and more evidence is found that even the most pristine ecosystems, such as
tropical forests, have already been transformed by human activities and at much
earlier times than was formerly believed. Ecosystems seem to be shaped in much
more accidental and random ways than traditional ecological theories may suggest.
They are constantly being remade by fire and flood, disease and the arrival of new

61 On novel ecosystems see generally, R. Hobbs et al., 'Novel ecosystems; implications for
    conservation and restoration' (2009) 24(11) *Trends in Ecology and Evolution* 599–605.
62 See, e.g., D. Macdonald et al., 'Rewilding' in D. Macdonald and K. Willis, *Key Topics
    in Conservation Biology 2* (Oxford University Press 2013) Chapter 23.

species. Climate change[63] and the arrival of alien species have sped up this process dramatically.[64] Some scholars are claiming that we need to revise traditional conservation and restoration priorities as a result of anthropogenic harm done to the functioning of Earth systems. Does it make sense then, or is it even possible, to try and return to a historical trajectory approach in light of scientific uncertainties? Ecosystems are dynamic entities, and some see restoration goals that are based on relatively static compositional or structural attributes as problematic.[65] Instead of looking backwards, and trying to re-create ecosystems as they were in the past, some scholars suggest that restoration and conservation goals must focus on building ecosystems that can cope with the future or provide human beings with desired benefits.[66] While this may still include the restoration of compositional or structural elements from the past, it is not limited to just these goals.[67] Recent literature suggests that a new term, 'novel ecosystems',[68] would better capture the ambitions of an alternative approach to the active recovery of an ecosystem that is not driven by fidelity to historical characteristics.

Novel ecosystems are defined in different ways but one more commonly used definition characterises them as:

> a system of abiotic, biotic and social components (and their interactions) that, by virtue of human influence, differ from those that prevailed historically, having a tendency to self-organize and manifest novel qualities without intensive human management. Novel ecosystems are distinguished from hybrid ecosystems by practical limitations (a combination of ecological, environmental and social thresholds) on the recovery of historical qualities.[69]

This definition seeks to recognise that 'historically rooted ideals' cannot always guide all kinds of restoration activities. Sometimes this is because of ecological complications, but also practical limitations due to social and economic influences on decision-makers and restoration practitioners. In the earlier discussions in this chapter, it was noted that where an ecosystem is unable to be assisted so that it can recover back to a historical trajectory of some kind, then restorationists could engage in some kind of rehabilitation work. Rehabilitation, however, was not seen as an alternative to ecological restoration but more simply as an approach to recovery, recognising that sometimes the damage done to ecosystems is almost fatal.

63 See also Chapter 11 on climate change.
64 F. Pearce, 'True nature: revising ideas on what is pristine and wild' (2013) *Yale Environment* 360.
65 See, e.g., E. Ellis et al., 'All is not loss: plant biodiversity in the Anthropocene' (2012) 7 *PLos One*, e30535.
66 See, e.g., R. Corlett, 'New approaches to novel ecosystems' (2014) 29 *Trends in Ecological Evolution* 137–138.
67 R. Hobbs and J. Harris, 'Restoration ecology: repairing the Earth's ecosystems in the new millennium' (2001) 9(20) *Restoration Ecology* 239–246.
68 See, e.g., Hobbs et al.
69 R. Hobbs et al., 'Defining novel ecosystems' in R. Hobbs et al. (eds), *Novel Ecosystems: Intervening in the New Ecological World Order* (Willey-Blackwell 2013) 58.

The roots of the novel ecosystem concept have emerged from what is known as 'interventionist ecology'.[70] This branch of ecology can help to clarify a more implicit feature in the above definition of novel ecosystems, which is that the concept can assist planners to move an ecosystem towards more desirable states, or assist it away from undesirable states.[71] That is, the recovery of an ecosystem can be engineered so it supports ecologically, socially or economically important dimensions of the habitat in concern. This element of choice reinforces two ideas in the concept of novel ecosystems. First, the definition gives the impression that novel ecosystems can be used to explain why historical authenticity is not necessary to the recovery of an ecosystem. Second, if the creation of a novel ecosystem is to meet recovery needs then a range of interest groups will need to articulate what they see as desirable features compared with what was historically significant in the ecosystem under consideration.

The same criticisms that are applied to the 'new conservationist' approach to nature can be applied to 'novel ecosystems'. The new conservationists argue that some conservation efforts are impossible or impracticable, given the state of the natural environment, and as such we should do what is desirable in terms of designing approaches to conservation that support or build on available ecosystem services.[72] However, criticisms of the new conservationists are often aimed at the extreme ways in which the ideology can be used to support short-term initiatives that do not leave any long-term ecological benefits for future generations but are rather aimed at increasing available ecosystem services for certain groups of people. That is, the approach is often criticised for its potential to support consumerism. The concept of novel ecosystems potentially faces similar problems in terms of whether it should or shouldn't supplement ecological restoration. Novel ecosystems could be useful if created or allowed to continue where rehabilitation is the only possible option for an ecosystem. However, where it is substituted in preference to returning an ecosystem to an historical trajectory, then novel ecosystems can have negative implications for the goals of ecological restoration.[73]

Rewilding

The concept of rewilding was introduced in the 1980s by the conservation biologist Michael Soule who created the Wildlands Project.[74] The project aimed at

70 R. Hobbs et al., 'Intervention ecology: applying ecological science in the twenty-first century' (2011) 61 *BioScience* 442–450.

71 C. Murcia, J. Aronson, G. Kattan, D. Moreno-Mateos, K. Dixon and D. Simberloff, 'A critique of the "novel ecosystem" concept' (2014) 29(1) *Trends in Ecology & Evolution* 548–553.

72 See, e.g., M. Marview, 'New conservation is true conservation' (2013) 28(1) *Conservation Biology* 1–3.

73 See also Chapter 11 where we argue for the negative implications of using novel ecosystems.

74 M. Soule and R. Noss, 'Rewilding and biodiversity: complementary goals for continental conservation' (1998) *Fall Wild Earth* 22; see <www.wildlandsnetwork.org/> for more on this project.

testing the idea that reintroducing a certain keystone species or large predator could help restore biodiversity and lost ecosystem functioning in very large and protected habitats. Rewilding is described as restoring 'missing or dysfunctional ecological processes and ecosystem function via a process of species reintroduction'.[75] Although rewilding is more popularly associated with the introduction of a keystone species into a habitat, the approach to restoration is seen as being built on three pillars:

1 large and protected reserves;
2 connectivity between habitats, ecosystems or regions;
3 species reintroduction to restore ecosystem functioning.[76]

For instance, there is research suggesting that the extinction of giant tortoises in the Mascarene Islands located in the Western Indian Ocean has contributed to the loss of native biota and the settling of significant numbers of invasive species on the island.[77] Scientists introduced a non-native tortoise into one of the three islands which helped with the recovery of the seed dispersal capacities of native plants.[78] It has also been hypothesised that the tortoise will also help with the recovery of native plants and controlling of invasive species.[79] The introduced tortoise interacted with native plants, which then helped with the restoration of an important ecosystem function on the islands.

Rewilding generally requires an analysis that goes through the following steps in order to identify the specific restoration goal and the approach that needs to be taken in terms of keystone species:

• identification of the issue of conservation concern (e.g. overgrazing by large ungulates);
• identification of the missing ecological processes (e.g. predation);
• identification of the functional characteristics required to restore the missing processes (e.g. large apex consumers);
• selection and reintroduction of the most suitable species to restore the missing or dysfunctional processes.[80]

The important issue in this process is that the historical information about the site is only necessary to the extent that it helps with decisions that need to be

75 C. Sandom, C. Donlan, J.-C. Svenning and D. Hansen, 'Rewilding' in D. Macdonald and K. Willis (eds), *Key Topics in Conservation Biology 2* (Oxford University Press 2013) 430–451.
76 Ibid., 431.
77 See C. Griffiths et al., 'The use of extant non-indigenous tortoises as a restoration tool to replace extinct ecosystem engineers' (2010) 18 *Restoration Ecology* 1–7.
78 C. Griffiths et al., 'Resurrecting extinct interactions with extant substitutes' (2011) 21 *Current Biology* 762–765.
79 Griffiths et al., 'The use of extant non-indigenous tortoises' 1–7.
80 Sandom et al., 'Rewilding' 430–451.

made when going through the above four steps. That is, historical information is used in terms of helping planners and decision-makers to learn about what needs to be done or if something is possible. Most often the historical information that is needed is likely to be sourced from a time when the habitat in question had not been influenced by anthropogenic factors. The importance of going through these steps – in terms of restoration – is that rewilding assists large and connected areas with the recovery of native biodiversity but without considering human connections within an ecosystem. The assumption is that by restoring ecosystem functions back into an area the need for human management of the area will also disappear.

Rewilding has been described as ecological restoration.[81] It is perhaps more accurate to describe it as an ecological healing process for ecosystems that are in large and connected protected areas. This is because the historical information that is taken into account is aimed simply at identifying dysfunction, interactions amongst species and the relative health of the habitats being studied. The historical information does not serve as a model that restorationists then seek to replicate. It is, however, a process that requires active involvement from human beings to assist the ecosystem to recover. The difficulty lies in how this judgement is exercised in that rewilding presumes that the human involvement in the identification of the problems to be solved and the solutions to be applied will lead to desirable outcomes for the habitats under consideration. While rewilding may achieve some of the same goals as ecological restoration, it does not take into consideration the human relationship factor that is fundamental to ecological restoration.

Conclusion

This chapter has sought to define and differentiate some important conceptual approaches to restoration. This is necessary to do because environmental law rarely defines the concept of restoration even when it seeks to deploy it for recovery efforts. One thing is clear from this chapter, which is that restoration has to be more than remediation, rehabilitation or mitigation in order to support recovery efforts. These terms are useful as part of the legal lexicon because of the ease with which dispute resolution processes can account for the costs involved when asking parties to carry them out. The value of definitions – particularly for international legal instruments – is that they enable a diverse range of actors to share information, technologies and approaches that are effective in supporting recovery efforts from around the world. This is one reason why the historical trajectory approach taken by the SER to ecological restoration has been discussed in some detail.

As was made evident from the above discussion a range of approaches to what is native or historically authentic can be supported by ecological restoration. Increasingly however, it seems that the outcomes of projects are being measured against standards that reflect the values of ecological restoration for specific communities including social and cultural values.

81 Ibid.

Concepts like novel ecosystems and rewilding continue to ignite a great deal of interest in what constitutes an appropriate restoration project. While there is no evidence to suggest that the concepts of novel ecosystems and rewilding are not rigorous enough to support recovery efforts, these concepts do support values that have to be made explicit when they are discussed as part of restoration planning efforts. The kind of engineering of ecosystem functioning that they support – in very different ways – has to be recognised and appreciated before they are adopted and used.

3 Why should we restore?

Introduction

The goals of restoration depend to a large degree on the motivation driving a project. There is, however, no one paradigm or context for how restoration goals are designed or set. Goals need to be developed appropriately for each project, relative to the scope and reasons for the restoration effort.[1] Depending on the habitat under consideration and the nature of the degradation, states may support restoration for a variety of reasons including the following:

a   building ecological resilience or the health of ecosystem processes;
b   increasing or maintaining available ecosystem services to support human welfare goals and those of the biosphere more generally;
c   supporting moral goals, including justice for the future and current generations in terms of their access to the natural world;
d   achieving the goals of environmental human rights, like the right to a clean environment;
e   removing fear, anxiety and guilt over the degradation of the natural environment;
f   conserving and preserving plant and animal species;
g   adapting to climate change.

Some of these reasons for engaging in restoration will be discussed in this chapter. In Chapters 1 and 2, this book discussed the importance of restoration for international environmental law and the meaning of ecological restoration. In this chapter we explore why it is necessary to support restoration activities, broadly speaking. This question is relevant as it helps with an inquiry into a more common question, which is whether current efforts in conservation and protection of the natural environment are sufficient to help with the recovery of ecosystems. Rehabilitation and remediation, as discussed in Chapter 2, are not always adequate strategies as they do not restore an ecosystem to a reference system such as its

---

1 J. Ehrenfeld, 'Defining the limits of restoration: the need for realistic goals' (2000) 8(1) *Restoration Ecology* 2–9.

historical trajectory. International law can encourage and require that states make decisions to assist with the recovery of ecosystems but sometimes states need to be persuaded to realise that what is immediately politically, social or economically inconvenient can be immensely beneficial for a variety of other important reasons. The cost and amount of time required for carrying out significant ecological restoration projects can be high and may not be convenient given the short political cycles that many democracies support.

This chapter starts with the assumption that the question of why we should restore deserves a major treatise in itself.[2] While this chapter will not be a major treatise, it puts forward and discusses three arguments that suggest why ecological restoration is important in strengthening and supporting the development of international environmental law and governance initiatives and outcomes. The first suggests that ecological restoration is essential for the survival of planet Earth given that human beings have potentially stopped symbiotically living on it. The second argues that ecological restoration will be essential for ensuring that the availability of ecosystem services and benefits do not become a direct or indirect source of global conflict. Lastly, this chapter makes the moral case for restoration in that it is an important way of achieving environmental and ecological justice whether it is for the current or future generations. It must be noted that these three arguments – namely survivability, ecosystem services and environmental justice – are intentionally focused on ecological restoration ultimately benefiting human beings. That is because the three arguments allow this chapter to examine important social, economic and cultural roots of the problems that underline the challenging human-nature relationship and the potential of ecological restoration to possibly influence it.

Symbiosis and the Anthropocene epoch

It is arguable that the current relationship that human beings have with their natural environment is not a symbiotic one, but something that could more easily be described as exploitative and even parasitic. Symbiosis has been defined in many ways but here it is described as a persistent and mutual beneficial relationship between human beings and the natural environment.[3] Symbiosis generally does not preclude the possibility of conflict amongst partners but rather it assumes that the resolution of problems and issues is part of the continued mutually beneficial relationship.[4] The metaphor of symbiosis describes the human-nature relationship during what is known as the Holocene epoch. The Holocene is a geological term describing the epoch during which planet Earth sustained a natural environment supportive of human societies.[5] In this context, symbiosis would not preclude the

---

2 E.g. Ehrenfeld, 'Defining the limits of restoration'; and S. Allison, *Ecological Restoration and Environmental Change: Renewing Damaged Ecosystems* (Routledge 2012) especially Chapter 3.
3 A. Douglas, *The Symbiotic Habit* (Princeton University Press 2010) 4.
4 Ibid., Chapter 1.
5 W. Steffen et al., 'Planetary boundaries: guiding human development on a changing planet' (2015) 347(6223) *Science* 1259855-1.

possibility of growth, consumption and development. However, the question that has been continuously asked since the 1970s is whether there are limits to growth and consumption. In this context, can the size of the human population and human consumption be reconciled with the natural world or have humans become exploitative and even parasitic on the natural world?

In 1972 the Club of Rome commissioned a report that was titled *The Limits to Growth* and pointed to the dangers of the nature of the relationship that humans have with the natural world.[6] One of the conclusions of this report was that:

> If the present growth trends in world population, industrialisation, pollution, food production, and resource depletion continue unchanged, the limits to growth on this planet will be reached sometime within the next one hundred years. The most probable result will be a rather sudden and uncontrollable decline in both population and industrial capacity.[7]

As this paragraph highlights, the nature of our relationship with nature has the potential to create testing conditions on planet Earth, which could then easily generate the 'sudden and uncontrollable decline' in the conditions that sustain the biosphere.

It would be easy to argue that little has changed since 1972. In 2004 the authors of the initial report completed a 30-year update that went further in terms of restating their dire predictions.[8] One of the comments in the update points to the significance of restoration in terms of the human-nature relationship when the authors stated 'we are drawing on the world's resources faster than they can be restored, and we are releasing wastes and pollutants faster than the Earth can absorb them or render then harmless'.[9] In a symbiotic relationship the human population would only consume to the extent that the natural world was capable of recovering from the harm that is done to it. In the 1980s the dumping of wastes at sea was commonly done on the assumption that the seas had unlimited carrying capacity.[10] Where ecosystems have stopped functioning when compared with their more natural baselines or reference point then ecological restoration would be an effective way of re-establishing symbiosis. The research on symbiosis does not presuppose that one partner in the relationship cannot be dominant, and it has been noted that in a lot of instances in the biological world this is actually

---

6 On the continued relevance of this work see K. Higgs, *Collision Course: Endless Growth on a Finite Planet* (MIT Press 2016).

7 D. Meadows, D. Meadows, J. Randers and W. Behrens III, *The Limits to Growth* (Universe Books 1972). This report used 'system dynamics theory and a computer model' to generate findings about whether there were limits to how we lived on this planet.

8 See D. Meadows, J. Randers and D. Meadows, *A Synopsis. Limits to Growth: The 30-Year Update* (Chelsea Green 2004).

9 Ibid., 3.

10 A. Akhtar-Khavari, 'Beyond compliance and the sea disposal of dredged and excavated materials' (1998) 1 *Maritime Studies* 23–33.

the case.[11] This is often necessary to resolve conflicts amongst symbionts. As such, ecological restoration is potentially a solution to the challenging conditions in which consumption results in the degradation of an ecosystem. The following discussion describes the concepts of planetary boundaries and the Anthropocene to illustrate why without ecological restoration the human population risks becoming parasites and moving away from what has so far been a mutually beneficial partnership with the natural world. It reinforces the argument as to why ecological restoration is needed in the Anthropocene.

Planetary boundaries and the limits to growth

The comments made by the update to *The Limits to Growth* report in 2004 suggested that human beings are drawing on the natural world faster than it can recover. Beyond the work done by the authors of this report, Earth systems scientists have developed the idea of planetary boundaries to examine what limits exist on the functioning of Earth systems. Although Earth systems scientists do not use terms like symbiosis and parasitic, the upshot of their work is that it provides evidence for how we can describe our relationship to the natural world.

A salient but critical question underpinning discussions of the planetary boundaries framework is whether the human population is placing too much pressure on Earth systems.[12] The idea of planetary boundaries refers to scientifically determined baselines that state the point at which anthropogenic activities or 'perturbations' can 'destabilise' the Earth systems. If these boundaries are crossed the human population will influence the 'safe operating space for global societal development'.[13] The idea of planetary boundaries is based on the assumption that the Earth systems are resilient to certain amounts of anthropogenic influences and 'perturbations', but at a particular point they begin to destabilise.[14] Scholars have identified nine Earth systems with planetary boundaries:

1   biosphere integrity;
2   climate change;
3   novel entities;
4   stratospheric ozone depletion;
5   atmospheric aerosol loading;
6   ocean acidification;
7   freshwater use;
8   land-system change; and
9   biogeochemical flows.

11 A. Douglas, *The Symbiotic Habit* (Princeton University Press 2010) Chapter 3.
12 Steffen et al., 'Planetary boundaries' 1259855–1.
13 Ibid.
14 Steffen et al. define resilience as the ability of Earth systems 'to persist in a Holocene-like state in the face of increasing human pressures and shocks' (ibid.).

Based on scientific research, for each of these planetary boundaries a certain threshold has been identified which, if exceeded as a result of human activities, would destabilise the relevant Earth systems. Exceeding planetary boundaries will have an impact on the welfare and prosperity of human beings as well as the rest of the biosphere. For instance, increases in the use of fertilisers in agriculture will elevate the availability of phosphorous in regional watersheds. This can then destabilise the biogeochemical flows that support ecosystems in areas with increased agricultural activities.[15]

Being able to identify a threshold beyond which human beings will fail to live symbiotically is significant. The idea of planetary boundaries provides scientific evidence for the limits to growth should we choose to avoid destabilising various Earth systems. Recent scholarly activities suggest that a breach of the planetary boundaries can change what has been the mild and supportive conditions of the Holocene epoch for the growth and healthy functioning of the human population and create the conditions for planet Earth to actually leave the Holocene epoch.[16] Is this the same as describing a situation whereby we no longer live in symbiosis but have become parasites on planet Earth? The concept of the Anthropocene is discussed next to shed more light on this distinction between our symbiotic and parasitic relationship to nature.

The Anthropocene epoch and planetary boundaries

In the past decade the idea of the Anthropocene epoch has been used to describe the general impact of human beings as a geological force on planet Earth.[17] Paul Crutzen in 2000 had suggested the term Anthropocene as a new geological epoch, describing the profound impact of human activities on planet Earth.[18] In the geological sciences, the Anthropocene epoch describes the variety of stratigraphical signals that human beings leave behind by building cities, dams, roads and so on. It has been suggested that the Anthropocene epoch may have possibly started from the industrial age or maybe even as far back as when humans cleared trees, cultivated land for the purposes of agriculture, or domesticated animals.[19] Some scientists have argued that from the 1850s onwards the levels of carbon dioxide in the atmosphere, as a result of increased fossil fuel usage in urbanised areas, made significant differences in terms of the human impact on Earth systems,

---

15 Steffen et al., 'Planetary boundaries' 1259855–1, 1259855–6.
16 A. Zlasiewic, M. Williams, W. Steffen and P. Crutzen, 'The new world of the Anthropocene' (2010) 44 *Environmental Science Technology* 2228–2231; the authors examine the scale of human modifications of planet Earth to conclude that we are in the Anthropocene.
17 F. Biermann et al., 'Down to Earth: contextualizing the Anthropocene' (2015) *Global Environmental Change*, <http://dx.doi.org/10.1016/j.gloenvcha.2015.11.004>.
18 P. Crutzen, 'Geology of mankind' (2002) 415 *Nature* 23.
19 Most commentary points to the start of the industrial age as the harbinger of the Anthropocene; e.g. see Biermann et al., 'Down to Earth'.

which explains the onset of the Anthropocene epoch. In 2016, the International Commission of Stratigraphy, which has set up a Working Group on the Anthropocene, is expected to decide whether planet Earth has officially entered into the Anthropocene epoch.[20]

During the Holocene epoch – the epoch we are currently leaving – and other geological periods, major changes to the geological conditions on planet Earth were through events like shifts in plate tectonics, meteoric impacts or significant melting of ice sheets. Human species in the Anthropocene epoch are significant in geological and morphological ways not just because of the size of the human population but in terms of the anthropogenic impact we have on Earth's subsystems. These impacts have come from increased carbon and nitrogen oxide levels in the atmosphere, biodiversity loss or a range of other conditions that place pressure on one of the Earth's subsystems. In terms of changes to the biosphere, for instance, humans are seen as being as powerful, through the release of carbon and nitrogen, if not more so than the forces of climate itself. The geologists look for actual human footprint whereas Earth system scientists look at the changes occurring in the functioning of planetary systems. The concept of the Anthropocene epoch creates a useful context for the discussion of planetary boundaries. It enables a scientific exploration of the conditions that are likely to lead human beings to have a significant geological impact on planet Earth.[21]

The significance of the concept of the Anthropocene for scientific fields, like ecology, but also the humanities and social sciences has yet to be adequately researched and discussed,[22] and scholars are only just beginning to explore its relevance for environmental law and justice issues.[23] The idea of the Anthropocene has, however, given Earth system scientists and a range of other scholars the ability to more easily represent what they observe as the 'radical anthropogenic alteration of the planet's natural cycles and systems'.[24] Scholars reviewing planetary boundaries note that already in relation to the Earth systems regulating biogeochemical flows and the biosphere, human beings have contributed to the breach of the safe threshold of these planetary boundaries.[25] Although this breach is only in relation to two out of the nine planetary boundaries the implications of this for ecosystem recovery and health is still likely to be fairly significant.

20 Zlasiewic et al., 'The new world of the Anthropocene'.
21 This point is made most aptly by Zlasiewic et al., 'The new world of the Anthropocene'.
22 For exceptions see V. Galaz, *Global Environmental Governance, Technology, and Politics: The Anthropocene Gap* (Edward Elgar 2014); and J. Baskin, 'Paradigm dressed as epoch: the ideology of the Anthropocene' (2015) 24 *Environmental Values* 9–29.
23 A. Akhtar-Khavari, 'Accessing ecological justice in the Anthropocene epoch!' in P. Keyzer, V. Popovski and C. Sampford (eds), *Access to International Justice* (Routledge 2015) 199–224.
24 See Biermann et al., 'Down to Earth'; and Zalasiewicz et al., 'The new world of the Anthropocene'.
25 Steffen et al., 'Planetary boundaries' 1259855-1, 1259855-6.

Symbiosis and ecological restoration in the Anthropocene: case of megafires

The idea of the Anthropocene poses the question of whether human beings already have left or will leave their geological footprint on planet Earth. Whether this geological footprint is good or bad ecologically is something that is difficult to predict, but if the planetary boundaries are used as a hallmark for what we can expect if human beings continue to dominate the functioning of Earth systems, then we need to think more critically about whether we are failing to symbiotically live with the natural world. A range of scholars have already thought about what the Anthropocene means in governance terms. It is arguable that some of these responses already presume that we live parasitically and that we have to mitigate against both potential future harm to the natural environment and welfare needs of the human population. Galaz for instance has used the term the 'Anthropocene gap',[26] to suggest that in this epoch – meaning the Anthropocene – human beings will increasingly have to deal with governance gaps because of the uncertainty and potential state of chaos that the 'alteration of the planet's natural cycles and systems' can create.[27] Another group of scholars have suggested that in the Anthropocene we have to conserve nature but only for human survival rather than for nature's sake.[28] In other words, where the Holocene created clement conditions for our survival, the Anthropocene now requires humans to focus only on what is necessary to ensure human survival and prosperity. Others have, however, argued that the Anthropocene requires that we make better use of established concepts, such as sustainability, to govern and limit growth.[29]

The governance challenges emerging from being in the Anthropocene require that states also make choices as to whether human beings respond to harm as mutual partners or as potential parasites. For instance, the term 'megafires' has been increasingly used to describe very large and severe fires that have devastated large ecosystems, such as the 2015–2016 fires in Tasmania, Australia.[30] Fire has always had an important role to play in nature's own management of its affairs, but as scholars have noted, 'Fire in the Anthropocene has … shifted dramatically, from an ecological phenomenon driven by natural factors to a spatially and temporally variable hazard strongly associated with humans.'[31] Megafires are now very common around the world. What makes the idea or the potential of megafires interesting is that it puts pressure on our governance structures to adapt in order to protect potentially affected human populations and ecosystems from the

26 Galaz, *Global Environmental Governance, Technology, and Politics.*
27 Ibid.
28 See the new conservationists represented by the following article: P. Kareiva, M. Marvier and R. Lalasz, 'Conservation in the Anthropocene: beyond solitude and fragility' (2012) *The Breakthrough*, <http://thebreakthrough.org/index.php/journal/past-issues/issue-2/conservation-in-the-anthropocene>.
29 L. Kotze, 'Rethinking global environmental law and governance in the Anthropocene' (2014) 32(2) *Journal of Energy and Natural Resources Law* 121–156.
30 See <www.abc.net.au/radionational/programs/scienceshow/urban-areas-threatened-by-megafires-in-early-days-of-climate-ch/6939756>.
31 Biermann et al., 'Down to Earth'.

immense magnitude of the fires. The problem with megafires is that they have severe impacts not just on the burnt-down trees and ecosystems, but also for ecosystems in other parts of the world, whether it be direct or indirect.[32] One issue with fires, including megafires, is that research suggests that over 90 per cent of all fires are human-lit, suggesting that human beings are likely to be responsible for lighting megafires. Further, land-use policies and other kinds of human activities in and around forests can account for these fires starting and spreading.[33] In other words megafires potentially provide us with visible signs of the Anthropocene in the same way as meteoric damage to planet Earth would leave geological footprints.

Megafires can help illustrate the potential significance of ecological restoration as an intentional and obvious strategy for human beings to engage symbiotically with the natural environment. Although a variety of regulatory responses can be adopted to stop megafires from happening in the first place, particularly by prohibiting human occupation of pristine forests and protected habitats, they are not always going to be effective. Ecological restoration as a strategy can in turn contribute significantly to our mutual partnership with nature by, for example:

- Restoring greater connectivity between important landscapes to ensure that ecological processes that can help prevent megafires are not lost.[34]
- Restoring well-established forests to an ecologically resilient historical trajectory. This can include removing dry materials that have accumulated in the forest or removing trees to increase the space in between more established trees. These simple restoration strategies seek to reverse what are sometimes caused by established land-use practices and fire-exclusion strategies that could now be the cause of megafires.[35]

These two ideas for avoiding or reducing the chances of megafires illustrate how in the Anthropocene a symbiotic rather than a parasitic relationship with nature will require that we turn to ecological restoration for strategies and goals. Megafires are a useful way of illustrating the importance of restoration for reducing the potentially adverse implications of being in the Anthropocene. Restoration is not only one of very few effective solutions to megafires but it is a proactive solution to what Galaz referred to as the Anthropocene gap. Ecological restoration can

32 See J. McConnell et al., '20th-century industrial black carbon emissions altered Arctic climate forcing' (2007) 17 *Science* 1381–1384; and Biermann et al., 'Down to Earth'.
33 Biermann et al., 'Down to Earth', referring to O. Dube, 'Challenges of wildland fire management in Botswana: towards a community inclusive fire management approach' (2013) 1 *Weather Climate Extremes* 26–41.
34 On the importance of where to restore for landscape connectivity, see B. McRae et al., 'Where to restore ecological connectivity? Detecting barriers and quantifying restoration benefits' (2012) 7(12) *PLOS One* e52604.
35 An example of this strategy is the four forest restoration initiatives in northern Arizona which involve significantly thinning the forest; see <http://4fri.org/>.

assist states to continue functioning by living symbiotically with nature or – and both possibilities seem to be on the table – to create options for them to reverse the parasitic and exploitative relationship that they have potentially ushered in through the Anthropocene.

Ecosystem services

The discussion above described how anthropogenic pressures that are put on Earth systems are likely to continue having a destabilising effect on a planetary scale. A consequence of the exploitative use of the natural world or a breach of planetary boundaries is that harm to functioning ecosystems will in turn reduce available services and benefits for human welfare and the biosphere more generally. The availability of fresh water, for instance, will continue to be a major source of conflict internationally.[36] A range of ecosystem services are needed to make fresh water available for human consumption. In the last decade, the concept of ecosystem services has become popular and is being increasingly used to better understand and to discuss the value of ecosystems for human welfare.[37] The concept was further popularised in the work of the *Millennium Ecosystem Assessment* (MA), a report commissioned by the then Secretary-General of the United Nations (UN).[38] The MA was completed in 2005, involving around 1360 experts worldwide, and concluded that 15 of the 24 ecosystems investigated were in decline and could no longer adequately support services essential for meeting the Millennium Development Goals of the UN and for the support of human beings more generally. A vast body of scholarship now exists on the concept of ecosystem services and the way in which it is drawn upon by a range of decision-makers and scholars to measure, value and understand the benefits we obtain from these services.[39]

Ecosystem services are financially valuable as illustrated by a study estimating that ecosystem services in 2011 contributed upwards of around US$125 trillion per year either directly or indirectly to human welfare and well-being.[40] The significance of ecosystem services is even more pronounced when viewed as a loss. A decline in services like pollination and pest control, as well as water and nutrient cycles, may reduce global food production by around 25 per cent by the year

---

36 See M. Klare, *Resource Wars: The New Landscape of Global Conflict* (Owl Books 2001).

37 For an influential and early book on this subject see, G. Daily, *Nature's Services: Societal Dependence on Natural Ecosystems* (4th edn, Island Press 1997).

38 Millennium Ecosystem Assessment, *A Framework for Assessment* (2005); see also the following United Nations Environment Programme (UNEP) website for more information on the MA: <www.unep.org/maweb/en/about.aspx>.

39 On this see the paper by B. Fisher, K. Turner and P. Morling, 'Defining and classifying ecosystem services for decision making' (2009) 68 *Ecological Economics* 643–653, showing on p. 644 the very significant exponential rise in the use of the term ecosystem services between 1998 and 2007.

40 R. Costanza et al., 'Changes in the global value of ecosystem services' (2014) 26 *Global Environmental Change* 152–158.

2050.[41] More critically, however, are the suggestions of an estimate that over 75 per cent of the ecosystems are already degraded, reducing their ability to support ecosystem services of some kind.[42] Some of these services, such as those that play a critical role in regulating elements in the biosphere (e.g. climate systems), are invaluable because of the high costs of reproducing them and the dangers in not having them.

As part of efforts to develop protective juridical regimes, countries are increasingly providing for rights such as those to a clean and healthy environment, or to water and sanitation.[43] Governments are thereby obligated to protect ecosystems in order to ensure that achieving such rights remains viable.[44] In addition to rights-based frameworks, states are legislating domestically to preserve and restore ecosystems with a view to ensuring that ecosystems continue to provide the services that benefit communities.[45] Increasingly, it is being recognised that ecological restoration can play a significant role in the holistic recovery of ecosystems and in turn assist in achieving human welfare needs and goals through the recovery of lost ecosystem services. For instance, soil formation is essential as an ecosystem service for agriculture, which is necessary for food production. Without ecological restoration to support the recovery of a range of services and benefits coming from ecosystems, states would put even more pressures on the international system when negotiating globally for access to scarce resources.[46] This discussion presumes that a greater availability of ecosystem services will in the long term reduce political and social pressures on individual states and global systems. It is common to argue that ecosystem services are a goal for restoration activities. This discussion takes that argument further and suggests that ecological restoration is a valuable goal internationally in ensuring that a range of ecosystem services continue to provide for human welfare.

41 C. Nellemann and E. Corcoran (eds), *Dead Planet, Living Planet: Biodiversity and Ecosystem Restoration for Sustainable Development – A Rapid Response Assessment* (United Nations Environment Programme 2010); referred to by Allison, *Ecological Restoration and Environmental Change* 51.

42 Nellemann and Corcoran (eds), *Dead Planet, Living Planet* 5.

43 For a visual presentation of the content of environmental human rights that countries have adopted see: <http://envirorightsmap.org/>.

44 For studies examining the role of law in protecting ecosystem services because they benefit human beings, see E. Blanco and J. Razzaque, 'Ecosystem services and human well-being in a globalized world: assessing the role of law' (2009) 31 *Human Rights Quarterly* 692–720; see also their book on the same subject: *Globalisation and Natural Resources Law: Challenges, Key Issues and Perspectives* (Edward Elgar 2011).

45 See K. Mertens, A. Cliquet and B. Vanheusden, 'Ecosystem services: what's in it for a lawyer?' (2012) 21(1) *European Environmental Law Review* 31–40, for more on this subject.

46 Aronson et al. have written that 'restoration of natural capital is the most direct and effective remedy for redressing the debilitating socioeconomic and political effects of its scarcity' (J. Aronson et al., 'Restoring natural capital: definitions and rationale' in J. Aronson, S. Milton and J. Blignaut (eds), *Restoring Natural Capital: Science, Business, and Practice* (Island Press 2007) 3–9, 3).

The concept of ecosystem services

The concept of ecosystem services has a long history.[47] Central to the idea of ecosystem services is the notion that ecological processes in an ecosystem support human welfare in a way that humans are incapable of doing without ecosystem support, or where the human performance of the services would be too costly. The concept has become important because it has been argued that the benefits of ecosystems for human welfare are not adequately understood or recognised by markets.[48] As a result, ecosystem services have been inadequately considered when making policy decisions. Increasingly, the concept and language of ecosystem services have gained popularity in a range of decision-making environments because this enables social and ecological factors to be integrated in decisions about human welfare.[49] Some have referred to ecosystem services as a 'communication device' in order to contrast it with other economic tools that are used to value nature.[50] The idea of ecosystem services also provides the language for discussions about the present generational footprint, in terms relative to the benefits that future generations can also derive from the same service provided by nature. Ecosystem services – in short – set up a user-based structure for recognising and establishing the means to engage with the range of human-welfare questions that arise from short and long-term benefits provided by ecosystems.[51] These benefits are diverse, highly significant and can often go unrecognised. Some examples include coastal dunes controlling storm surges and wetlands of various kinds helping to mitigate against floods. The services from a single ecosystem can be diverse; a forest can provide timber, fruit and other forest products, but can also assist with carbon sequestration, water conservation, flood control and erosion.[52]

One of the earliest common expressions of the idea of ecosystem services is provided by Daily, who in 1997 defined it as the 'conditions and processes through which natural ecosystems, and the species that are part of them, help

---

47 See H. Mooney, P. Ehrlich and G. Daily, 'Ecosystem services: a fragmentary history' in G. Daily (ed) *Nature's Services: Societal Dependence on Natural Ecosystems* (Island Press 1997) 11–19.

48 For the original version of this argument, see R. Costanza and C. Folke, 'Valuing ecosystem services with efficiency, fairness and sustainability as goals' in G. Daily (ed) *Nature's Services: Societal Dependence on Natural Ecosystems* (Island Press 1997) 49–68.

49 In this regard see the recent report commissioned for an Australian Department on the use and benefits of the concept of ecosystem services in public policy. Australia 21 Limited, *Discussion Paper on Ecosystem Services for the Department of Agriculture, Fisheries and Forestry (Final Report)*, 5 July 2012, <www.australia21.org.au/publica tion-archive/discussion-paper-on-ecosystem-services-for-the-department-of-agricul ture-final-report-for-the-department-of-agriculturefisheries-and-forestry/#. VPRAKk2zXcc>.

50 Australia 21 Limited, *Discussion Paper on Ecosystem Services* 18–19.

51 See J. Salzman, B. Thompson and G. Daily, 'Protecting ecosystem services: science, economics, and law' (2001) 20 *Stanford Environmental Law Journal* 309.

52 See M. Balick and L. Mendelsohn, 'Assessing the economic value of traditional medians from tropical rain forest' (1992) 6 *Conservation Biology* 128–130; and D. Pearce and D. Moran, *The Economic Value of Biodiversity* (Earthscan 1994).

sustain and fulfil human life'.[53] The closely related idea of drawing on the language of markets and economics to sell conservation and preservation goes back even further in the United States where it can be traced to the activism of Gifford Pinchot in the early 1900s.[54] More recently, a report by an influential organisation assisting countries with the evaluation of ecosystem services, The Economics of Ecosystems and Biodiversity (TEEB), defined ecosystem services more simply as 'the direct and indirect contributions of ecosystems to human well-being'.[55] This is more explicit than the earlier and even simpler definition in 2005 by the Millennium Ecosystem Assessment (MA), which referred to ecosystem services as 'the benefits people obtain from ecosystems'.[56]

Much scientific work has been done in terms of quantifying how landscapes, freshwater systems and other natural ecosystems add value to or provide for basic human welfare needs.[57] The varieties of services provided by ecosystems are classified into categories to recognise and accurately specify the kinds of benefits that they provide. In particular scholars often adopt the classification of ecosystem services by the report of the MA to discuss the benefits that ecosystem services provide.[58] The categories or typologies of ecosystem services in the report are: provisioning,[59] regulating,[60] cultural[61] and supporting services.[62] Supporting services underpin other kinds of ecosystem services in that they help make them possible. Of particular significance is the inclusion and recognition of the cultural benefits that ecosystems provide for people and their characterisation as a separate feature of the functioning of ecosystems. This could include, for instance, the

---

53 G. Daily, 'Introduction: what are ecosystem services?' in G. Daily (ed) *Nature's Services: Societal Dependence on Natural Ecosystems* (Island Press 1997) 1–10, 3.

54 J.G. Lewis, 'The Pinchot family and the battle to establish American forestry' (1999) 66(2) *Pennsylvania History* 143–165.

55 TEEB, *The Economics of Ecosystems and Biodiversity: Mainstreaming the Economics of Nature: A Synthesis of the Approach, Conclusions and Recommendations of TEEB* (2010) 33.

56 *Millennium Ecosystem Assessment* (Island Press 2005) 53.

57 For an example of research developing categories for the benefits of freshwater ecosystems see S. Postel and S. Carpenter, *Nature's Services: Societal Dependence of Natural Ecosystems* (Island Press 1997).

58 *Millennium Ecosystem Assessment* 56–57; on the idea of classification of ecosystem services and the use made of the MA ones, see Fisher et al., 'Defining and classifying ecosystem services for decision making'; and A. Nahlik, M. Kentula, S. Fennessy and D. Landers, 'Where is the consensus? A proposed foundation for moving ecosystem service concepts into practice' (2012) 77 *Ecological Economics* 27–35.

59 Provisioning services refer to 'products' that ecosystems help produce. These could include, for instance, food or fresh water.

60 Ecosystem processes are regulated which results in a range of benefits for people. These include, for instance, climate regulation.

61 Ecosystems provide us with a range of non-material benefits that include, for instance, recreation or ecotourism. For a useful critique of the problematic nature of this when it comes to practice see K. Chan, T. Satterfield and J. Goldstein, 'Rethinking ecosystem services to better address and navigate cultural values' (2012) 75 *Ecological Economics* 8–18.

62 Without the supporting services of ecosystems other services would not exist. For instance, soil formation is essential for a range of provisioning services by ecosystems.

beauty of an inland lake that local tourists have become accustomed to enjoying regularly. Returning an ecosystem to its historical trajectory could have cultural implications for some communities around the world. A turn to culture broadens the notion of consumption and utility deployed in the valuation of ecosystem services and enables decision-makers expressly to discuss ecosystems as having value for people and communities.

Ecosystem services and ecological restoration

Scholars tend to agree that the reduced availability of natural capital is a 'limiting factor for human well-being and economic sustainability'.[63] If the services and benefits of certain ecosystems become scarce because of overconsumption, for instance, then conflicts can arise within nations and internationally to secure access to other available resources. Dealing with scarcity could hinge on reducing consumption and waste or more generally conserving what is available. However, restoration in times of scarcity would also be an effective way of supporting the availability of a broad range of ecosystems services. For instance, restoration can increase the provisioning services available through forests by increasing tree counts. This initiative would also assist with carbon sequestration and farming sustainable timber for a range of construction activities.

In this simple sense, restoration efforts are needed globally to avoid the problems that come with scarcity. Ecological restoration is quickly becoming a management strategy for the accumulation and development of ecosystem services because ecological restoration efforts may increase available ecosystem services.[64] A meta-analytical study suggests that the correlation between engaging restoration and being able to observe recovered ecosystem services and functions is fairly solid.[65] Restoration efforts appear to be most successful in increasing available ecosystem services when conducted in degraded ecosystems.[66] Restoration can return around 144 per cent in biodiversity and 125 per cent for ecosystem services in comparison with what is available on degraded lands. Restoration efforts also produced 86 per cent of the biodiversity and 80 per cent of ecosystem services when recovery efforts were compared with what is expected from the historical or reference ecosystems.[67] A different research project analysing over 89 different studies has identified a different set of results. Ecological restoration can only increase biodiversity and ecosystem services by around 44 and 25 per cent respectively.[68] Notwithstanding, researchers suggests that at the right scale 'ecological restoration is

63 Aronson et al., 'Restoring natural capital: definitions and rationale' 3.
64 E.g. Nellemann and Corcoran (eds), *Dead Planet, Living Planet*; and Aronson et al. (eds), *Restoring Natural Capital*.
65 See J. Rey Benayas, A. Newton, A. Diaz and J. Bullock, 'Enhancement of biodiversity and ecosystem services by ecological restoration: a meta-analysis' (2009) 325(5944) *Science* 1121–1124.
66 Ibid., 1122.
67 Ibid.
68 Nelleman and Corcoran (eds), *Dead Planet, Living Planet* 22.

likely to lead to large increases in biodiversity and provision of ecosystem services, offering the potential of a win-win solution in terms of combining biodiversity conservation with socio-economic development objectives'.[69]

The discussions so far have coalesced all restoration activities as equivalent to ecological restoration. What makes ecological restoration, as discussed in Chapter 2, different from other restoration activities is the return of an ecosystem to some historical trajectory. Restoration initiatives can be undertaken to improve soil fertility for intensive agriculture. The argument in this chapter is not that general restoration goals – as opposed to those influenced by the principles of ecological restoration – aimed at securing a particular ecosystem service is inherently bad or even problematic, but that to avoid the global governance challenges that come with scarcity, states have to remain vigilant about the choices that they make and support. For instance, novel ecosystems as described in Chapter 2 are increasingly seen as an alternative and holistic approach to restoration, which can support and develop a wide range of ecosystems services. A novel ecosystem can range from a reforestation programme and extend to other types of commercially oriented restoration initiatives. The idea of restoring and creating a novel ecosystem can be used in support of a wide range of goals if it is assumed that a return to a historical trajectory cannot be achieved or is costly, socially problematic or politically challenging to achieve.[70]

The challenge with novel ecosystems is not just that they can be used by anyone as a commercial restoration initiative, but that states and regulators can become complacent with what would be the natural condition and state of an ecosystem, or its historical trajectory. The additional problem, however, is that sometimes novel ecosystems do not deliver on ecosystem services because of the uncertainty that is generally associated with predicting how ecosystems will respond to being designed by human beings or for human welfare purposes. For instance, the Grain to Green Project in China sought to solve soil erosion by planting non-native trees on agricultural lands rather than shrubs or grass.[71] These unintended consequences of the project are that the non-native trees are now contributing to the loss of native vegetation and reducing available water in some regions.[72] The term 'ecosystem disservices' has emerged as a way of capturing the potential harm that some ecosystem services can do to ecosystems and human beings.[73] Ecosystem disservices can

69 Rey Benayas et al., 'Enhancement of Biodiversity and ecosystem services by ecological restoration' 1123.

70 See, e.g., M. Perring et al., 'Novel urban ecosystems and ecosystem services' in R. Hobbs et al. (eds), *Novel Ecosystems: Intervening in the New Ecological World Order* (John Wiley & Sons 2013) 310–325.

71 See S. Cao et al., 'Impact of China's grain for green project on the landscape of vulnerable arid and semi-arid agricultural regions: a case study in northern Shaanxi province' (2009) 46 *Journal of Applied Ecology* 536–543; and C. Deland and Z. Yuan, *China's Grain for Green Program: A Review of the Largest Ecological Restoration and Rural Development Program in the World* (Springer 2014).

72 Deland and Yuan, *China's Grain for Green Program* Chapter 9.

73 P. Dohren and D. Haase, 'Ecosystem disservices research: a review of the state of the art with a focus on cities' (2015) 52 *Ecological Indicators* 490–497; see below.

be 'effects or side-effects of deliberate manipulation of ecosystems'.[74] There is nothing deterministic about novel ecosystems, but states need to remain aware of why ecological restoration can be a preferred strategy for maintaining and reviving ecosystem services as compared with other restoration initiatives.

In some situations ecological restoration initiatives cannot be used to support desired ecosystem services and benefits that need to be realised by the state. Given that agricultural or 'human-dominated' lands and ecosystems have cleared a vast area of natural habitats, resulting in significant biodiversity losses, it is critical that at least general or targeted restoration initiatives continue to support agriculture and food production. This will ensure that other natural habitats are not increasingly modified for food production or grazing.[75] This is not a general critique of ecological restoration which takes into accounts human activities that define how an ecosystem functions. In fact, research suggests that sometimes rare plants and animals or highly diverse ecosystems emerge because of human activities.[76]

However, simply restoring certain ecosystem services for economic purposes is not always seen as being compatible with supporting the diversity and depth of biodiversity that can exist in an ecosystem. For example a tree plantation can enhance the provisioning services available through an ecosystem but at the cost of reducing diversity that would have otherwise been available had the recovery focused on historical ecosystems. This is potentially problematic in some regions of the world where restoration is viewed entirely as a strategy to support ecosystem services rather than aiming for the recovery of historical ecosystems, its historical trajectory or ones that aim to increase biodiversity. Allison through his analysis of a range of research papers discussing restoration projects has pointed out that Asian restorationists almost exclusively engage in restoration whose goal is the recovery of ecosystem services and functions.[77] This contrasts with, for instance, their counterparts in North American and Oceania who engage in restoration mostly for the purposes of recovering historical ecosystems and also biodiversity.[78] The problem with restoration goals that are focused entirely on ecosystem services is that biodiversity loss, for example, can sometimes have adverse impacts on the actual ecosystem services that are being restored.[79]

The conflict or tensions between the choices that have to be made between restoring ecosystems purely for economic purposes – even if they support human welfare – or returning an ecosystem to a historical trajectory is likely to create challenges for states at the global level. This is because states want to continue,

74 J. Lyytimaki, 'Ecosystem disservices: embrace the catchword' (2015) 12 *Ecosystem Services* 1, 36.

75 Nellemann and Corcoran (eds), *Dead Planet, Living Planet* 17–19.

76 Ehrenfeld, 'Defining the limits of restoration' 7.

77 Allison, *Ecological Restoration and Environmental Change* Chapter 6 in particular.

78 Ibid.

79 For potential failures of restoring for ecosystem services see, e.g., F. Chapin et al., 'Consequences of changing biodiversity' (2000) 405 *Nature* 234–242; B. Bond, 'Trends in the state of nature and their implications of human well-being' (2005) 8 *Ecological Letters* 1218–1234.

like they have been with issues such as climate change, thinking about how to support their growth agendas and the direct and immediate welfare needs of their populations. These tensions, however, are not as stark as they might appear. This is because the concept of ecosystem services – as is argued by a recent discussion paper for an Australian government department – is going through a transition: definitions of ecosystem services are being redefined from 'benefits to people from ecosystems' to ones that define ecosystem services as 'ecological phenomena' and 'benefits' as things that 'flow from services as a result of human inputs'.[80] For instance, ecosystems help with soil fertility, but it is people who sow the seeds and reap the benefits from the process.[81] The involvement of people in ecosystems resulting in benefits to them is the source of the social welfare outcome. In this shift, more emphasis is being placed on the intermediate and final services provided by ecosystems in order to separate them from the benefits that people get through their own interventions. Several ecosystem services, both intermediate and final, could, for example, be involved before clean water is actually drawn from a source. The benefits from an ecosystem service in this case results from people building the necessary infrastructure to utilise clean water. What is important about this shift is that restoring to the historical trajectory of an ecosystem can still enable human interventions – beyond the restoration process – to capitalise on the services for productive ends. Ecological restoration can be used to restore the ecological features of the historical ecosystem while not compromising on the availability of biodiversity and also ecosystem services. By taking out of the equation the required human interventions, it is even possible that alternative and more productive ways of benefitting from the ecosystem in question could be imagined once the ecosystem has been restored.

The benefits of turning to ecological restoration could be that the process also assists in reducing ecosystem disservices. An ecosystem disservice has been described as 'functions or properties of ecosystems that cause effects that are perceived as harmful, unpleasant or unwanted'.[82] One of the important observed ecosystem disservices is the effect that they can have on 'ecosystem structures, processes and/ or the services that they consequently provide'.[83] An example of this kind of ecosystem disservice is increased water consumption by vegetation that is not suited or native to the area. The long-term impacts of ecosystem disservices can create challenges for communities and need to be avoided. This is where ecological restoration becomes necessary and inherently conservative in its approach to how we assist with the recovery of ecosystems. It remains a useful way to avoid conflicts that could arise in how states manage the long-term availability of natural capital.

---

80 Australia 21 Limited, Discussion Paper on Ecosystem Services 6; on this, see also E. Lugo, 'Ecosystem services, the millennium ecosystem assessment, and the conceptual difference between benefits provided by ecosystems and benefits provided by people' (2008) 23(1) *Journal of Land Use and Environmental Law* 243–261.
81 On this distinction and associated analysis, see Fisher et al., 'Defining and classifying ecosystem services for decision making'.
82 Lyytimaki, 'Ecosystem Disservices' 136.
83 Dohren and Haase, 'Ecosystem Disservices Research' 492.

Environmental justice and ecological restoration

The last argument in this chapter will discuss why the stability and clement conditions in the natural world are essential, not just for the survival of human beings – as discussed in the section above dealing with planetary boundaries and the Anthropocene – but for social justice concerns. In particular it argues that ecological restoration is necessary to secure environmental justice for current and future generations of human beings.[84] This argument follows from the discussions of ecosystem services to the extent that the availability or distribution of how nature is used can turn into a socio-environmental problem. The degradation of ecosystems, or ecosystem disservices can be responsible for a range of problems which, if not managed properly, will raise social and economic problems for people and communities. In this section we take the discussion of human welfare further and examine whether people can be victims of environmental harm or injustice. In particular it assesses if ecological restoration can extend or develop the conceptual framework of environmental justice.

Environmental justice has been a concern of academics and social activists for some time and has traditionally served as a conceptual frame for describing and analysing how environmental benefits and harms can create social advantages and disadvantages amongst human populations and communities around the world.[85] Markets have not fairly distributed the costs of environmental harm, making the environment an important social justice issue for the public at large. For instance, poor households most often live near areas with factories, polluting industries, and fewer parks and environmental amenities. This is despite governments taxing everyone at the same levels for environmental pollution and other externalities. Environmental justice concerns have extended to what the environment means; the underlying causes that created communities which are then subjected to environmental externalities or harms; and the demands that justice imposes on markets and the state itself.[86]

A more robust discussion of environmental justice is outside the scope of this book. It is important to note, however, that although the field began with activists in the United States pursuing and fighting for justice for communities who had been exposed to environmental harms, such as hazardous wastes and toxins, there is now a more robust and academically rigorous engagement with the subject. In the past decade, environmental justice issues and concerns have expanded to an extent that scholars have been suggesting that the discipline is at an important

---

84 See D. Schlosberg, *Defining Environmental Justice: Theories, Movements, and Nature* (Oxford University Press 2007); and L. Westra, *Environmental Justice and the Rights of Unborn and Future Generations: Law, Environmental Harm and the Right to Health* (Taylor & Francis 2008). One of the classic texts in this field is the original work by L. Cole and S. Foster, *From the Ground Up: Environmental Racism and the Rise of the Environmental Justice Movement* (NYU Press 2001).

85 See B. Holland, *Allocating the Earth* (Oxford University Press 2014).

86 D. Schlosberg, 'Theorising environmental justice: the expanding sphere of a discourse' (2013) 22(1) *Environmental Politics* 37–55, 38–40.

crossroad.[87] For instance, the ability of multinational companies to create serious environmental damage in developing countries has led to the expansion of the environmental justice frame to include within its scope the socio-environmental concerns that span legal jurisdictions.[88] The expansion of the environmental justice frame is seen as important for it has been argued as having 'the potential to be an integrative and empowering framework for a variety of movements and concerns'.[89]

The question that is more difficult to answer in relation to environmental justice is if addressing environmental conditions themselves is necessary in order to deal with the social justice concerns. It has been argued that concerns with environmental justice initially developed in the United States at a time when environmentalists did not seem to be concerned with the victims of environmental injustice.[90] The work of the Sierra Club has been criticised, for instance, for not connecting social justice concerns with those of conservation in the 1970s.[91]

The tide is turning again and it has been suggested that environmental justice is already on its way to achieving a 'conceptual shift – that a working environment is necessary for justice, and that justice entails creating practices and material flows that do not undermine environmental processes and systems'.[92] Extending environmental justice concerns to include ecological issues is done on the assumption that all living matter is potentially vulnerable when ecosystems fail to function properly and ideally. The term ecological justice recognises that the natural environment has intrinsic value, which needs to be socially, politically and legally recognised.[93] Concerns around ecological justice have, for instance, inspired the rights of the nature movement internationally.[94] This was a necessary counterpart to arguments that suggest that only human beings are capable of being protected, using the justice lens.[95] This concern with the biotic features

87 J. Sze and J. London, 'Environmental justice at the crossroads' (2008) 2(4) *Sociological Compass* 1331–1354.
88 For the argument on how environmental justice has expanded, see G. Walker, 'Globalizing environmental justice: the geography and politics of frame contextualization and evolution' (2009) 9(3) *Global Social Policy* 355–382.
89 Schlosberg, *Defining Environmental Justice*; J. Sze et al., 'Defining and contesting environmental justice: socio-natures and the politics of scale in the Delta' in R. Holifield, M. Porter and G. Walker (eds), *Spaces of Environmental Justice* (Wiley-Blackwell 2010) 219–256.
90 K. Shrader-Frechette, *Environmental Justice: Creating Equality, Reclaiming Democracy* (Oxford University Press 2002) in particular Chapter 1.
91 Ibid., 34.
92 Schlosberg, 'Theorising environmental justice' 37–55, 45–46.
93 E.g. K. Bosselmann, *The Principle of Sustainability: Transforming Law and Governance* (Ashgate 2014).
94 E.g. R. Nash, *The Rights of Nature: A History of Environmental Ethics* (University of Wisconsin Press 1989).
95 See A. Dobson, *Fairness and Futurity: Essays on Environmental Sustainability and Social Justice* (Oxford University Press 1999).

of an ecosystem does not take away from the social justice focus of environmental justice. It just deepens and broadens it so that it is holistically concerned about human beings.

A great deal of scholarship now exists on the idea of climate justice, generally arguing that ecological damage caused by global warming has the potential to harm and devastate people's lives, livelihood and communities around the world.[96] One consequence of the developments in momentum behind climate justice has been the ability of non-governmental communities to shift debates from the scientific and technical issues and focus on the human impacts and vulnerabilities of communities to potential harm.[97] Environmental justice has the potential to encompass not just climate-related issues, but any activity that contributes to the potential breach of planetary boundaries. As such it is a rather larger and ecologically focused conceptual frame than just climate justice itself. Critical to the environmental justice movement has been the momentum that individuals, non-government organisations and other entities have been able to generate to advocate for issues. A critical example is the case that the Urgenda Foundation brought against the State of the Netherlands where the Hague District Court in June of 2015 ordered the government to reduce by at least 25 per cent within five years the carbon emissions in that country.[98] This case is supportive of the idea that the extended conceptual frame of environmental justice discussed here is essential in the Anthropocene.

Ecological restorationists and the practice itself can contribute to deepening the concerns and also approaches that are taken to environmental justice issues. Given the heritage of the environmental justice advocacy movement, it is fair to suggest that it has emerged out of concerns to stop construction activities; to provoke the state into delivering on law reforms and compel them to engage in clean-up activities; and to pressure the state and its registered companies to stop engaging in activities that can directly harm the environment and the people living in them. Restoration ecology can also be potentially reactive in that restoration initiatives and activities start as a response to some kind of degradation. Although ecological restoration aims to restore an ecosystem to its historical trajectory, that does not mean that human cultures cannot be a part or central to the healthy recovery of the ecosystem. Restoration can support to reinstate sustainable cultural practices that require certain ecological processes for them to take place. Environmental justice initiatives would support a return to culturally sustainable practices as well as help ecosystem processes to recover.

96 E.g. H. Shue, *Climate Justice: Vulnerability and Protection* (Oxford University Press 2014); and B. Tokar, *Toward Climate Justice: Perspectives on the Climate Crisis and Social Change* (New Compass Press 2014).

97 J. Agyeman, B. Doppelt and K. Lynn, 'The climate-justice link: communicating risk with low-income and minority audiences' in S. Moser and L. Dilling (eds), *Communicating a Climate for Change: Communicating Climate Change and Facilitating Social Change* (Cambridge University Press 2007) 119–138, 125.

98 For an unofficial but English translation of the judgment see: <http://edigest.elaw. org/sites/default/files/urgenda.pdf>.

Principles and practices of ecological restoration can support development in environmental justice because advocacy in this field cannot always account for long-term environmental degradation. One of the concerns with some deleterious environmental challenges is that they take a long time to manifest. In reality, it could be argued that climate change started with the industrial age and is only beginning to manifest itself and its most severe impacts on populations is yet to emerge. Biodiversity loss can be subtle, gradual and almost undetectable in some instances. Advocacy with respect to biodiversity loss, through the conceptual lens of environmental justice, is likely to focus on short-term and immediate health or subsistence concerns. The trigger for ecological restoration, however, is degradation, damage or the transformation of an ecosystem and can often be used to help with the recovery of ecosystems.

Environmental justice has been a useful conceptual framework that enables social issues and concerns to be addressed. Ecological restoration can support the social justice and moral ambitions of societies in new ways.

Conclusion

This chapter has aimed at examining a simple but complex question, which is why we need to concern ourselves with ecological restoration. This is a different but somewhat connected question to asking about the goals that ecological restoration practitioners should pursue when helping to recover damaged ecosystems. They are somewhat connected because concerns with why we should do something ultimately have to comment on goals that we need to pursue. In answering this question, however, this chapter concerned itself with the institutional issues that would matter to someone who is thinking about international law and global environmental governance. From that perspective this chapter has dealt with the cultural, social and economic dimensions of issues that ecological restoration can contribute something to. It examined: (1) how ecological restoration can help us think about the cultural dimensions of the human species survival on planet Earth; (2) the potential economic benefits of ecological restoration for supporting human welfare and well-being; and (3) enriching social-environmental justice issues with concerns and approaches that ecological restoration can enable us to think about. Given these discussions the rest of this book will explore what role ecological restoration currently plays in international environmental law.

Part II
Law of ecological restoration

4 International soft law and ecological restoration

Restoration as a national activity, particularly forest restoration is not a new historical phenomenon,[1] but ecological restoration as a legal obligation has only become a legal concept in the last 50 years.[2] The next three chapters in this book will describe the evolution of the international legal framework for ecological restoration that began to emerge in the early 1970s with the first multilateral international environmental agreements. This chapter will describe and analyse the international legal context for restoration beginning with five non-binding but multilaterally negotiated documents spanning several decades: the Stockholm Declaration, the World Charter for Nature, the Rio Declaration, The Future We Want and the Sendai Declaration. Based on these declarations and the actions plans that have emerged from a number of international meetings (Stockholm Action Plan, Agenda 21, Johannesburg Implementation Plan and the Sustainable Development Goals), this chapter will identify a number of principles of international environmental law that might inform the implementation of national and international ecological restoration efforts. While this chapter and the following two chapters will identify a number of legal texts that include the term 'restoration' in the operative language, it is worth noting at the outset that the concept of restoration is rarely defined in any of the legal texts.[3]

---

1 Since 1700, Japan has been the site of three major reforestation efforts; see L. Walker and P. Bellingham, *Island Environments in a Changing World* (Cambridge University Press 2012) 201.

2 Ecological restoration as a scientific practice was already being pursued by environmental leaders such as Professor Aldo Leopold at the University of Wisconsin and had been eloquently described as an ethical concept by George Perkins Marsh in *Man and Nature: The Earth as Modified by Human Action* (1868) who wrote the book according to its preface to 'indicate the character and, approximately, the extent of the changes produced by human action in the physical conditions of the globe we inhabit; to point out the dangers of imprudence and the necessity of caution in all operations which, on a large scale, interfere with the spontaneous arrangements of the organic or the inorganic world; to suggest the possibility and the importance of the restoration of disturbed harmonies and the material improvement of waste and exhausted regions ...' (Preface:iii).

3 One of the only treaties to define 'restoration' is the Protocol on Conservation and Sustainable Use of Biological Diversity to the Framework Convention on the Protection and the Sustainable Development of the Carpathians; see below Chapter 5.

Even though there have been no disputes requiring the interpretation of restoration in any treaty or other international document, the international community would benefit from refining its understanding of what constitutes 'restoration' for legal purposes if 'restoration' is to be a meaningful obligation.[4]

The current international movement for ecological restoration, particularly landscape restoration is in part a response to the industrial revolution and the current forces of economic globalisation. The industrial revolution transformed global landscapes. In the name of social progress, old-growth forests were cut with abandon, fishing fleets pillaged the seas and mining industries ravaged the earth. When transboundary interests in exploitation were eventually threatened by over-exploitation, states exercised their sovereignty and responded with a variety of bilateral or regional conservation agreements such as the Fur Seals Convention or the International Convention for the Regulation of Whaling.[5] In spite of the environmental legacy of the industrial revolution, the concept of environmental protection was given no lip service by states when the United Nations became a formal institution.[6] Ecological restoration was not on the minds of high-level policy-makers in part because there was a lack of collective understanding of the extent of the global loss of habitat and species.

The tide of indifference to the environment began to change in the decades after the creation of the United Nations. Over the course of the 1960s, environmental values were mainstreamed into political values culminating in the emergence of new national laws and institutions.[7] Recognising the transboundary nature of a number of environmental issues, particularly airborne pollution issues, states invested substantial political capital to attempt to collectively respond to environmental threats. The product of this political effort was the first modern multilateral document to address global environmental concerns including natural resource loss.[8]

---

4 See Chapter 2 for a discussion of why the term 'restoration' when used as a legal term should be regarded as ecological restoration for the purposes of at least reviving ecosystem functions that would be part of the historical trajectory of ecological development at a given location.

5 Convention between the United States, Great Britain, Japan and Russia providing for the Preservation and Protection of the Fur Seals, 7 July 1911, 37 Stat. 1542; International Convention for the Regulation of Whaling, 2 December 1946, 161 *UNTS* 361.

6 United Nations, Charter of the United Nations, 24 October 1945, 1 *UNTS* XVI.

7 For example, the US National Environmental Protection Act of 1969, 42 USC §§4321–4370h. The US Environmental Protection Agency was created on 2 December 1970.

8 J. McCormick, *The Global Environmental Movement* (2nd edn, John Wiley & Sons 1995) 127: 'The Stockholm conference was the single most influential event in the evolution of the global environmental movement, and of a global environmental consciousness.' Environmental programmes had been part of regional policy responses with, for example, the Organisation of Economic Co-operation and Development launching environmental programmes in the 1960s.

International soft law documents

Stockholm Declaration (1972) and Stockholm Recommendations

> The capacity of the earth to produce vital renewable resources must be maintained and wherever practicable, restored, or improved.
>
> (Principle 3, Stockholm Declaration)

One of the earliest direct references to the concept of 'restoration' is in the Stockholm Declaration concluded at the United Nations Conference on the Human Environment.[9] While non-binding in terms of its legal precedent, the 26 negotiated principles of the Declaration reflect the consensus of 114 states and today provide international evidence of state practice to support the application of several international environmental law principles.[10] As a follow-up to the Biosphere Conference organised by UNESCO in 1968,[11] the Stockholm Conference was the product of four years of preparatory meetings designed to integrate environmental concerns into human decision-making.[12] In 1969, the UN General Assembly directed the Conference 'to serve as a practical means to encourage, and to provide guidelines for, action by Governments and international organizations designed to protect and improve the human environment and to remedy and prevent its impairment'.[13]

Statements peppered throughout the negotiating history covering the convening of the Stockholm Conference support a desire for states to pursue restoration as a priority remedial action. Some of these statements represent extreme optimism about the potential for restoration. For example, in November 1969, the diplomatic representative from Byelorussian Socialist Soviet Republic commented that his country attached 'great importance to the problem of protecting and rehabilitating the environment'.[14] The representative went on to indicate that 'Natural resources

---

9 Declaration of Principles for the Preservation and Enhancement of the Human Environment, Report of the UN Conference on the Human Environment, Stockholm 5–16 June 1972, UN Doc. A/CONF.48/14/Rev.1 (1973), 3, reprinted in (1972) 11 *ILM* 1416 ('Stockholm Declaration').

10 See below on international environmental law principles; see also Ibid., J.A. Beesly, United Nations Representative of Canada, stated after the adoption of the Stockholm Principles that the Principles were 'a first step towards the development of international environmental law'. UN DOC. A/CONF.48/14, 115.

11 UNESCO, Intergovernmental Conference of Experts on the Scientific Basis for Rational Use and Conservation of the Resources of the Biosphere (1968), p. 31, Recommendation 17, <http://unesdoc.unesco.org/images/0001/000172/017269eb.pdf> (Included participation of 63 states plus the Food and Agriculture Organization, World Health Organization, the International Union for the Conservation of Nature, the International Biological Programme, United Nations Development Programme, World Meteorological Organization and the International Labour Organization.)

12 L. Caldwell, *International Environmental Policy: Emergence and Dimensions* (Duke University Press 1984) 47.

13 UNGA Resolution 2581 (XXIV) (1969).

14 UN General Assembly, A-C.2-SR.1276 (November 1969) para. 35.

were unlimited, but they must be protected, used rationally, increased and constantly replenished'.[15] For at least this state representative, restoration seemed to be linked directly to preserving a capacity for resource production.

This idea of 'restoration' and improvement of natural resources found favour with the Preparatory Committee that was formed in advance of the Stockholm Conference. This Committee understood the Draft Declaration to be a document that:

> [W]ould serve to stimulate public opinion and community participation for the protection and betterment of the human environment and, where appropriate, for the restoration of its primitive harmony etc. in the interest of present and future generations. It would also provide guiding principles for Governments in their formulation of policy and set objectives for future international cooperation.[16]

At the Second Preparatory Meeting, state representatives indicated that the Declaration was intended to be formulated as 'a document of universally recognised fundamental principles recommended for action by individuals, states and the international community'.[17] An Intergovernmental Working Group was formed which eventually produced a set of draft principles for the Stockholm Declaration. The original version of Principle 3, the only principle to mention restoration, as proposed by Sweden did not include an explicit reference to restoration but instead read 'The productive basis of renewable resources of the earth such as farmland, forests, crops and fish, which in many cases and places have been threatened or destroyed, must be maintained and enhanced'.[18] The Netherlands proposed the following alternative that, 'Each State shall do its utmost to restore and improve the productive capacity of renewable resources of the earth, such as farmland, forests, crops and fish for the proper supply of future generations with food and other material products.'[19] Brazil, Egypt and Yugoslavia narrowed the proposal by suggesting that states should 'restore, wherever possible, the productive capacity of those renewable resources that have been unnecessarily depleted'.[20] A later version of the text called on states to 'maintain and wherever practicable, restore or improve, the capacity of the earth to produce vital renewable resources'.[21]

15 Ibid.
16 Report of the Preparatory Committee for the United Nations Conference on the Human Environment, First Session, UN Doc. A.CONF.48/PC/6 (10–20 March 1970) p. 19, para. 36.
17 Report of the Preparatory Committee for the United Nations Conference on the Human Environment, Second Session, UN Doc. A/CONF.48/PC/9 (8–19 February 1971) p. 16, para. 30.
18 UN Doc.A/CONF.48/WG.1(II)/CRP.2 (1971) 3.
19 UN Doc.A/CONF.48/PC/WG.1(II)/CRP.5 (1972) 3.
20 UN Doc.A/CONF.48/PC/WG.1(II)/CRP.3/Rev.3 (1972) 3.
21 UN Doc.A/CONF.48/PC/WG.1(II)/CRP.3/Rev.4 (1972) 3.

The final draft by the Intergovernmental Working Group included the following language for Principle 3 providing that states should share a 'common conviction' that 'The capacity of the earth to produce vital renewable resources must be maintained and wherever practicable, restored, or improved.'[22] This language was ultimately adopted by states as one of the principles of the Stockholm Declaration. Arguably, this principle can be read to stand for the proposition that 'restoration' of vital renewable resources that humans need for survival such as food is a common concern for the international community. The final version of Principle 3 included the word 'must' suggesting a recognition by states that investments in restoration of ecosystem services may be essential for long-term human well-being.

While there is no negotiated set of definitions from the 1972 Conference,[23] one can assume that 'renewable resources' referred to those components of nature that states considered to have some inherent economic values to humans. At the time, this would have been tradable commodities such as fish stocks and timber since the focus of the Declaration was squarely on the 'enhancement of the human environment'.[24] This reading is supported by previous proposed versions of the Principle 3 text. This emphasis on restoring resources of value is not surprising given the focus of the Conference on directly acknowledging the interconnectedness between humanity and nature with man as 'creature and molder of his environment'.[25] As the preamble observed the 'improvement of the human environment is a major issue which affects the well-being of peoples and economic development throughout the world'.[26] Restoration as a form of enhancement of the human environment is offered, then, as one antidote for 'massive and irreversible harm to the earthly environment on which our life and well-being depend'.[27]

The Declaration is informative in terms of contemporary attitudes towards restoration, many of which continue to define the field of restoration as understood by national and international policy-makers. Limited to natural resource management, Principle 3 suggests that restoration in 1972 was understood to be an effort intended to serve primarily human production ends. As the Principle also indicates, restoration will only be pursued 'wherever practicable' suggesting that states were aware of both the financial obstacles to reviving certain 'vital renewable resources' as well as the technical challenges of restoring degraded sites. The difference between the term 'restored' and 'improved' in Principle 3 is also worth noting because the 1972 Declaration is one of the only international legal

---

22 United Nations Conference on the Human Environment, A/CONF.48/4 Annex Draft Text of a Preamble and Principles of the Declaration of the Human Environment. The same non-controversial language was included in the second Intergovernmental Working Group reports at UN Doc. A/CONF.48/PC/16 (1972).

23 'Renewable resources' is not defined in the Declaration. 'Natural resources' are defined to include 'air, water, land, flora and fauna and especially representative samples of natural ecosystems' (Stockholm Declaration, Principle 2).

24 Ibid., Preamble.

25 Ibid., Preamble, para. 1.

26 Ibid., Preamble, para. 2.

27 Ibid., Preamble, para. 6.

documents to really make a distinction between these two terms. The use of the term 'improved' may have been included in the text in order to allay state concerns over how much they might be expect to do to achieve a condition of restoration, because the plain use of the term 'restored' in Principle 3 would have set a high bar for state performance at the time the Declaration was negotiated. As defined in the 1971 *Oxford English Dictionary*, restoration referred to 'the act of restoring to a former state or position … or to an unimpaired or perfect condition' and restore meant 'to bring back to the original state … or to a healthy or vigorous state'.[28] Given this definition, it suggests that state negotiators would have used the term 'restored' as a reference to bringing back a resource to some former condition of plentitude or perhaps ecological autonomy. For those resources where this would simply not be possible such as salmon runs in a river whose headwaters had been dammed, the states could instead more realistically seek to create 'improved' resource capacity through programmes such as aquaculture or timber plantations.

Principle 3, like several other Stockholm principles, is drafted in the passive voice. The question is, then, who is responsible for achieving the goals of Principle 3? In the Preamble, states agreed that the 'protection and improvement of the human environment is … the duty of all Governments'.[29] In particular 'Local and national governments will bear the greatest burden for large-scale environmental policy and action within their jurisdictions.'[30] But the Preamble also suggests that restoration triggers 'the acceptance and responsibility by citizens and communities and by enterprises and institutions at every level, all sharing equitably in common efforts'.[31] This Preambular language signals the contributions to be made by both state and non-state actors in 'environmental protection and improvement'. While the state representatives were unlikely to have been contemplating ecological restoration practice as it has subsequently developed in the twenty-first-century, the 1972 language reinforces the significant role for non-state actors in achieving environmental principles of the Stockholm Declaration including restoration of renewable natural resources.[32]

Based on the language from the Stockholm Declaration, states appear to have committed at least to the idea that they may have some obligation to participate in 'restoration' of vital natural commodities as long as such efforts are financially and technically practical. While this hardly reflects a legally cognisable restoration ethos, it does signal the potential start of a larger international dialogue about humanity's role in mediating nature. Having collectively rejected the idea that humanity serves as an overseer of nature, the states who attended the Stockholm Conference understood that humanity depended on nature and to some extent nature depended on humanity. As expressed in the Declaration, humans must 'For

---

28 *Oxford English Dictionary* (Oxford University Press 1971).
29 Stockholm Declaration, Preamble, para. 2.
30 Ibid., Preamble, para. 7.
31 Ibid.
32 See Chapter 9 on non-state actors.

the purpose of attaining freedom in the world of nature ... use knowledge to build, in collaboration with nature, a better environment.'[33]

Principle 3 is the only principle in the Stockholm Declaration to explicitly refer to restoration but other principles provide context for the type of restoration work that is currently being pursued through organisations such as the Global Partnership for Forest Landscape Restoration.[34] For example, Principle 24 offers international cooperation as a levelling mechanism where 'International matters concerning the protection and improvement of the environment should be handled in a cooperative spirit by all countries, big and small, on an equal footing.'[35] States should enter into bilateral and multilateral cooperative arrangement 'to effectively control, prevent, reduce and eliminate adverse environmental effects resulting from activities conducted in all spheres'.[36] States can also act collectively through their active membership in international organisations. The Stockholm Declaration suggests that states should ensure international organisations 'play a coordinated, efficient and dynamic role for the protection and improvement of the environment'.[37]

States understood that 'political will' was essential to achieving the norms embodied in the principles.[38] In the spirit of cooperation and the recognition of the power of 'political will' that can be accessed at a world conference, the states attending the Conference on the Human Environment also negotiated a number of recommendations for an action plan. While there was no specific effort to highlight broader concepts of ecological restoration in the hundred plus recommendations that eventually formed the action plan, states did agree to promote 'soil restoration' by recommending that the UN Secretary General in conjunction with appropriate UN bodies provide capacity training in support of soil 'restoration'.[39]

When the state representatives made the decisions to move forward with the Stockholm Conference, the representative from Sweden observed that 'the preparations for the Conference, the Conference itself, and the follow-up to the Conference' should be considered as parts of one single continuing process'.[40] While this observation was intended to broadly apply to the emerging practice of international environmental policy-making, it can be applied more specifically to emerging attitudes about 'restoration' as part of 'one single continuing process' in

---

33 Stockholm Declaration, Preamble, para. 6.
34 See Chapter 5.
35 Stockholm Declaration, Principle 24.
36 Ibid.
37 Ibid., Principle 25.
38 Stockholm 1972, Brief Summary of the General Debate, para. 64. (Speakers at the Stockholm Conference 'often expressed' the view 'that man possessed the skills to foresee and avert ecological misfortunes and to create a much happier and richer world, but that no positive advances could be made without the political will'.)
39 Recommendations for action at the International Level adopted by the United Nations Conference on the Human Environment 1972, Recommendation 15, <www.unep.org/documents.multilingual/default.asp?DocumentID=97&ArticleID=1506&l=en>.
40 UN General Assembly, A/PV.1834 (December 1969) para. 87.

mediating how state and non-state actors interact with the natural environment and with each other. Ultimately, the Stockholm Declaration's Principle 3 can be regarded as setting the stage for subsequent multilateral negotiations that would recognise restoration as an appropriate policy choice where resource management through conservation was no longer viable.

World Charter for Nature (1982)

> [A]reas degraded by human activities shall be rehabilitated for purposes in accord with their natural potential and compatible with the well-being of affected populations.
>
> (Article 11(e), World Charter for Nature)

After the adoption of the Stockholm Declaration and the creation of the United Nations Environmental Programme, efforts for environmental protection and particularly conservation were mainstreamed. Restoration was also becoming popularised as a last resort option to address degraded or impacted areas. The World Charter for Nature reflected an effort by states and civil society to enhance cooperation on protecting the natural environment. In 1975, the President of Zaire made the proposal for a Charter and requested the International Union for the Conservation of Nature and Natural Resources (IUCN) to draft a Charter for negotiation by the UN General Assembly. The result was a document modelled in part on the Universal Declaration of Human Rights insofar as the Charter assigns 'each person' a duty to act in accordance with the terms of the Charter and 'strive to ensure that the objectives and requirements of the present Charter are met'.[41] One hundred and twelve members of the 154 voting members of the United Nations endorsed the Charter including all of the European Community States and all of the NATO nations except the United States. The Charter is not legally binding today but reflects an important joint political statement and a potential basis for the evolution of customary international law.

The term restoration is never used in the Charter. Instead, the drafters referred to rehabilitation as one of many 'principles of conservation'.[42] Specifically, the drafters urged persons to control activities with impacts on nature and provided that 'areas degraded by human activities shall be rehabilitated for purposes in accord with their natural potential and compatible with the well-being of affected populations'.[43] The language in this section raises some questions about what the states' intent might have been in adopting this section of the Charter. First, there is no indication of whether the 'well-being of affected populations' refers to populations of animals and plants or merely to human communities. It is fair to assume that the term 'populations' is intended to refer to human communities

---

41 World Charter for Nature, United Nations General Assembly, A/RES/37/7 (28 October 1982).
42 Ibid., Preamble.
43 Ibid., Article 11(e).

because the term 'well-being' is not typically used to refer to the non-human environment.

While we can assume an interest in safeguarding human community values, the concept of rehabilitation 'for purposes in accord with their natural potential' in Article 11 remains vague. One possible reading of the term 'rehabilitation' suggests a twenty-first-century concept of ecological restoration where actors re-establish environmental values for a given ecosystem that restore the development of an ecosystem to a historical trajectory. A second interpretation of this language is more pragmatic and focuses instead on what potential there is after human degradation to recover any 'natural potential'. Under this second interpretation, a formerly forested region that is heavily quarried might be appropriately rehabilitated into an area with lakes that may or may not have forest values. The first interpretation captures a progressive understanding of human obligations in relation to environmental interventions. The second interpretation more closely reflects practice at the time the Charter was endorsed as evidenced by national laws and practices such as the United Kingdom National Parks and Access to the Country Act 1949 to transform land made derelict by underground mining operations for potential reuse.[44]

While the World Charter for Nature is still an aspirational document rather than an operational agreement, the Charter was important in establishing general principles focused on conserving the ecological functions of natural systems. In particular, the World Charter formally introduced to the international community an environmental interaction hierarchy with the emphasis on avoiding environmental impacts, minimising environmental impacts, and, only as a final resort, rehabilitating environmental impacts.[45]

Rio Declaration (1992) and Agenda 21

Twenty years after the Stockholm Conference on the Human Environment, states in 1992 convened the United Nations Conference on Environment and Development (nicknamed 'the Earth Summit') to further refine principles that states should apply to 'respect the interests of all and protect the integrity of the global environmental and developmental system'.[46] The final meeting was a public relations success drawing hundreds of heads of state and active participation from a large sector of civil society. During the meeting, two treaties – the UN Framework Convention on Climate Change and the Convention on Biological Diversity (also

---

44 National Parks and Access to the Country Act (1949), Section 89, <www.legislation. gov.uk/ukpga/Geo6/12-13-14/97> (providing that local authorities for land that is derelict or likely to become derelict may 'carry out, for the purpose of reclaiming or improving that land or of enabling it to be brought into use, such works on that land or any other land as appear to them expedient').
45 World Charter for Nature, Article 11(a), (c) and (e).
46 UN Conference on Environment and Development, Rio de Janeiro, Rio Declaration on Environment and Development, 13 June 1992, UN Doc. A/CONF.151/26 (Vol. 1) (1992), reprinted in (1992) 31 *ILM* 874 ('Rio Declaration').

known as the Biodiversity Convention or CBD) – with substantive obligations involving restoration were adopted. The content of these treaties as it relates to restoration will be discussed in subsequent chapters.[47] The other major output of the UN Conference on Environment and Development was the Rio Declaration, negotiated as an extension to the Stockholm Declaration, to reflect over mainstreaming human development into environmental management.

The negotiating history of the Rio Declaration was markedly different than the Stockholm Declaration and bringing together the disparate thinking of the various state participants proved challenging. At an early meeting, 136 separate proposals were made by states with potential principles.[48] At the final Preparatory Committee, there was still no working draft leading to seven representatives from the Global North and seven representatives from the Global South collaborating on a text that was ultimately adopted at the Rio Conference.[49]

Out of the 27 principles that became the 1992 Rio Declaration only Principle 7 specifically mentions restoration:[50]

> States shall cooperate in a spirit of global partnership to conserve, protect and restore the health and integrity of the Earth's ecosystem. In view of the different contributions to global environmental degradation, States have common but differentiated responsibilities. The developed countries acknowledge the responsibility that they bear in the international pursuit to sustainable development in view of the pressures their societies place on the global environment and of the technologies and financial resources they command.

Principle 7 reflects an amalgam of ideas. The first portion of the Principle provides that 'States shall cooperate in a spirit of global partnership to conserve, protect, and restore the health and integrity of the Earth's ecosystem'. The remainder of Principle 7 introduced the concept of common but differentiated responsibilities. Linking these two parts of Principle 7 together suggests that 'developed countries' have special obligations to help other states to 'conserve, protect and restore' by providing technological assistance and financial resources. While it is never stated in Principle 7, the combination of the two parts of Principle 7 suggests that developed countries may have a 'differentiated responsibility' to support restoration efforts particularly within former colonies where large amounts of timber, biodiversity, or

47 See Chapter 6 for a discussion of the Biodiversity Convention and Chapter 11 for a discussion of the UN Framework Convention on Climate Change.

48 Principles on General Rights and Obligations: Chairman's Consolidated Draft, UN Doc. A/CONF.151/PC/WG.III/L.8/Rev.1 (1991).

49 H. Mann, 'The Rio Declaration' (1992) 86 *Proc. American Society of International Law* 405–411.

50 The initial text of Principle 3 proposed by Japan included a specific reference to 'restoration' as an obligation of intergenerational equity. The text originally read: 'Today and in the future, the individual has both a fundamental right to benefit from the common resources of humankind, which constitute the global environment, and at the same time a responsibility to protect, restore and improve for present and future generations.' U.N. Doc. A/CONF.151/PC/WG.III/L.22, Principle 4.

minerals were extracted in the pre-colonial era. Even if legitimate post-colonial concerns to address past environmental wrongs motivated the drafting of Principle 7, the final language in Principle 7 still reflects curious drafting choices.

For example, the Principle refers to 'the Earth's ecosystem' which is an incorrect usage of the term 'ecosystem' as understood by ecologists. Using the definition of ecosystem from the Biodiversity Convention which was drafted contemporaneously with the Rio Declaration, an ecosystem is understood to be 'a dynamic complex of plant, animal and micro-organism communities and their non-living environment interacting as a functional unit'.[51] The Earth does not have a single 'ecosystem' but rather a multitude of ecosystems. It may have been that the drafters wanted to emphasise a 'one world', 'all for one–one for all' approach in order to bolster the common but differentiated responsibility principle,[52] but the choice of the Earth as the unit to be restored ultimately provided no helpful guidance to states seeking new environmental policy directions. If anything, this word choice may have diluted the potential impact of this principle as guidance for states.

The choice of the phrase 'conserve … and restore' rather than 'conserve … or restore' also reflects an interesting choice of words. It probably should be read to mean that states are expected to seek to implement parallel conservation and restoration strategies because some areas are in need of conservation attention while other areas that have already experienced serious degradation now need restoration. This seems to be both a logical and practical reading of what the Principle requires of states.

Because 'the Earth's ecosystem' is the singular grammatical object of Principle 7, however, the verbs 'conserve', 'protect' and 'restore' might also be interpreted to be read in a sequence. Under this reading, a state would endeavour to first conserve and then protect 'the Earth's ecosystem' up to a point before 'the Earth's ecosystem' would be deemed to have an inadequate level of 'health and integrity'. Once the singular 'Earth's ecosystem' has been exhausted by the various states, then the states should collectively and cooperatively restore it to some condition of 'health and integrity'. While this is a grammatical reading of the text, it leads to a less logical outcome. A less confusing reading of the first part of Principle 7 that corresponds with the need for states to achieve improved health and integrity for all of Earth's ecosystems would be to read the text 'conserve … and restore' as 'conserve … or restore' and apply the obligation to all of Earth's ecosystems.[53]

Even though the Principles were adopted as aspirational measures rather than binding obligations, the lack of attention given to the drafting of the first half of Principle 7 matters. Governments in the Global South regularly cite and rely upon a now legalised concept of 'common but differentiated responsibilities' to explain

---

51 Convention on Biological Diversity, 5 June 1992, 1760 *UNTS* 79, Article 2.
52 Rio Declaration, Preamble. (The Conferences agreed to 'Recognizing the integral and interdependent nature of the Earth, our home'.)
53 See EU Habitats Directive, Chapters 7 and 10 (text of the Directive refers to 'maintain or restore').

why their states should be given additional time to comply with an international obligation or to urge Global North states to provide adequate resources to support Global South states in pursuing sustainable development.[54] In contrast to the attention given to the second half of Principle 7, there has been far less effort over the past two decades dedicated to pursuing cooperation through global partnership for conservation, protection and restoration of ecosystems.[55] Yet, some of the most at-risk ecosystems in the world are ecosystems from the Global South where resources are lacking for conservation and restoration.

In spite of the confusing language in Principle 7, the evolution of the Principle from the initial reference to restoration in the Stockholm Declaration is remarkable. In 20 years, the emphasis on restoration has shifted from a narrow focus on restoring ecosystem provisioning services to a focus on much broader holistic concepts of undertaking restoration for the purpose of ecosystem health and integrity. This is a notable normative change in light of the criticisms by some that the Rio Conference was taking the international community's attention away from environmental protection.[56]

In addition to the Rio Declaration, states agreed to a similar action plan to the one formulated at the Stockholm Conference. Entitled 'Agenda 21', the 40-chapter document was drafted on the basis of 'global consensus and political commitment at the highest level' to further development and environmental cooperation.[57] Designed to be a practical document, the Agenda includes specific shared objectives, proposals for activities and means for implementation. The concept of ecological restoration is threaded throughout various Agenda chapters as an action item. For example, in the Agenda chapter on combatting deforestation, states share the goal of 'enhancing the protection, sustainable management and conservation of all forests, and the greening of degraded areas, through forest rehabilitation, afforestation, reforestation and other rehabilitative means'.[58] In order to achieve this goal, 'rehabilitation' activities should be designed to 'restore productivity and

54 See, e.g., Statement by China on Behalf of Brazil, India, South Africa and China at the Opening Plenary of the Durban Platform, Warsaw, Poland (12 November 2013) ('The process and outcome of the Durban Platform [to develop a new climate protocol] shall be under the [Climate Change] Convention and guided by its principles and provisions, in particular the principles of equity and common but differentiated responsibilities and respective capabilities'), <http://unfccc.int/files/documentation/submissions_from_parties/adp/application/pdf/adp2.3_basic_20131112.pdf>.

55 Since the negotiation of the UN Convention on the Law of the Sea in 1982, there still remain areas of the high seas that lack basic conservation and management measures. A paradigm shift may be beginning as individual countries recognise the need for greater external technical or financial involvement in conservation or restoration efforts and join multi-stakeholder conservation and restoration projects. Cf. Global Partnership on Forest Landscape Restoration (launched in 2003 by UNEP) and the Global Partnership for Oceans (launched in 2012 by the World Bank).

56 D. Wirth, 'The Rio Declaration on environment and development: two steps forward and one back, or vice versa' (1995) 29 *Georgia Law Review* 599–653, 650.

57 United Nations Agenda 21 (1992), Chapter 1, para. 1.3, <https://sustainabledevelopment.un.org/content/documents/Agenda21.pdf>.

58 Ibid., Chapter 11(B).

environmental contributions, giving particular attention to human needs for economic and ecological services, wood-based energy, agroforestry, non-timber forest products and services, watershed and soil protection, wildlife management, and forest genetic resources'. While one can take issue with this strongly anthropocentric perspective of the goals for forest rehabilitation under the Agenda, the same ecosystem service motivations continue to drive many contemporary projects today that are labelled instead as forest restoration.[59]

In the chapter on combatting desertification and drought, states agreed to the goal of 'combating land degradation through, inter alia, intensified soil conservation, afforestation and reforestation activities'.[60] To achieve this goal, states 'at the appropriate level and with the support of the relevant international and regional organizations, should take corrective measures to restore the productivity of drylands'.[61] Governments should promote 'integrated research programmes on the protection, restoration and conservation of water and land resources'.[62] Both of these Agenda examples demonstrate that a core driver for restoration is our dependency on the environment. Typical restoration proposals are largely investments to revive resource productivity for human development ends. A similar type of commitment to productivity for human ends is made in the chapter on oceans calling for states on the high seas and within waters under their national jurisdiction to 'restore populations of marine species at levels that can produce the maximum sustainable yield as qualified by relevant environmental and economic factors'.[63]

In the chapter on conservation of biodiversity, the States called upon each other to 'promote the rehabilitation and restoration of damaged ecosystems'.[64] This is nearly verbatim language from the Biodiversity Convention.[65] But unlike the CBD which speaks primarily to states, the Agenda provides that the state 'with the support of indigenous people and their communities, non-governmental organizations and other groups, including the business and scientific communities' should pursue restoration of damaged ecosystems.[66] This multi-stakeholder approach to restoration embedded in the Agenda 21 document remains critical to the success of effective restoration for any country where there is a patchwork of land tenure.

A similar multi-stakeholder approach is proposed in the chapter on the protection of the oceans. Coastal states are encouraged to develop integrated management for the sustainable development of the coastal regions drawing on support from

---

59 See, e.g., US Forest Service Collaborative Forest Landscape Restoration Program Overview, <www.fs.fed.us/restoration/CFLRP/overview.shtml> (describing one of the goals of the landscape restoration to be managing woods for local community economic development and to reduce wildfire management costs).
60 United Nations Agenda 21 (1992), Chapter 12(B).
61 Ibid., Chapter 12, para. 12.18.
62 Ibid., Chapter 12, para. 12.23(b).
63 Ibid., Chapter 17, para. 17.46(b) and para. 17.74(c).
64 Ibid., Chapter 15.5(h).
65 CBD, Article 8.
66 United Nations Agenda 21 (1992), Chapter 15.5.

'the academic and private sectors, non-governmental organizations, local communities, resource user groups, and indigenous peoples'.[67] Part of any proposed integrated coastal zone plan is 'restoration of altered critical habitats'.[68]

As will be detailed further in Chapter 6 on the Biodiversity Convention, 1992 was a landmark year for the recognition of restoration of ecosystems. Agenda 21 took special notice of restoration in its chapter detailing the role of science to support environmental management and sustainable development. Specifically, Agenda 21 called upon states to 'develop further restoration ecology' in order to enhance scientific understanding to better respond to 'short- and long-term perturbations' to terrestrial, freshwater, coastal and marine ecosystems.[69]

The Rio Declaration in conjunction with Agenda 21 suggests dual policy attitudes to restoration. On the one hand, restoration is regarded as a utilitarian exercise to recover vital human commodities as indicated in the forest and ocean chapters of Agenda 21. On the other hand, restoration is designated as a therapeutic exercise for purposes of recovering global 'health and integrity'. These dual justifications for restoration persist in contemporary discourses on state-directed ecological restoration.

Johannesburg Plan of Implementation of the World Summit on Sustainable Development (2002)

Ten years after the Rio Declaration, states convened in Johannesburg, South Africa for the World Summit on Sustainable Development. This meeting did not produce sets of principles like the Stockholm and Rio meetings but did include the negotiation of a plan of implementation.[70] Focused on reducing poverty, changing consumption and production patterns, and protecting natural resource bases for economic and social development, the Implementation Plan did not set any new goals. Instead, the implementation plan reinforced Agenda 21 goals with additional specificity. For example, states agreed working 'at all levels' to 'maintain or restore [marine] stocks to levels that can produce the maximum sustainable yield' by 2015.[71]

The Plan identified two additional developments that have become increasingly relevant in assessing the costs and benefits of restoration. First, in relation to freshwater allocation the Plan suggested that states should develop integrated water resources management plans and water efficiency plans by 2005 that balance 'the requirement of preserving or restoring ecosystems and their functions, in particular in fragile environments, with human domestic, industrial and agriculture needs, including safeguarding drinking water quality'.[72] The issue of restoring instream riparian rights in over-allocated water basins remains a sensitive issue.

67 Agenda 21, Chapter 17, para. 17.6.
68 Ibid., para. 17.6(h).
69 Ibid., Chapter 35, para. 35.12(e).
70 UN Doc. A/CONF.199/20 (2002).
71 Ibid., para. 31(a).
72 Ibid., para. 26(c).

Second, the Plan proposes ecosystem restoration work to reduce disaster risks for communities where environmental degradation has increased vulnerability. Under the Implementation Plan, states 'at all levels' are expected to take action to 'Reduce the risks of flooding and drought in vulnerable countries by, inter alia, promoting wetland and watershed protection and restoration.'[73] The Implementation Plan was intended to accelerate the efforts of states in achieving sustainable development. While many of the implementation targets were missed, the Implementation Plan is still an interesting document in terms of highlighting the global need to restore not just individually useful species such as commercial fish stocks but also complex systems such as wetlands.

Rio+20 The Future We Want (2012) and Sustainable Development Goals (2015)

> Protect, restore and promote sustainable use of terrestrial ecosystems, sustainably manage forests, combat desertification, and halt and reverse land degradation and halt biodiversity loss.
>
> (Sustainable Development Goal 15)

Twenty years after the UN Conference on Environment and Development and ten years after the UN Summit on Sustainable Development, states organised another multilateral meeting to take stock of the progress made and the challenges remaining in greening the economy and implementing sustainable development.[74] The outcome of the meeting was a single document, The Future We Want, which was generally regarded as a disappointment because it failed to elicit any new commitments from states beyond an agreement to develop sustainable development goals to be adopted by the UN General Assembly. Perhaps the failure to take on new commitments should not have been surprising in light of the number of unmet commitments under existing treaty regimes and other international agreements such as the Johannesburg Implementation Plan. Even so, The Future We Want reflects important normative developments in how restoration is regarded.

Even more so than previous international negotiated documents, The Future We Want explicitly recognised restoration as a key environmental management strategy that could be mainstreamed into economic decision-making. On the first page of the document, the authors wrote about the need to achieve sustainable development by 'promoting integrated and sustainable management of natural resources and ecosystems that supports *inter alia* economic, social and human development while facilitating ecosystem conservation, regeneration and restoration and resilience in the face of new and emerging challenges'.[75] This language indicates a shared commitment to 'ecosystem restoration' and not just restoration of vital commodities.

---

73 Ibid., para. 37(d).
74 This meeting was convened as the United Nations Conference on Sustainable Development.
75 The Future We Want (2012) para. 4, <www.uncsd2012.org/content/documents/727The%20Future%20We%20Want%2019%20June%201230pm.pdf>.

The term regeneration has replaced the former use of 'improve' from the Stockholm Declaration and might have been included to distinguish holistic restoration efforts from ecological rehabilitation efforts such as afforestation. States recognise a new condition for environmental management; now conservation, regeneration and restoration efforts are expected to be resilient. While the term 'resilience' is never explained in the document, it can be assumed that the States intended for restoration projects to be capable of responding to or adapting to external stressors posing 'new and emerging challenges' such as climate change.

With a direct reference to Rio Declaration Principle 7, sustainable development was identified as a complex process requiring 'holistic and integrated approaches ... which will ... lead to efforts to restore the health and integrity of the Earth's ecosystem'.[76] Restoration activities were also recognised by states as a public works job generator and as essential for the recovery of certain industries such as the fishing industry.[77] While most of The Future We Want document is drafted in general aspirational language, the section on restoration of fish stocks is specific in its expectations. Building on the commitments under the Johannesburg Implementation Plan, the Rio+20 outcome document defines specific action items for states including implementing science-based management plans, reducing or suspending fishing catch and effort, better managing bycatch and discards, and effectively using impact assessments.[78]

Three years after the conclusion of Rio+20, states through the UN General Assembly adopted the Sustainable Development Goals, as a set of goals and targets to continue measuring progress towards internationally shared objectives.[79,80] Like the Millennium Development Goals, the Sustainable Development Goals are not mandatory but reflect instead an effort for states in cooperation with stakeholders to identify chronic challenges and devise national strategies to address these challenges. One of the 17 goals and 6 of the 169 targets explicitly mention restoration. Goal 15 provides that states must 'Protect, restore and promote sustainable use of terrestrial ecosystems, sustainably manage forests, combat desertification, and halt and reverse land degradation and halt biodiversity loss.[81] In order to achieve this goal states agreed to three short-range targets involving restoration as an environmental management strategy:

> 15.1 By 2020, ensure the conservation, restoration and sustainable use of terrestrial and inland freshwater ecosystems and their services, in particular forests, wetlands, mountains and drylands, in line with obligations under international agreements

76 Ibid., para. 40.
77 Ibid., paras 154 and 158.
78 Ibid., para. 168.
79 Transforming Our World: the 2030 Agenda for Sustainable Development, UN Doc. A/Res/70/1 (21 October 2015).
80 Ibid., Preamble ('All countries and all stakeholders, acting in collaborative partnership, will implement this plan').
81 Ibid.

15.2 By 2020, promote the implementation of sustainable management of all types of forests, halt deforestation, restore degraded forests and substantially increase afforestation and reforestation globally

15.3 By 2030, combat desertification, restore degraded land and soil, including land affected by desertification, drought and floods, and strive to achieve a land degradation-neutral world.[82]

Two additional goals include restoration targets. Goal 6 provides that states must ensure availability and sustainable management of water and sanitation for all. Target 6.6 calls upon states by 2020 to 'protect and restore water-related ecosystems, including mountains, forests, wetlands, rivers, aquifers and lakes'. Finally Goal 14 on the conservation and sustainable use of the oceans, seas and marine resources includes one target to take action by 2020 for the restoration of coastal and marine areas 'in order to achieve healthy and productive oceans' and a second target to 'restore fish stocks in the shortest time feasible, at least to levels that can produce maximum sustainable yield as determined by their biological characteristics'.[83]

In order to measure achievement towards these goals and targets, States adopted a set of indicators. For Target 6.6 on the restoration of water-related ecosystems, the indicator of progress is "change in the extent of water-related ecosystems over time."[84] This could prove problematic where states such as the United States at one point in its measurement of wetland restoration identified an increase in wetlands acreage by counting man-made features such as golf-course features that were not designed to increase wetland ecosystem functions.[85] For Target 14.2 on restoration of coastal and marine areas, the indicator of progress is whether a State within its exclusive economic zone uses "ecosystem-based approaches."[86] It is unclear that an ecosystem-based approach will by definition include restoration efforts so this indicator does not seem to be a useful proxy for measuring progress towards ecological restoration of degraded areas.[87] Finally, the indicators for terrestrial restoration address forests and deserts.[88] The indicators are vague and include forest area as a proportion of the total land area, progress towards sustainable

---

82 Ibid.

83 Ibid., Targets 14.2 and 14.4.

84 Report of the Inter-Agency and Expert Group on Sustainable Development Goal Indicators, E/CN.3/2016/2/Rev.1 (2016).

85 T.E. Dahl, *Status and Trends of Wetlands in the Conterminous United States 2004 to 2009*, Report to Congress, Department of Interior, US Fish and Wildlife Service (2011) 42 (noting that between 2004 and 2009 freshwater ponds increased by over 200,000 acres but that 'the functional characteristic of these water bodies continues to be debated'); F. Barringer, 'Fewer marshes + more man-made ponds = increased wetlands', *New York Times* (31 March 2006), <www.nytimes.com/2006/03/31/washington/31wetlands.html?_r=0>.

86 Report of the Inter-Agency and Expert Group on Sustainable Development Goal Indicators.

87 Ibid.

88 Ibid.

management and the proportion of land characterized as degraded over the total land area. Taken as a whole, there indicators provide little guidance to States regarding the "quality" of a restoration project. To the extent that the existence of the indicators might shape State behaviour, there may be States who will seek to pursue a "quantity" rehabilitation strategy over "quality" ecological restoration.

Sendai Framework

In 2015, states negotiated a non-binding but normative framework for disaster risk reduction.[89] The document was intended to continue the progress that had been made under the 2005 Hyogo Framework for Action 2005–2015: Building the Resilience of Nations and Communities to Disasters, which encouraged states to develop effective disaster response plans.[90] The word 'restoration' is never mentioned in the text. But the concept of ecological restoration as a strategy for disaster risk reduction is implicitly included in the framework with the frequent references in the framework to 'environmental assets' and the need for investing in disaster risk reduction to protect these assets. States recognise that it is 'urgent and critical to anticipate, plan for and reduce disaster risk in order to more effectively protect persons, communities and countries, their livelihoods, health, cultural heritage, socioeconomic assets and ecosystems'.[91] States have committed to achieving by 2030, 'The substantial reduction of disaster risk and losses in lives, livelihoods and health and in the economic, physical, social, cultural and environmental assets of persons, businesses, communities and countries.'[92] One internationally agreed upon means of achieving disaster risk reduction objectives is ecological restoration. Nationally funded projects such as mangrove restoration or mountain forest restoration are often approached as disaster risk reduction projects.[93]

Taken together, the Stockholm Declaration, Rio Declaration, Johannesburg Implementation Plan, The Future We Want and the Sendai Framework reflect a gradual transition in community thinking on the various roles for ecological restoration in international law. While there has been a recognition since 1972 that society at large has a role to play in restoration, it is only since 1992 that holistic ecological restoration has become a shared objective. The relationship between conservation and restoration has become clearer over the decades as the concept of restoration has become mainstreamed. Restoration is regarded as a legitimate but last resort environmental management strategy to support the broader goals of conservation. The relationship between sustainable development and restoration has

---

89 Sendai Framework for Disaster Risk Reduction 2015–2030, UN Doc. A/CONF.224/L.2 (7 April 2015).
90 Hyogo Framework for Action, UN Doc. A/CONF.206/6 (22 January 2005).
91 Sendai Framework, para. 5.
92 Ibid., para. 16.
93 Philippines Department of Environment and Natural Resources, Philippines National Greening Program, <http://ngp.denr.gov.ph> (describing a programme of reforestation designed for both economic benefits and reduction of disaster risks).

also been clarified with restoration playing an increasingly important role in jump-starting the recovery of socio-ecological systems that benefit human development needs.

Assuming that the five international documents previously discussed reflect political will for states to engage in ecological restoration, the remainder of this chapter examines what general principles of international environmental law might apply to the national and international restoration efforts that states have committed themselves to pursue.

Principles of international environmental law to guide implementation of state restoration commitments

In addition to treaties that will be discussed in the next two chapters, there are also international legal principles that may apply in how states develop and implement national and transnational restoration policies. Principles are not the same as rules of law that require a particular outcome,[94] but provide guidelines for actions and offer justifications for specific policy choices by decision-makers.[95] Principles are more than simply ideas. They are normative concepts. Over time, principles may mature into customary obligations or may be referenced in binding conventions where states choose to create more precise directions of how to apply a principle. While in theory many different principles that have been articulated by states through declarations such as the Stockholm Declaration or the Rio Declaration could be applied in implementing restoration commitments, this section will only deal with the 'directing principles' of the precautionary principle, the polluter pays principle and the duty to prevent, which each have the potential to 'assume an autonomous normative value'.[96]

Precautionary principle

Probably, the most significant principle for restoration efforts is the precautionary principle. As restoration ecologist Stuart Allison notes, 'ecological restoration is a crisis discipline and as such we often have to make decisions before we know enough to be able to predict outcomes of our efforts with a high degree of precision'.[97] The precautionary principle, as reflected in Principle 15 of the Rio Declaration, provides that 'where there are threats of serious or irreversible damage, lack of full scientific certainty shall not be used as a reason for postponing

---

94 U. Beyerlin, 'Different types of norms in international environmental law policies, principles, and rules' in D. Bodansky, J. Brunee and E. Hey (eds), *Oxford Handbook of International Environmental Law* (Oxford University Press 2008) 433.

95 R. Dworkin, *Taking Rights Seriously* (Harvard University Press 1977) 25 (principles that 'do not set out legal consequences that follow automatically when the conditions provided are met' but should be taken into consideration when appropriate).

96 N. de Sadeleer, *Environmental Principles: From Political Slogans to Legal Rules* (Oxford University Press 2002) 305.

97 S. Allison, *Ecological Restoration and Environmental Change* (Routledge 2012) 71.

cost-effective measures to prevent environmental degradation'.[98] Ideally in taking State-based action, a state would have already applied the precautionary principle to avoid environmental degradation by either avoiding a harm or minimising a harm. To the extent that a harm has occurred or there is a threat of harm causing serious or irreversible damages, States must then consider restoration strategies. What this might mean in practice is that a state may need to undertake restoration efforts on a given wetland not because the ecological functions of the given wetland are likely to decline further but rather because an adjacent wetland site will be further degraded due to a delay in restoration.

The lack of full scientific uncertainty comes into play for a restoration project when a restoration practitioner must decide what reference model will become the target for a restoration project. Should a state insist on a reference model based in part on past systems or can a state offer a reference model based wholly on future projections there are contemporary controversies among restoration ecologists about how to approach restoration baselines. Recognising that it is impossible to achieve historical fidelity because of so many changes in land use and climate, some ecologists ardently advocate for the use of reference models including historic reference models.[99] For these ecologists, this means a good faith effort on the part of restoration professionals to work with species that are likely to have formerly resided on the restoration site and to revive ecological structures and functions based on a historic ecological trajectory that should be self-sustaining and self-organising.[100] For other ecologists, the historical trajectory or reference model approach appears to be too limiting. These ecologists argue that so much has changed in the last decades that ecologists should strive instead to foster appropriate novel ecosystems that are capable of surviving under current conditions while also contributing ecological value.[101]

The debate among ecologists is more than just a theoretical debate because the outcome of the debate determines how financial and labour resources will be deployed. Without weighing in on the scientific positions that favour the two positions in the debate, it appears from a purely legal perspective that an application of the precautionary principle favours the approach of using 'reference models' rather than actively pursuing restoration efforts that are likely to reinforce 'novel ecosystems'.[102] In choosing between two situations that both involve scientific uncertainty, the application of the precautionary principle should favour the approach that is more likely to avoid future environmental degradation. In theory,

98 Rio Declaration, Principle 15.

99 A. Clewell and J. Aronson, *Ecological Restoration: Principles, Values and Structure of an Emerging Profession* (Island Press 2008) 80–87.

100 Ibid., 84. (Proponents of a reference model system recognize that the concept of a 'historical trajectory' seems to suggest some degree of environmental stability which cannot be assumed since ecological restoration projects do not involve single project factors but rather a multitude of factors.)

101 See generally R. Hobbs, E. Higgs and C. Hall (eds), *Novel Ecosystems: Intervening in the New Ecological World Order* (Wiley 2013); see also Chapter 11 on climate change.

102 See Chapter 2 for a discussion of these competing models of ecological restoration.

work based on a historic reference model should reverse trends towards environmental degradation and restore some pre-disturbance ecological values. Work based on bolstering novel ecosystems may also reverse the state of environmental degradation but may also inadvertently contribute to further environmental degradation by, for example, introducing novel ecological disturbances into a neighbouring system.

Restoration projects involve a great deal of decision-making. To ensure that decision-makers employ a precautionary approach, states or other governing bodies may require some form of pre-restoration project impact assessment in order to identify actions that may impact environmental resources. From a practical perspective, States might require restoration projects of a certain size or restoration of sensitive habitats to obtain permits so that the state can require certain minimum conditions for restoration work in order to avoid risks of environmental damage (e.g. the use of hand tools rather than mechanical tools in areas with known threatened or endangered species).

Polluter pays principle

The polluter pays principle provides that 'National authorities should endeavour to promote the internalization of environmental costs and the use of economic instruments, taking into account the approach that the polluter should, in principle, bear the costs of pollution.'[103] In the context of restoration, a modified version of the 'polluter pays principle' could apply that might be labelled the 'degrader pays principle' or 'user pays principle'. When an entity under the control and jurisdiction of a state undertakes a certain type of habitat intensive activity such as mining or farming and damage occurs within the boundaries of the legal obligations owed by the industry to the community, then industry may be required by the state to invest in restoration efforts. This understanding of the 'degrader pays principle' or 'user pays principle' is uncontroversial and depends on shifting the restoration burden to the entity responsible for the damage. It is a matter of fairness that an entity that fails to control its activities and damages property must be prepared to offer restitution. This concept of burden-shifting is increasingly captured in national soil remediation statutes requiring industries to clean up contaminated sites.[104]

The more controversial portion of an application of the 'polluter pays principle' is whether industries should be expected to engage in restoration efforts when they are engaged in a socially useful enterprise that damages the environment in order to deliver a socially acceptable good. Would it be adequate for the industry to simply restore some ecological functions and who would decide which functions – the industry or the community? If the industry must provide restoration, is it acceptable for the polluter to pay for a restoration project that does not actually

103 Rio Declaration, Principle 16.
104 Comprehensive Environmental Response, Compensation and Liability 42 USC §§ 9601 to 9628 (statute designed to identify Potential Responsible Parties and hold them liable for clean-up costs associated with contaminated sites).

reflect historical values but creates new ecosystems – e.g. creating lakes in former quarries which were previously forested land? Should this qualify as restoration?

Principle of prevention

States are expected to prevent transboundary harm.[105] Most restoration projects do not involve cross-border projects or have cross-border impacts unless the restoration projects involve a transboundary watercourse or other shared resource. Some scholars suggest that states apply a broader 'principle of prevention' requiring a state to exercise a certain level of vigilance on the part of the government to ensure that states limit or control activities that are likely to cause harm to the environment.[106]

In the case of a needed restoration project, there has already been an environmental harm. It can be argued that where there is harm and the state may have had an obligation of due diligence to create, implement and enforce laws to prevent harm, the duty to prevent harm has already been breached. Unfortunately, the only fair conclusion is that every state has violated this principle of prevention.

A better reading of the principle of prevention in the context of restoration is that a state breaches the principle of prevention when it has a large number of potential restoration sites requiring active intervention. For example, there may be a site with extensive land degradation where soil continues to be lost or a damaged wetland site where effluent continues to be released. Applying the principle of prevention, a State should use its resources, even if they are limited resources, to prioritise restoration efforts for those sites where a continued failure to restore will lead to ongoing or future harm.

This chapter has covered the 'soft law' context for the international development of state commitments to undertake ecological restoration. In the twenty-first-century, states understand that they have obligations to restore marine, coastal and terrestrial resources, and have formulated these general obligations most recently into a series of goals and targets to achieve by 2020. What is less clear from the various declarations is what constitutes ecological restoration or whether each state simply defines the terms for its own ends. The next chapter continues this inquiry by examining a selection of treaties that create binding commitments for restoration.

---

105 Stockholm Declaration, Principle 21.
106 P. Sands, *Principles of International Environmental Law* (2nd edn, Cambridge University Press 2003) 194.

5 International conventions and ecological restoration

In addition to the numerous non-binding declarations and action plans negotiated by states described in Chapter 4, states have entered into numerous multilateral treaties requiring them to engage in restoration activities. The treaties that will be discussed in this chapter fall into two general categories. The first category includes restoration treaties focused on the recovery of certain species. The second category features treaties that endeavour to revive particular habitats. The chapter concludes with a discussion of a recent soft-law multi-stakeholder agreement designed to scale up habitat restoration to the landscape level and to enhance technical cooperation. The Biodiversity Convention that is the landmark international treaty on restoration will not be discussed here, but in Chapter 6. The United Nations Framework Convention on Climate Change that catalyses ecological restoration work under the Reduction Emissions from Deforestation and forest Degradation initiative is discussed in Chapter 11.

In almost every treaty described below, there is no definition of restoration. What this means in practice is that the requirement to undertake restoration is regarded largely as an obligation of conduct or obligation of means rather than an obligation of result. An obligation of conduct as applied to restoration provides greater latitude for a state to act since it can define for itself through its domestic government branches what will constitute adequate restoration for purposes of fulfilling its international obligation. In contrast, an obligation of result requires parties to achieve a specific negotiated result. Obligations of result related to restoration obligations feature in only two of the multilateral treaties discussed in this chapter, the Law of the Sea Convention and the Straddling Stocks Agreement, and in EU law, discussed in Chapter 7. More recent international implementation efforts like those under the Biodiversity Convention's Aichi Targets, discussed in the next chapter, do not create obligations of results but employ language that closely correlates with obligations of result.

Whether ecological restoration in the context of international law should be an obligation of conduct or an obligation of result is a significant question that will be discussed further in the conclusion to this book in Chapter 12. For purposes of this question, it is worth considering whether it is adequate from an environmental protection perspective for states to enter into broad conduct commitments where it is difficult to determine compliance with these commitments and even more difficult to ascertain whether compliance with the commitments eventually leads to optimal ecological outcomes.

Restoration conventions for the recovery of species

Many of the earliest international environmental law treaties were wildlife treaties, focused on the recovery of a species threatened by over-exploitation or by a lack of coordinated governance. This section highlights restoration obligations within two species oriented treaties focused on creating transboundary solutions to wildlife management: bilateral migratory bird treaties and the various Law of the Sea treaties addressing marine conservation of living resources. These treaties reflect a functional approach to restoration focused on recovering species deemed to be of commercial importance. Compliance with these treaties largely does not require a holistic view of ecosystem restoration.[1]

Migratory bird treaties

In the nineteenth century, farmers regarded most birds in their fields as pests and the world of high fashion prized bird plumage. As a result bird populations plummeted, leading to the extinction of some species such as Labrador ducks. Conservation groups such as the Audubon Society formed in response to this tragedy and pressure was placed on governments to stem the losses. Over the course of the next century, several bilateral treaties were signed to protect migratory birds. While the earliest versions of some of these treaties assumed that conservation efforts by closing hunting seasons would lead to the passive restoration of bird populations, later versions focused on the need for a more intensive regulatory framework to actively recover dwindling migratory bird populations. For example, when a Protocol to the 1916 Migratory Bird Protection Convention was negotiated between Canada and the United States in 1999, the parties agreed that one of the shared objectives would be 'to restore depleted populations of migratory birds'.[2] Significantly, the Protocol went beyond simply curtailing the hunting season and began to also require an early ecosystem approach to protecting birds. States were specifically called upon to prevent damage to 'birds and their environments ... resulting from pollution' and 'to take such measures as may be necessary to control the importation of live animals and plants ... hazardous to the preservation' of migratory birds.[3] While concepts of preservation, conservation and

---

1 In recent decades, Parties to the Law of the Sea Treaties have been promoting the implementation of ecosystem-based fisheries management which reflects a more holistic view of ecological restoration than simply restoring commercially significant stocks.
2 Protocol Between the Government of Canada and the Government of the United States of America Amending the 1916 Convention Between the United Kingdom and the United States of America for the Protection of Migratory Birds in Canada and the United States, Article II, <www.treaty-accord.gc.ca/text-texte.aspx?id=101589>.
3 Ibid., Article IV.

restoration are used sometimes in interchangeable fashion when they appear in the migratory bird treaties, it is clear that current restoration work undertaken to comply with the bird treaties is focused largely on the recovery of a subset of species. Any additional environmental benefits that flow from the work undertaken in response to the treaties such as recovery of non-migratory birds or insect populations are side benefits.

One advantage of the species restoration approach is that it provides a clear focus for policy-makers. While wildlife scientists may find themselves needing to confront complex issues in the field regarding the viability of maintaining certain types of habitat in a changing landscape, the migratory bird treaties are clear about their objective: to restore a subset of migratory birds. Restoration metrics are relatively simple under the species recovery treaties. In terms of measuring restoration success, policy-makers can agree to a baseline number of birds and then determine the growth in number of migratory birds across their migration path. The same single-factor species restoration approach based on population growth applies to the restoration commitments under the Convention on Migratory Species[4].

Law of the Sea: UNCLOS and the Straddling Stocks Agreement

In 1982, after decades of negotiations, states adopted the United Nations Convention on the Law of the Sea (UNCLOS).[5] Encompassing topics as diverse as marine pollution and piracy, the UNCLOS was designed to rationalise management of marine resources. Parties to UNCLOS agreed that part of fisheries management would include restoration. The obligation to restore appears in two sections of the treaty concerning the total allowable catch for commercially harvested species: Article 61 and Article 119. First in Article 61, states agreed within the exclusive economic zone, a jurisdictional zone of 200 nautical miles as measured from specified territorial basepoints, that coastal states must take:

> Measures ... to maintain or restore populations of harvested species at levels which can produce the maximum sustainable yields, qualified by relevant environmental and economic factors, including the economic needs of coastal fishing communities and the special requirements of developing States, and taking into account fishing patterns, the interdependence of stocks and any generally recommended international minimum standards, whether subregional, regional or global.[6]

Regarding this obligation states are expected to take into account 'effects on species associated with or dependent upon harvested species with a view to maintaining or

---

4 See below.
5 United Nations Convention on the Law of the Sea, Montego Bay, 10 December 1982 (1982) 21 *ILM* 1261.
6 Ibid., Article 61(2).

restoring populations of such associated or dependent species above levels at which their reproduction may become seriously threatened'.[7] This language offers an early ecosystem approach by recognising a wider marine ecology beyond target commercial species. Second in Article 119, the identical restoration language from Article 61 is applied to the high seas, the jurisdictional area designated as the waters beyond the 200-nautical-mile exclusive economic zones. Read collectively, Article 61 and Article 119 provide an obligation for parties to undertake restoration activities in all ocean waters.

This treaty language marks one of the rare instances where states have agreed to an obligation of result related to restoration – namely in setting total allowable catch, states must 1) take measures to restore harvested species to maximum sustainable yield levels and 2) factor into decision-making the possibility of restoring species that are 'associated or dependent' on harvest species. The restoration obligation under the Law of the Sea Convention is an obligation of result driven by a functional restoration scheme designed to ensure adequate commodity levels for commercial harvest. The drafters of Articles 61 and 119 understood that in practice restoration could not be limited to simply target harvested species but must include a broader web of species that would include species that form essential habitat for harvested species such as coral or species that are part of a target species' food chain.

It was noted earlier that obligations of conduct and means can offer greater latitude for states to interpret how they will comply with a text that requires them to engage in 'restoration of stocks'. Article 61 illustrates that an obligation of result depending on how it is negotiated can also provide a great deal of latitude for parties. The concept of 'maximum sustainable yield' (MSY) as an outcome of restoration is controversial among fishery managers.[8] In many instances, maximum sustainable yields are set on the basis of political considerations rather than ecological concerns and become difficult to ratchet down even in the face of data about fishery declines.

States have flexibility in what measures they choose to implement in order to further restoration goals for both target species and associated/dependent species. For example, states can choose to create marine protected areas or close fisheries during particular seasons to allow for passive restoration. States may also chose to pursue an active restocking of a fishery through hatcheries as in the case of salmon in the United States and Canada. While restocking is prevalent in the fisheries world, there is some question as to whether this alone qualifies as best practices in restoration since the mortality rate of restocked fish is unusually high.[9]

---

7 Ibid., Article 61(4).
8 See generally C. Finley, *All The Fish in the Sea; Maximum Sustainable Yield and the Failure of Fisheries Management* (University of Chicago Press 2011).
9 C. Brown and R. Day, 'The future of stock enhancements: lessons for hatchery practice from conservation biology' (2002) 3 *Fish and Fisheries* 82–83 (distinguishing between restocking and reintroducing of species with the difference being that restocked fish exhibit few to no 'wild' behaviours such as dispersal behaviour so that individual fish do not have to intensely compete for food sources).

For both of the obligations under Articles 61 and 119, there is an important distinction made between a general unconditional obligation to restore and an obligation based on 'the special requirements of developing States'. Both Articles imply an application of the common but differentiated responsibility (CBDR) principle.[10] Given food security concerns, developing States with limited resources may be able to simultaneously increase total allowable catch to meet fishers' immediate needs while also investing fewer resources in restoration efforts than other states. Increasingly, setting a total allowable catch without also desgining and implementing a restoration plan may lead to the unintended consequence of depleting fishery resources. Applying a CBDR principle to national fishery man-agement, developing coastal states may also continue to acquiesce to the operation of distant water flagged vessels in their coastal waters in order to generate revenue even when a given state lacks the capacity to detect unreported fishing activity or to engage in ecosystem-based fishery management research.

In 1995, states negotiated a post-UNCLOS implementing agreement on the conservation and management of straddling fish stocks and high migratory fish stocks.[11] Drawing directly on UNCLOS authority and the language adopted in Article 119, the coastal states and distant water fishing states reaffirmed principles intended to maintain and restore both target stocks and associated/dependent stocks.[12] The Straddling Stocks Agreement went much further than UNCLOS in terms of applying the precautionary principle to stock recovery. The Agreement called upon states to identify reference points for various stocks that would 'correspond to the state of the resource'.[13] These reference points would be calculated based on 'best scientific information available' and would be 'stock-specific'.[14] Points needed to be stock-specific because stocks exist across a spec-trum of production states.[15] The Agreement contemplated states collectively and

10 Rio Declaration, Article 7.
11 Agreement for the Implementation of the Provisions of the United Nations Conven-tions on the Law of the Sea of 10 December 1982 Relating to the Conservation and Management of Straddling Fish Stocks and Highly Migratory Fish Stocks, 4 August 1995, S. Treaty Doc. No. 104–24, 2167 *UNTS* 88 ('Straddling Stocks Agreement').
12 Ibid., Article 3(b) (measures should be 'designed to maintain or restore stocks at levels capable of producing maximum sustainable yield, as qualified by relevant environmental and economic factors, including the special requirements of developing states, and taking into account fishing patterns, the interdependence of stocks and any generally recom-mended international minimum standards, whether subregional, regional or global'); ibid., Article 3(e) (states should 'adopt, where necessary, conservation and management measures for species belonging to the same ecosystem or associated with or dependent upon the target stocks, with a view to maintaining or restoring populations of such species above levels at which their reproduction may become seriously threatened').
13 Ibid., Article 6(3)(b) and Annex II(1).
14 Ibid., Article 6(3)(b).
15 D. Spencer and J.S. Collie, 'Patterns of population variability in marine fish stocks' (1997) 6 *Fisheries Oceanography* 188–204 (describing a spectrum of stock production types including category I [steady state], category II [low variation and frequency], category III [low frequency, cyclic], category IV [irregular], category V [high variation and frequency] and category VI [spasmodic]).

cooperatively setting two reference points for the straddling stocks and highly migratory stocks: a conservation reference point (also called a limit reference point) 'to constrain harvesting within safe biological limits within which the stocks can produce maximum sustainable yield' and a target reference point for purposes of management.[16]

States are expected on the basis of the precautionary reference points 'to maintain or restore populations of harvested stocks, and where necessary associated or dependent species, at levels consistent with previously agreed precautionary reference points'.[17] When a harvested stock or dependent stock falls below a limit reference point or is at risk of falling below such a reference point, states should immediately deploy conservation and management measures to reverse the decline in a species.[18] Specifically the Agreement contemplates that states undertake pre-agreed conservation and management actions 'to restore the stocks'.[19] Subsequent non-binding state agreements have reinforced this approach.[20]

On paper, this seems to be a good quantitative approach to determining when restoration obligations are triggered for a party. Yet, as with so much of restoration practice, there is a great deal of discretion remaining with any institutional actors. As one group of fisheries experts observed the precautionary reference points required by the Agreement 'reflect individual organizations' interpretations and implementations of precautionary management'.[21] This comment suggests that precautionary reference points might be concluded as a political matter without adequate regard for the ecological management of the resource in question.

The existing legal regime embodied in UNCLOS and the Straddling Stocks Agreement centres on reviving fish to be commodities. The priority for restoration management has been restoring stocks to some population level that continues to permit harvesting according to a maximum sustainable yield formula. Only focusing on reducing fishing effort by creating no-fish areas or limiting fishing seasons may not be enough to restore stocks in an increasingly polluted or low-productivity regime (e.g. physically damaged coral reefs).[22] There may also be a need for

16 Straddling Stocks Agreement, Annex II(2).
17 Ibid., Annex II(4).
18 Ibid., Annex II(5).
19 Ibid., Article 6(3)(b), 6(4) and Annex II(4).
20 Food and Agriculture Organization, FAO Code of Conduct for Responsible Fisheries 1995 (Section 6.5.4 'When precautionary or limit reference points are approached, measures should be taken to ensure that they will not be exceeded. These measures should where possible be pre-negotiated. If such reference points are exceeded, recovery plans should be implemented immediately to restore the stocks').
21 W. Gabriel and P. Mace, *A Review of Biological Reference Points in the Context of the Precautionary Approach*, Proceedings, 5th NMFS NSAW, 1999. NOAA Tech. Memo. NMFS-F/SPO-40, <www.st.nmfs.noaa.gov/Assets/stock/documents/workshops/nsaw_5/gabriel_.pdf>.
22 J. Caddy and J. Sejio, 'This is more difficult than we thought! The responsibility of scientists, managers and stakeholders to mitigate the unsustainability of marine fisheries' (2005) 360(1453) *Philosophical Transactions B Royal Society London Biological Sciences* 59–75.

judicious interventions to restore habitat conditions. The following section describes a number of agreements negotiated to address habitat restoration.

Restoration treaties for reviving damaged habitats

Investing in the restoration of specific migratory birds and fish stocks offers an easy legal strategy where states create hatchery or nursery industries for breeding new generations of birds and fish. Yet, the survival of this next generation outside of the hatchery or nursery will depend on access to suitable habitat. Early in the history of contemporary international environmental law, states recognised that collective responses to support domestic efforts to recover habitat would be critical to restoring species abundance.

The treaties discussed below are mostly described in chronological order even though the habitat-related treaties reflect two very different models of international cooperation. The World Heritage Convention and the Ramsar Convention rely on each state to nominate sites and then create management plans for the specific sites that may include implementing restoration strategies (see also Chapter 10 on protected areas). The Convention on Migratory Species and the Convention on Desertification require domestic actions but focus on generating regionally focused responses either through the negotiation of additional agreements among states where migratory species reside or transit or through the production of national action programmes designed to support regional implementation.

World Heritage Convention

When the World Heritage Convention was negotiated in 1972, it was designed so that states would propose sites with particular universal values and agree to provide adequate resources to protect these sites. There is no language in the Convention text that specifically provides for ecological restoration but the Convention did contemplate that states may ask for assistance in the 'rehabilitation' of a site.

In practice, the World Heritage Committee has frequently invoked the concept of ecological restoration in its decisions related to natural heritage sites. For example in a 2010 Decision regarding relisting the Everglades National Park on the World Heritage List in Danger, the Commission offered a number of observations including its approval of a project which if executed 'should restore historical water flow volumes and pathways through the property and secure long-term ecosystem functions'.[23] In response to the Wenchuan Earthquake in Central China, the Commission called upon the Chinese government in its work in the Sichuan Giant Panda Reserve to 'implement ecosystem restoration aspects of the post-earthquake recovery plan'.[24]

---

23 World Heritage Committee Decision, 34 COM 7B.29 (2010) Everglades National Park (United States of America).
24 World Heritage Committee Decision, 34 COM 7B.11 (2010) Sichuan Giant Panda Sanctuaries – Wolong, Mt Siguniang and Jiajin Mountains (China).

One role that the Committee plays is to offer informed input on the feasibility of restoration and to provide an intergovernmental support network for ecological restoration activities. For example, in 1992, the Committee expressed concerns that the wetland restoration plans for the Srebarna Biosphere Reserve in Bulgaria may not be technically feasible to 'enable a full restoration of Srebarna as a naturally functioning wetland ecosystem'. The Committee gave Bulgaria one year to report back with a viable plan or the World Heritage site would be removed from the World Heritage List.[25] In 1993, Bulgaria had prepared a restoration plan and the Committee decided to defer its decision on delisting and instead to request the International Union for the Conservation of Nature, in cooperation with the Ramsar Convention, to monitor 'the extent to which the project(s) implemented by the Bulgarian authorities are restoring the ecological integrity of Srebarna'.[26] In 1995, the IUCN representative informed the Committee that steps had been taken but that the 'integrity of the site' had 'not yet been adequately restored'.[27] In 2004, after additional restoration efforts, the Srebarna site was removed from the World Heritage in Danger List and is back on the 'normal' World Heritage List.[28]

Many of the sites that are presently listed on the World Heritage in Danger List are sites that require ecosystem restoration. For example, the Belize Barrier Reef System requires restoration of areas where unauthorised oil exploration and extraction activity has been taking place.[29] The Tropical Rainforest Heritage of Sumatra site requires 'ecosystem restoration programmes' to control invasive species to support species recovery efforts.[30] The Committee has also called for the restoration of 'Outstanding Universal Value' for a number of parks including the Los Katios National Park in Colombia,[31] the Manovo-Gounda St Floris National Park in the Central African Republic[32] and Kahuzi-Biega National Park in the Democratic Republic of the Congo.[33] Based on the criteria of 'outstanding universal value', the restoration standards for these parks will require the articulation of restoration objectives that will support the recovery of 'significant on-going ecological and biological processes in the evolution and development of terrestrial, fresh water, coastal and marine ecosystems and communities of plants and animals'

25 WHC-92 CONF 002 VIII (1992) Srebarna Biosphere Reserve (Bulgaria), <http://whc.unesco.org/document/940>.
26 WHC-93 CONF 002 VIII.2 (1993) pp. 8–9, <http://whc.unesco.org/archive/1993/whc-93-conf002-2e1.pdf>.
27 WHC-95 CONF 203 VII.A.1.3 (1995) p. 8, <http://whc.unesco.org/document/799>.
28 WHC-03/27.COM/24 27 (2003) p. 15, <http://whc.unesco.org/archive/2003/whc03-27com-24e.pdf>.
29 WHC-13/37.COM/20 (2013) 37 COM 7A.16, p. 25, <http://whc.unesco.org/archive/2013/whc13-37com-20-en.pdf>.
30 WHC-14/38.COM/7A (2014) 38 COM 7A.28, <http://whc.unesco.org/archive/2014/whc14-38com-7A-en.pdf>.
31 WHC-14/38.COM/7A (2014) 38 COM 7A.32.
32 Ibid., 38 COM 7A.34.
33 Ibid., 38 COM 7A.38.

and the revival of 'important and significant natural habitats for in-situ conservation of biological diversity'.[34]

Convention on the Conservation of Migratory Species of Wild Animals (Bonn Convention)

The Convention on Migratory Species was negotiated in 1979 in recognition that species that crossed geographical boundaries needed coordinated management plans for their long-term conservation and protection.[35] For purposes of this Convention, states negotiated obligations to restore habitats that were critical to endangered species and species with unfavourable conservation status as a means for achieving conservation objectives. The Convention provides that range states for endangered migratory species should 'where feasible and appropriate, restore those habitats of the species which are of importance in removing the species from danger of extinction'.[36] This obligation is a qualified obligation leaving a large degree of discretion on the part of range states to determine what is 'feasible and appropriate'. Does this refer to economic feasibility or environmental feasibility? Does this refer to political appropriateness or ecological appropriateness?

The language in Article III contains an important ambiguity leaving the text open to at least two interpretations. First, it could be read narrowly to provide that range states must protect habitats, but only those habitats 'which are of importance in removing' the endangered species from the list. Second, it can be read broadly to endow protection not just for the habitat occupied by listed endangered species, but also for the habitat occupied by non-endangered species that 'are of importance in removing the [endangered] species from danger of extinction'. This second reading would then encompass habitats for those species that endangered species depend upon and would presumably cover a far greater range. The practical consequences of reading the text broadly and narrowly matter in terms of policy interventions. For example, if a given endangered species such as a panda eats a particularly tree or bush such as mountain bamboo, should a state simply restore the handful of habitats where the panda is known to currently reside? Or should the state undertake a much broader landscape restoration project to also restore bamboo groves where pandas might reside if there was a larger wild population of pandas?

There are no clear answers offered in the Convention text about whether the scope of restoration obligations are intended to be broad or narrow. Given that the Convention was negotiated in an era before there was diplomatic acknowledgement of the value of the ecosystem-based approach, it is more likely that the

---

34 UNESCO, Criteria for Selection, <http://whc.unesco.org/en/criteria/> (see in particular criteria IX and X).
35 Convention on the Conservation of Migratory Species of Wild Animals, Bonn, 23 June 1979 (1979) 19 *ILM* 15.
36 Ibid., Article III(4).

states were only committing to restore habitat for the listed endangered migratory species and not for dependent non-listed species. The Convention does not provide any sanctions where states fail to restore habitats for endangered migratory species.

In addition to the unilateral requirements for states to act under Article III to restore habitat for endangered species, the Convention also provides under Article V that parties should negotiate agreements 'to restore' listed migratory species 'to a favourable conservation status or to maintain it in such a status'.[37] The Article V obligation, like the Article III obligation, leaves a great deal of discretion for a State Party to decide when an agreement regarding a migratory species is 'appropriate and feasible'.[38] Restoration is just one of many conservation strategies that parties are expected to incorporate into an agreement. The language in the Convention instructs states 'where required and feasible' to engage in 'restoration of habitats of importance in maintaining a favourable conservation status, and protection of such habitats from disturbances, including strict control of the introduction of, or control of already introduced, exotic species detrimental to the migratory species'.[39] In addition to restoring habitats by, for example, removal of invasive species, parties are expected to coordinate 'conservation and management plans'.[40] By implication, parties should coordinate their restoration efforts as part of these plans which might include 'elimination of … or compensation for activities and obstacles which hinder or impede migration'[41] as well as 'prevention, reduction or control of the release into the habitat of the migratory species of substances harmful to that migratory species'.[42]

In 2005, states submitted national reports in a harmonised form for publication on the Convention website indicating restoration work that they were engaged in for various listed species. For example, Chad reported habitat restoration work for gazelles and Senegal reported habitat restoration efforts for a species of falcon.[43] While the documented mapping of effort is notable, the reports, in general, provide extremely minimal information about what activities have been undertaken by the states.[44] While the reporting has continued until present as an obligation of conduct, the reports, with a few exceptions, remain at a level of generality that do not seem to be facilitating either communication or coordination on restoration

---

37 Ibid., Article V(1).
38 Ibid., Article V(5).
39 Ibid., Article V(5)(e).
40 Ibid., Article V(5)(b).
41 Ibid., Article V(5)(h).
42 Ibid., Article V(5)i).
43 Convention on Migratory Species National Reports from 2002 are available at <www.cms.int/en/meeting/seventh-meeting-conference-parties-cms>.
44 Most reports provide no indication of what steps the State Party has taken. A few of the reports such as the Slovakian report provide a description of the activities and locations for restoration activities (e.g. installation of artificial nests for white-tailed eagles near the Oravska water reservoir); Slovakia, 2002 National Report, <www.cms.int/sites/default/files/document/national_report_slovakia_0.pdf>.

efforts between states.[45] The Convention Secretariat has requested more comprehensive responses, but most states provide little information that describes progress towards national or regional goals. The reports could provide an important vehicle for states to share both the successes and challenges associated with national restoration efforts and notable non-state actor-based efforts for restoration.

Since the Convention went into effect, states have negotiated several Article V agreements under the auspices of the Convention on Migratory Species. Many of these agreements contain explicit references to restoration duties. What is interesting to observe in the history of the drafting of these agreements is an increasing trend towards specificity of state duties under these agreements. In the Agreement on the Conservation of Seals in the Wadden Sea in 1990, parties simply agreed to 'explore the possibility of restoring degraded habitats and of creating new ones'.[46] Written so broadly, this obligation required very little of the states beyond a brief consultation.

In 1995, States concluded the African-Eurasian Migratory Waterbirds Agreement, the largest geographical agreement of the agreements negotiated under the Convention on Migratory Species covering 119 range states. Parties are expected 'to take coordinated measures to maintain migratory waterbird species in a favourable conservation status or to restore them to such a status'[47] and to apply the precautionary principle when making decisions about species maintenance or restoration.[48] Unlike the Conservation of Seals Agreement, the African-Eurasian Agreement contemplates specific verifiable actions. First, the parties must 'identify sites and habitats for migratory waterbirds occurring within their territory and encourage the protection, management, rehabilitation and restoration of these sites'.[49] Second, the parties are expected to 'investigate problems that are posed or are likely to be posed by human activities and endeavour to implement remedial measures, including habitat rehabilitation and restoration, and compensatory measures for loss of habitat'.[50] The African-Eurasian Agreement reflects a new development where restoration projects are the outcome of coordinated environmental planning.

In 2007, range states negotiated the Agreement on the Conservation of Gorillas and their Habitat and agreed 'to take co-ordinated measures to maintain gorillas in a favourable conservation status or to restore them to such a

45 Some of the reports do offer interesting insights in regards to unintended consequences. For example in the 2010 German Report, the authors observed that the less-white-fronted goose might have lower levels in the German coastal region 'owing to the restoration project in Scandanavia': 2011 German National Report, UNEP/CMS/Inf.10.12.28, p. 16, <www.cms.int/sites/default/files/document/028_germany_e_0.pdf>.

46 Agreement on the Conservation of Seals in the Wadden Sea (1990), Article VII, <www.cms.int/species/wadden_seals/sea_text.htm>.

47 Agreement on the Conservation of African-Eurasian Migratory Waterbirds (1995), Article II (1), <www.cms.int/species/aewa/aew_text.htm>.

48 Ibid., Article II(2).

49 Ibid., Article III(2)(c).

50 Ibid., Article III(2)(e).

status'.[51] In order to achieve species restoration, each Party to the treaty agreed to 'identify sites and habitats for gorillas occurring within their territory and ensure the protection, management, rehabilitation and restoration of these sites'.[52] The 2007 obligation language from the Agreement on the Conservation of Gorillas parallels the language from the 1995 African-Eurasian Agreement and offers a significant hierarchy of action for states. In this hierarchy of action, protection and management of essential habitat are prioritised over rehabilitation and restoration of those habitat values. Restoration is correctly identified as the option of last resort. In a world of finite resources, this hierarchy of action makes sense, though it does raise some issues about planning for conservation. At what point are protection and maintenance efforts not adequate and rehabilitation and restoration efforts thereby become necessary? Given the costs in initiating rehabilitation and restoration work, who gets to make that decision for migratory species?

In 2012 states negotiated an amended version of the Agreement on the Conservation of Albatrosses and Petrels.[53] Parties agreed to 'progressively undertake' efforts under an action plan including 'habitat conservation and restoration' in order to achieve and maintain a 'favourable conservation status' for albatrosses and petrels.[54] This Agreement marked a clear effort by its drafters to be specific about what measures parties must undertake and included by international standards a relatively detailed restoration plan.[55] The plan called for parties to introduce legislation 'and other controls' that will either maintain or restore the birds to favourable conservation status.[56] Under the Agreement, parties are expected to undertake specific restoration practices including eliminating non-native animals, plants, hybrids or disease-causing organisms[57] and potentially remediating a critical habitat area to remove pollution sources.[58]

While habitat restoration appears to be a primary means for achieving the goals of the Convention on Migratory Species, success under the terms of the Convention depends not upon restoration of habitat or inter-related species but rather upon the recovery of the listed migratory species to levels of historic abundance.[59] While the historical abundance and coverage standard is desirable from a 'corrective justice' perspective, whereby states make amends through restoration efforts

---

51 Agreement on the Conservation of Gorillas and their Habitat (2007), Article II(1), <www.cms.int/species/gorillas/agrmt_text/Scanned_Agreement_text_E.pdf>.
52 Ibid., Article III(2)(b).
53 Agreement on the Conservation of Albatrosses and Petrels (2012), Article III(1)(a), <www.cms.int/species/acap/acap_text_of_agreement/agreement_2012_e.pdf>.
54 Ibid., Article VI.
55 Ibid., Annex B.
56 Ibid., Annex B (2.1).
57 Ibid., Annex B (2.2.1).
58 Ibid., Annex B (2.3.1).
59 Convention on Migratory Species at Article I(1)(c)(4) (assigning favourable conservation status when 'the distribution and abundance of the migratory species approach historic coverage and levels to the extent that potentially suitable ecosystems exist and to the extent consistent with wise wildlife management').

for poor historic management, achieving this standard is increasingly unlikely. Two law professors offer an intriguing benchmark in lieu of unachievable historical abundance numbers. They propose protecting migratory species at a level of abundance providing for 'ecological viability'.[60] If such a benchmark was to be agreed upon by the parties and introduced into the operational portions of the Convention, the ecological 'abundance' standard would broaden existing restoration efforts by functioning as an obligation of result that states could rely upon for structuring their restoration interventions. Rethinking what restoration means under the Convention on Migratory Species as a conservation strategy may ensure that restoration efforts under the various Article V agreements remain relevant to achieving the underlying treaty objectives of the Convention on Migratory Species.

In 2016, Convention Parties agreed to set for themselves several long-term goals and targets as part of their Strategic Plan.[61] Target 11 requires states to identify migratory species and their habitats which provide important ecosystem services and either maintain or restore these species and their habitats 'to favourable conservation status'.[62] The structure of these goals parallels the Aichi Targets described in Chapter 6 but are specific to the concerns of migratory species.[63] Opportunities for States to achieve ecological restoration under multiple treaties is welcome in light of the common goals of the Convention on Migratory Species and the Convention on Biological Diversity and limitations on human and financial resources.

Ramsar Convention

Given the global pressure to convert wetlands into uplands and its impacts on particularly migratory birds, states negotiated in 1971 the Convention on Wetlands of International Importance especially as Waterfowl Habitat in Ramsar, Iran.[64] States agreed to identify wetlands deemed to be of 'international importance' and then manage these wetlands using 'wise use' values that would ensure conservation of both wetlands and waterfowl. The original Convention had no express mention of restoration. While the gold standard for the Ramsar Convention is the designation and protection of wetlands of international importance as protected areas, this book will discuss the Ramsar Convention both here as well as in Chapter 10 on protected areas because the Ramsar Convention is one of the

60 R. Fischman and J. Hyman, 'The legal challenges of protecting animal migrations as phenomena of abundance' (2010) 28 *Va. Envt. L.J.* 173.
61 Convention on Migratory Species, Strategic Plan for Migratory Species 2015–2023, Resolution 11–02, UNEP/CMS/COP11/Doc.15.2 (2016), <www.cms.int/sites/default/files/document/Res_11_02_Strategic_Plan_for_MS_2015_2023_E_0.pdf>.
62 Ibid., p. 13.
63 The Convention on Migratory Species Target 11 is based on Aichi Target 14.
64 Convention on Wetlands of International Importance especially as Waterfowl Habitat, Ramsar, 2 February 1971 (1971) 996 *UNTS* 245.

leading conventions shaping the discussion of how states should design and implement effective habitat restoration efforts.

Article 4 (2) implicitly provides for the possibility of restoration. When a party 'in its urgent national interest' takes action to remove protection from a wetland on the list, the state is expected to 'as far as possible compensate for any loss of wetland resource' by creating 'additional nature reserves ... either in the same area or elsewhere, of an adequate portion of the original habitat'. In practice, where there is no additional similarly situated habitat, a state may have to restore another type of degraded habitat to serve as a reserve.[65]

Over 20 years passed after the original treaty negotiation before restoration was considered to be a priority for the Ramsar Convention framework, as demonstrated by a series of recommendations and resolutions that are regarded by some states as authoritative interpretations of obligations under the Convention.[66] In 1990, the 4th Conference of the Parties recommended that every state 'examine the possibility' of undertaking wetland restoration projects particularly to restore degraded wetlands.[67] In addition, parties asked the Standing Committee to consider whether a technical guide for restoration of wetlands should be prepared with a focus on the impact of global warming and sea level rise on wetlands. In 1993 at the 5th Conference of the Parties, the parties observed that in spite of the Convention being in force, there were still an unacceptable number of wetlands being converted to uplands. Parties agreed in Resolution 5.1 to focus their attention on the 'conservation and management of wetlands of international importance' which would include an effort to 'restore degraded wetlands and compensate for lost wetlands'.[68] In 1996 at the 6th Conference of the Parties, the parties agreed to focus on 'wetland restoration' as a form of 'nature restoration' that can also provide important ecosystem services to humans as well as to waterfowl.[69] Parties recommended that Contracting Parties give a 'higher priority to

65 See also Chapter 10 on protected areas.
66 Article 6(2)(d) and Article 6(2)(f) of the Convention provides the Conference of the Parties with the competency respectively to 'make general or specific recommendations to the Contracting Parties regarding the conservation, management and wise use of wetlands and their flora and fauna' and 'to adopt other recommendations, or resolutions, to promote the functioning of this Convention'. Article 6(3) provides that Parties to the Convention 'shall ensure that those responsible at all levels for wetlands management shall be informed of, and take into consideration, recommendations of such Conferences concerning the conservation, management and wise use of wetlands and their flora and fauna'; the Dutch government has annulled project leases on the basis of the project not satisfying recommendations and resolutions issued by the Ramsar Convention on Parties; see J. Verschuuren, *Ramsar Soft Law is Not Soft at All* (2008), <http://archive.ramsar. org/pdf/wurc/wurc_verschuuren_bonaire.pdf> (describing the Netherlands Crown Decision of 11 September 2007 in the Case Lodged by the Competent Authority for the Island of Bonaire on the annulment of two of its decisions on the Lac Wetland by the Governor of the Netherlands Antilles).
67 Recommendation 4.1. Wetland Restoration (1990).
68 Resolution 5.1. The Kushiro Statement and the framework for the implementation of the Convention (1993), Annex 1.
69 Recommendation 6.15. Restoration of Wetlands (1996).

restoration of wetlands' and 'integrate wetland restoration into their national nature conservation, land and water management policies'.[70]

This recommendation was elaborated on in 1999 at the 7th Conference of the Parties in Resolution 7.17, calling for each Ramsar Convention Party to establish a national programme of wetland restoration. Parties were requested to undertake a research programme that would provide an 'assessment of the lost processes, functions, composition and values of wetland areas'.[71] Based on these assessments, parties should identify priority sites for restoration and then identify in their national reports to the 8th Conference of the Parties what ecological, social, and economic constraints there are to pursuing restoration and what solutions there may be to overcome the constraints.[72] While states are left to define restoration in their own legislation, the resolution suggested that states at least define their 'restoration objectives and priorities … with reference to lost wetland functions, processes and components'.[73]

Importantly, Resolution 7.17 recommends parties to undertake some degree of environmental assessment before embarking on a restoration programme.[74] The inclusion of this recommendation is important because it makes explicit that while restoration is often a remedial mitigation strategy for environmental impact, restoration efforts may also have environmental impacts. Unsupervised active restoration, for example, with heavy machinery designed to open a river channel in a sensitive wetland environment could end up causing more environmental degradation in the long term by removing endangered plants than a less intensive intervention. A robust government and public dialogue is needed to ensure that stakeholder input is taken into consideration in order to legitimise the decisions to restore.

The Ramsar Convention documentation on strategic environmental assessment introduces a significant standard for evaluating potential restoration.[75] A Conference of the Parties resolution makes it clear that the 'baselines' for evaluating restoration of Ramsar sites are the 'target condition (ecological character) described in the objectives of the management plan' and not necessarily the condition of the site at the time that the wetland was listed.[76] This interpretation offers a novel perspective on what can constitute a baseline with a 'baseline' potentially being an aspirational baseline for purposes of evaluating future project impacts. What this means in practice is that even if a restoration objective has not yet been achieved, such as a water quality target, the unachieved target can still for purposes of environmental assessment be the baseline condition against which to measure impacts.

---

70 Ibid.
71 Resolution 7.17. Restoration as an element of national planning for wetland conservation and wise use (1999).
72 Ibid., paras 11 and 15.
73 Ibid., Annex 1, para. 1.
74 Ibid., Annex 1, para. 6.
75 Environmental Impact Assessment and Strategic Environmental Assessment: Updated Scientific and Technical Guidance, Resolution X.17 (2008).
76 Ibid., p. 15, para. 25.

States submitted national reports for the 8th Conference of the Parties but they did not necessarily conform with the content on inventories requested under Resolution 7.17. Rather, the reports offered useful insight into the complexity of prioritising. As Austria's report to the 8th Conference of the Parties indicated, prioritisation of restoration was hampered by the need to coordinate the priorities of nine different administrative authorities operating from different budgetary resources.[77] The authors of the report noted that EC-funded projects tended to receive priority attention.[78] Some countries such as Algeria indicated that inventorying wetlands was not a priority.[79] The national reporting efforts have varied greatly depending on the State. By 2015, 68 per cent of Ramsar Parties had identified priority sites for restoration.[80]

At the 8th Conference of the Parties in 2002, there was an additional burst of energy around identifying restoration programmes. Parties were urged to undertake restoration activities for peatland and mangrove ecosystems.[81] Most importantly, for the Convention implementation, the parties negotiated a document detailing 'Principles and Guidelines for Wetland Restoration' that concludes that restoration cannot be a substitute for protection of wetlands.[82]

The principles and guidelines are appended to the Resolution 8.16 in a 33-paragraph annex and parties are expected to integrate the substance of the annex into their individual national wetland policy and administrative guidelines.[83] Significantly, the principles and guidelines note that the terms 'restoration' and 'rehabilitation' are used interchangeably throughout Ramsar Convention documentation. While restoration might refer to a 'return to pre-disturbance conditions', the term is given a broader meaning in the resolution. 'Restoration' is defined as 'projects that promote a return to original conditions and projects that improve wetland functions without necessarily promoting a return to pre-disturbance conditions'.[84]

77 National Planning Tool for the Implementation of the Ramsar Convention on Wetlands, Austria (2002) 12, <www.ramsar.org/sites/default/files/documents/library/a ustria_nr2002_1_2.pdf>.
78 Ibid.
79 Instrument de planification nationale pour l'application des dispositions de la Convention de Ramsar sur les zones humides, Algeria (2002) 12.
80 The Ramsar Strategic Plan 2016–2024, Resolution XII.2 (2015) p. 26, Target 12 Baseline.
81 Resolution 8.17. Guidelines for global action on peatlands (2002) para. 10 (defining the wise use of peatlands as 'including restoration and rehabilitation); Resolution 8.32. Conservation, integrated management and sustainable use of mangrove ecosystems and their resources (2002) (Requesting parties to 'implement measures to protect and restore [mangrove ecosystem] values and functions for human populations … and the maintenance of biodiversity …').
82 Resolution 8.16. Principles and Guidelines for Wetland Restoration (2002).
83 Ibid., Annex 1, paras 5–6.
84 Ibid., Annex 1, para. 3.

The primary principle for parties to incorporate into their restoration work is an emphasis on 'careful planning' for restoration work so that 'ecological engineering' (e.g. living shorelines) might be able to be incorporated into wetland restoration efforts rather than 'hard structures or extensive excavation'.[85] In terms of appropriate planning levels, states should focus their planning efforts on catchment basins that include both wetlands and uplands and on reviving ecosystem functions.[86] Additional principles call for open community engagement, ongoing management and monitoring, and if available, the incorporation of traditional management that also prioritises adaptable management.[87] While there has been no public evaluation of whether the principles and guidelines are making a difference in restoration projects, there is some question as to whether the content of the resolution is even reaching the national wetlands managers, as the frontline decision-makers on restoration projects.[88]

Of all of the international regimes, the Ramsar framework is perhaps the most robust in terms of creating a shared understanding of how states might set appropriate goals and objectives for restoration and then implement restoration strategies. The Ramsar states in recent meetings have analysed how their efforts can support ecological efforts adopted under other international legal frameworks. For example, the Ramsar states in 2012 evaluated the Biodiversity Convention's Aichi Targets to ascertain how Ramsar projects could help achieve these targets. The parties determined that restoration work of wetlands restoration under the Ramsar Strategic Plan would contribute to Aichi Target Goal 5 to at least halve the loss of all natural habitats by significantly reducing habitat degradation and fragmentation as well as Goals 14 and 15 (see Chapter 6).[89]

While Resolutions 7.17 and 8.16 are the pivotal documents in terms of defining state restoration obligations under the Ramsar Convention, subsequent Conferences of the Parties have continued to address restoration work. At the 10th Conference of the Parties in 2008, states recognised in the Changwon Declaration on human

---

85 Ibid., Annex 1, paras 10–11.
86 Ibid., Annex 1, paras 13–14.
87 Ibid., Annex 1, paras 17–18, see also para. 32 (recognising the significance of adaptable management since 'in almost all cases restoration projects should be considered experimental in nature').
88 G. van Boven, *An Evaluation of the Use & Utility of Ramsar Guidance: A Report to Ramsar Scientific & Technical Review Panel and Ramsar Secretariat* (2007) para. 8. One third of wetlands site managers polled in a Ramsar Convention sponsored review stated that they had no access to the various resolutions and another third responded that they were unaware of the existence of the resolutions. Interviewed parties offered a number of suggestions for improving the effectiveness of resolutions including tailoring the language to practitioners and covering more specific topics using clear language.
89 Resolution 11.3. Adjustments to the Strategic Plan 2009–2015 for the 2013–2015 triennium (2012) Appendix 1, page 7 (Ramsar Strategy 1.8 provides that states should 'Identify priority wetlands and wetland systems where restoration or rehabilitation would be beneficial and yield long-term environmental, social, or economic benefits, and implement the necessary measures to recover these sites and systems').

well-being and wetlands that restoration of wetlands was key to improving liveli-hoods and water quality, sustaining agriculture and fisheries, protecting biodi-versity and adapting to climate change.[90] In particular, states have highlighted the function of wetlands in alleviating some of the expected flooding impacts of cli-mate change.[91] Parties have been further encouraged to restore river basins, lake basins, aquifer basins, wetlands associated with these basins, lowland wetlands, coastal wetlands and peatlands as priority policies to respond to climate change.[92] At the 11th Conference of the Parties in 2012, the parties recognised that gov-ernment intervention to restore wetlands may not be sufficient and urged public and private actors to work in partnership.[93]

In the 2016–2024 Strategic Plan proposed in 2015, the parties noted 'the modesty of the efforts invested in restoring wetlands' and committed to improving efforts.[94] As part of the eight-year plan, parties agreed that in order to achieve the mission of the Ramsar Convention for 'conservation and wise use of all wetlands' 'it is essential that vital ecosystem functions and the ecosystem services they pro-vide to people and nature are fully recognized, maintained, restored and wisely used'.[95] Specifically, the parties agreed to a goal of 'effectively conserving and managing the Ramsar site network' with a target of either maintaining or restoring 'the ecological character of Ramsar sites'.[96] As part of the wise use of wetlands, states are expected to address wetlands beyond the Ramsar site network and to engage in restoration of degraded wetlands in order to enhance 'biodiversity con-servation, disaster risk reduction, livelihoods and/or climate change mitigation and adaptation'.[97] The Ramsar parties recognise these targets as complementary to the Biodiversity Convention's Aichi Targets described in Chapter 6.

The efforts of the Ramsar parties since 1999 have generated a great deal of momentum around restoration as a long-term conservation strategy for wetlands. Most states now have focused domestic implementation programmes to support Ramsar efforts to protect wetland values for a variety of management objectives including both biodiversity and human use. While the Convention has largely left implementation of restoration standards to each party, there may be additional coordination and collaboration opportunities under the Convention, particularly

90 Resolution 10.3. The Changwon Declaration on human well-being and wetlands (2008) Annex.
91 Resolution 10.24. Wetlands and Climate Change (2008) para. 18 (observing that wise use and restoration of wetlands contributes to building the resilience of human populations to climate change impacts and can attenuate natural disasters expected with climate change, such as the use of restored floodplain wetlands to reduce risks from flooding).
92 Ibid., paras 30–32.
93 Resolution 11.20. Promoting sustainable investment by the public and private sectors to ensure the maintenance of the benefits people and nature gain from wetlands (2012) para. 10.
94 The Ramsar Strategic Plan 2016–2024, Resolution XII.2 (2015) para. 24.
95 Ibid., Annex, p. 5.
96 Ibid., p. 14, Target 5.
97 Ibid., p. 14, Target 12.

in the form of assisting least-developed states through the transfer of technical knowhow and financing for restoration efforts. The parties in 2015 agreed to support regional initiatives including training and capacity building centres to improve Convention implementation.[98]

The evolution of the Ramsar Convention over 40 years offers a significant illustration of the role that restoration is playing in habitat conservation efforts. For the Ramsar parties, restoration is a conservation strategy. Where sites are degraded but still have environmental values, restoration activities may provide opportunities for reviving damaged ecological functions. One area that has been unexplored within the Convention context is whether shared management efforts such as restoration work on transboundary Wetlands of International Importance might offer new opportunities for state to state cooperation.[99]

The Ramsar Convention, however, faces the same challenge as the World Heritage Convention: implementation depends entirely on good faith efforts of parties with no obvious legal accountability framework. As such, both of these Conventions depend on the treaty framework continuing to offer 'carrots' to keep parties engaged. One such 'carrot' is being offered to cities through accreditation as a 'Wetland City of the Ramsar Convention' for cities that have set water quality and sanitation standards, implemented policies for sustainable production systems and are protecting ecosystem services.[100]

While in recent years there seems to be a genuine good faith effort to implement restoration, there is often the question when one is facing multiple declining ecosystems of whether the existing efforts are too little too late. States with Ramsar sites simply self-report on the ecological character of sites without any system of verification.[101] Little is known about whether existing restoration efforts will be 'effective restoration or rehabilitation projects'.[102] A Ramsar Site Management Effectiveness Tracking Tool has been designed but this rapid site assessment tool depends also on self-monitoring which can, depending on who is charged with filling out the form, be compromised by political concerns such as internal

---

98 Regional initiatives 2016–2018 in the framework of the Ramsar Convention, Resolution 12.8 (2015).

99 P. Griffin, *The Ramsar Convention: A New Window for Environmental Diplomacy* (Institute for Environmental Diplomacy and Security, University of Vermont 2012) 25–92, <www.uvm.edu/ieds/sites/default/files/Ramsar_IEDSResearchSeries.pdf> (discussing 75 transboundary Ramsar wetlands that could be jointly managed by neighbouring states and suggesting that trust between neighbouring states might be built through cooperative wetland management to meet Ramsar objectives).

100 Wetland City Accreditation of the Ramsar Convention, Resolution XII.10 (2015).

101 Ramsar Convention, Article 3.2 (parties must 'arrange to be informed at the earliest possible time if the ecological character of any wetland in its territory and included in the List has changed, is changing or is likely to change as the result of technological developments, pollution or other human interference'. This information is expected to be 'passed without delay' to the Ramsar Secretariat and is included in a register of wetlands referred to as the Montreux Record); see also Chapter 10.

102 The Ramsar Strategic Plan 2016–2024, Annex 1, Ramsar Goals and Targets, Target 12 Baseline and Indicators (indicating that 70 per cent of Ramsar Convention parties have implemented restoration or rehabilitation programmes).

political conflicts among domestic agencies.[103] While there are many civil society groups supporting the implementation efforts of the Ramsar Convention, there may be a formal role for third-party verification efforts to support state monitoring of the ecological health of both their Ramsar sites and other wetlands.

UN Convention on Combatting Desertification (UNCCD)

Negotiated to address the loss of arable land to forces of erosion, the Convention on Combatting Desertification does not include the term 'restoration' in its text. Even so, the Secretariat of the Convention was eager to join the Hyderabad Call for a Concerted Effort of Ecosystem Restoration to bolster the UNCCD Secretariat's effort to achieve zero net land degradation.[104] For UNCCD parties, ecological restoration is perceived as a strategy for slowing and potentially reversing existing trends of desertification and reviving productivity. The parties to UNCCD have agreed to combat desertification which they defined as the 'rehabilitation of partly degraded land' and the 'reclamation of desertified land'.[105] Based on parties' efforts to recover functionality from the degraded land, the parties to the Convention seem to implicitly understand the Convention's word choice of 'rehabilitation' as incorporating broader concepts of ecological 'restoration' since Article 2 of the Convention requires parties to promote 'long-term integrated strategies that focus simultaneously, in affected areas, on improved productivity of land, and the rehabilitation, conservation and sustainable management of land and water resources'. The connection between 'rehabilitation' and 'productivity' in this Article suggests an intent to do more than simply rehabilitate land. While rehabilitation of land in practice is usually associated with reducing hazards created by a particular land use such as mining and is not ordinarily associated with reviving the 'productivity of land', Article 2 links the restoration objective of recovering productivity with rehabilitation practices.

In proposing a zero net land degradation objective, the Secretariat of the Convention on Combatting Desertification recognised that states would need to undertake more than traditional rehabilitation activities of the nature legally required in post-mine operations if the parties are to meet the objectives of the Convention to 'combat desertification and mitigate the effects of drought … through effective action at all levels … with a view to contributing to the achievement of sustainable development in affected areas'.[106] It would not be enough to simply secure soils in place from windstorms; the soil that was protected would need to be nutrient rich to help communities sustain their livelihoods as herders, agriculturalists or hunter-gatherers.

---

103 Evaluation of the Management and Conservation Effectiveness of Ramsar Sites, Resolution 12. 15 (2015) Annex 1.

104 Hyderabad Call for Concerted Effort on Ecosystem Services.

105 United Nations Convention on Combating Desertification, Paris, 17 June 1994 (1994) 33 *ILM* 1328.

106 Ibid., Article 2.

In May 2012, the Secretariat published a policy brief on 'Zero Net Land Degradation' (ZNLD) that introduces the concept of ZNLD as 'the achievement of land degradation neutrality, whereby land degradation is either avoided or offset by land restoration'.[107] Land restoration is defined as 'reversing land degradation processes by applying soil amendments to enhance land resilience and restoring soil functions and ecosystem services'.[108]

The Secretariat's somewhat radical proposal arises from a frustration that it will not be enough to avoid expansion into non-desertified lands to reverse the trends towards desertification. Too many lands have already been degraded through conversion to cropland and timberland usages. Instead, parties will also need to invest in reviving ecological values on existing degraded lands since, by some estimates, up to 40 per cent of global cropland is subject to severe soil erosion or has reduced fertility.[109] To ensure that these lands are not abandoned, more attention needs to be given to restoration of lands through some combination of better land-management policies, community-based management and payments for ecosystem services.[110] The Secretariat envisions the payments for ecosystem services being delivered to private parties, not only to prevent degradation of healthy land but also to manage degraded land for the 'primary purpose' of 'restoring the natural functions of the land's ecosystems' in order to return productivity.[111] The Secretariat hopes that parties to the Convention will negotiate a protocol to the Convention to provide legal support for payment for ecosystem services programmes designed to promote restoration.[112]

The UNCCD as a framework treaty has struggled with adequate implementation, given the scope of its subject matter. In many respects, it is a far more ambitious treaty than many of the treaties previously discussed in this chapter with multiple regional implementation annexes recognising a variety of desertification drivers ranging from wildfires to steep slopes to poverty. Calling for zero net degradation and a return to land productivity, the UNCCD requires substantial national investments in restoration in the form of both policies that reduce pressure of desertification drivers and financial resources. Whether these types of regulatory and financial investments are being made on a state by state basis is unclear. State Parties in Africa, Asia, Latin America and the Northern Mediterranean are expected to submit national action plans and participate in the formation of regional action plans to designate priority areas for actions.[113] These plans fulfil

---

107 United Nations Convention to Combat Desertification Secretariat Policy Brief (May 2012) 7.
108 Ibid., 6.
109 J.A. Foley, 'Global consequences of land use' (2005) 309 *Science* 570–574.
110 United Nations Convention to Combat Desertification Secretariat Policy Brief (May 2012) 22–23.
111 Ibid., 23.
112 Ibid., 25.
113 See, e.g., United Nations Convention to Combat Desertification, Annex I, Articles 8, 11 and 13.

core obligations under the Convention because they detail domestic commitments to meet international objectives. Unfortunately the UNCCD suffers from the same shortfall as the Ramsar Convention with a lack of accountability measures to ensure that fulfilling an obligation of conduct (e.g. submissions of the national plans) will ultimately produce results that support the achievement of the Convention's object and purpose.

It is unclear what role the national action plans are serving in improving governance. For example, in 2007, the UNCCD negotiated 'the Strategy' which was a ten-year strategic plan and framework for enhancing implementation of the UNCCD. One of the expected outcomes from the Strategy was an increase in knowledge regarding the interaction among climate change adaptation, drought mitigation and restoration of degraded land.[114] In 2014, several countries indicated that their national action programme was either not being implemented or that the plan was not aligned with the 2008–2018 Strategic Plan.[115] While restoration of lands to improve productivity is desired by the Secretariat of the UNCCD, it is unclear that the State Parties are collectively taking policy steps and making investments necessary to achieve this type of restoration.

Transboundary Watercourses Convention

The Convention on the Protection and Use of Transboundary Watercourses is a regional convention negotiated by the United Nations Economic Commission on Europe and focused on joint protection and management of transboundary surface waters and groundwater.[116] Focused on integrated water management, the Convention sets out the general obligation to 'ensure conservation and, where necessary, restoration of ecosystems'.[117]

To help states with the implementation of the Convention, conservation is distinguished from restoration. Conservation is considered to be the maintenance of ecosystem structures, functions and species composition.[118] Restoration refers to

114 UN Convention to Combat Desertification, The ten-year Strategic Plan and Framework to Enhance the Implementation of the Convention (2008–2018) Outcome 3.4, <www.unccd.int/Lists/SiteDocumentLibrary/10YearStrategy/Strategy-leaflet-eng.pdf>.

115 See generally, UNFCCC Submitted Reports for 2014, <http://prais2.unccd-prais.com/node/508> (indicating that, for example, as of 2014, Azerbaijan and Yemen had national action programmes that were not being implemented; Argentina, China and Egypt did not have a plan aligned with the ten-year strategic plan even though the strategy period is almost at an end).

116 Convention on the Protection and Use of Transboundary Watercourses and International Lakes, Helsinki, 17 March 1992 (1992) 31 *ILM* 1312; due to a later amendment, the parties to the treaty now include states who are not within the geographical region of the United Nations Economic Commission for Europe but are using the treaty to guide their relationships with other states with whom they share transboundary waters.

117 Ibid., Article 2(2)(d).

118 United Nations Economic Commission for Europe, Guide to Implementing the Water Convention, ECE/MP.WAT/39 (2013) p. 26, para. 111.

measures to 'improve ecosystems' and 'return (damaged) ecosystems to a former viable or "natural" condition (or, as this cannot always be achieved, to a close approximation of its condition prior to disturbance)'.[119] Restoration focuses on proactive steps that may include 'chemical methods for cleanup and restoration' or 'biological manipulation'.[120] In terms of standards, there are no specific measures that states must undertake, but there is a wide variety of supporting documentation under the transboundary watercourse regime that provides baselines against which to measure whether a given aquatic ecosystem will be 'able to maintain viable structures, functions and species compositions' depending on the amount of oxygen and the amount of harmful substances found within a waterbody.[121] At a minimum, legal obligations to conserve and restore ecosystems associated with watercourses seem to include the implementation of some water-quality standards, sediment-quality standards and water-quantity measures.[122]

Carpathian Convention

In addition to the numerous multilateral environmental agreements, there are also intergovernmental restoration agreements specific to a single geographical region.[123] This section discusses one example of such an agreement. In 2003, the states bordering the Carpathian mountain range agreed to a Framework Convention on the Protection and the Sustainable Development of the Carpathians.[124] In 2008, the parties negotiated a Protocol on Conservation and Sustainable Use of Biological Diversity to the Carpathians to further implement the Convention ('Carpathian Protocol'). Notably, the Protocol specifically obliges parties to engage in restoration work within the Carpathians.

All of the states engaged in the Carpathian Convention or Protocol were conversant with the concept of ecological restoration since they were already parties to the Ramsar Convention, the World Heritage Convention and the Biodiversity Convention. Yet how the concept of ecological restoration is handled in the Protocol is very different from the other multilateral agreements. What is interesting about the Protocol was the decision by the parties to explicitly define restoration in a highly restrictive manner. Unlike the Watercourses Convention described above, which understands the concept of restoration to include efforts that are a close approximation of pre-disturbance conditions, the Carpathian Protocol defines restoration as 'the return of an ecosystem or habitat to its original structure,

---

119 Ibid.
120 Ibid., para. 112.
121 Ibid., p. 27, paras 117–118.
122 Ibid., paras 119–120.
123 See, e.g., Bern Convention on the Conservation of European Wildlife and Natural Habitat, partly discussed in Chapter 10.
124 Framework Convention on the Protection and the Sustainable Development of the Carpathians, May 2003, <www.carpathianconvention.org/text-of-the-convention.html> (including as parties Czech Republic, Hungary, Poland, Romania, Serbia, Slovak Republic and Ukraine).

natural composition of species, and natural functions'.[125] This is a very narrow definition that seems to contemplate a return to a particular point along an environmental time line to a state where there is 'original structure' (i.e. certain ratios of species diversity, certain distributions of nutrients, certain climatic conditions) and 'original natural functions'. Whether a return to what is 'original' is technically possible is uncertain. The parties do not decide in the Protocol what constitutes 'original'. It is left to the discretion of the parties to determine the time frame of what qualifies as 'original'.

The nature of this obligation to restore to 'original structure' and 'original natural functions' becomes more complex when read within the larger context of the Convention. Each party is expected to undertake measures within its own territory 'with the objective to restore degraded habitats in the Carpathians' which may include both natural habitats and semi-natural habitats such as agricultural fields.[126] What does it mean to restore an agricultural field to an 'original structure'? This question matters because parties must have some ability to measure success in achieving Convention and Protocol objectives. Subsequent Conferences of the Parties have largely not addressed restoration matters. In 2014, the parties did emphasise, as part of their communication about spatial planning efforts, the need to restore river beds and flood plans 'as much as possible to original/natural state'.[127]

Even with the challenges of teasing out what it is the parties are attempting to achieve in the Protocol, the Protocol reflects an important development in international negotiations regarding the obligation to engage in restoration. Unlike previous instruments, it takes a bold step in defining what constitutes restoration for purposes of the agreement. Yet in the very act of doing so, it potentially raises the bar so high as to invite failure on the part of parties. Even so, the Carpathian Protocol's restoration definition provides an important first textual bridge between ecology and law. It is the first example of an international treaty where parties are explicit about their intention to use ecological structures and function as a basis for evaluating whether a party meets its legal obligations. This is important because it reflects a policy awareness that is absent from so many of the other internationally negotiated texts. To the extent that there has been some appreciation that restoration work must be done to enhance ecological standards, this

125 Protocol on Conservation and Sustainable Use of Biological and Landscape Diversity to the Framework Convention on the Protection and Sustainable Development of the Carpathians, 2008, Article 3(r), <www.carpathianconvention.org/tl_files/carpathia ncon/Downloads/01%20The%20Convention/1.1.2.1%20BiodiversityProtocolFina lsigned.pdf>.

126 Ibid., Articles 8 and 10; see also Articles 3(n) and 3(s). Natural habitats are defined as areas 'where an organism or population naturally occurs' while semi-natural habitats are defined as habitats that have been 'modified and maintained by human activities' but still include species naturally occurring in the area.

127 Decisions Fourth Meeting of the Conference of the Parties to the Framework Convention on Protection and Sustainable Development of the Carpathians, Spatial Development, UNEP/CC/COP4/DOC10/REV1, Decision COP 4/2, para. 4.

understanding is usually only found in the workings of the Conferences of Parties through recommendations or resolutions.

Global Partnership on Forest and Landscape Restoration

As will be illustrated in Chapter 6, the concept of ecological restoration has continued to gain momentum among states with a great deal of attention on how they can fulfil their ecological restoration obligations. This chapter closes with an introduction to a multi-stakeholder development that includes contributions from a number of governments focused on the operational challenges of doing large-scale restoration work related to the various treaties described in this chapter.[128] While not an international treaty like the other documents discussed in this chapter that binds State Parties, the Global Partnership has been instrumental in building knowledge networks across national organisations, international organisations and non-governmental organisations to support community practice of restoration efforts. The Partnership reflects a new direction in connecting international law with practices of restoration.

Initiated around 2005, the Global Partnership was the product of ongoing dialogues among forestry experts working at the IUCN, World Wildlife Fund, the Biodiversity Convention, the International Tropical Timber Organization, the United Nations Environmental Programme, the Food and Agriculture Organization, the UK Forestry Commission, the US Forest Service, and many other inter-governmental and government organisations concerned with forest restoration. Members of the Partnership were concerned with the sharing of information among resource managers, policy-makers, researchers and other stakeholders on how to most effectively restore forest landscapes.

The Global Partnership has played a critical role in creating a normative framework for implementing international commitments under other treaties particularly the UNFCCC REDD+ targets through its learning networks.[129] In 2011, the Bonn Challenge catalysed a series of voluntary commitments towards restoring 150 million hectares of degraded and deforested land by 2020.[130] The 150 million hectare target has been increased to 350 million hectares of restored forests.[131] In addition to states, subnational governments, companies, indigenous people and non-governmental organisations have committed to supporting restoration efforts in their various capacities. In order to help states and others achieve their

---

128 Originally launched by the IUCN, WWF and the Forestry Commission of Great Britain, the partnership initiative has grown to include as participating governments China, El Salvador, Finland, Ghana, Italy, Japan, Kenya, Lebanon, the Netherlands, South Africa, Switzerland, the United Kingdom and the United States.

129 See also Chapter 11 on climate change.

130 N. Sizer et al., 'Bonn Challenge 2.0: Forest and landscape restoration emerges as a key climate solution' (Insights Blog, World Resources Institute 2015); see also Chapter 11.

131 New York Declaration on Forests, 2014, <www.un.org/climatechange/summit/wp-content/uploads/sites/2/2014/07/New-York-Declaration-on-Forest-%E2%80%93-Action-Statement-and-Action-Plan.pdf>.

commitments, the Partnership has been working outside of any treaty body to design tools such as restoration assessments to produce maps with priority restoration sites and promote guidelines to assist decision-makers, stakeholders and investors in restoring forest landscapes.[132]

What is perhaps most interesting from a legal perspective is the effort by Global Partnership members to create for themselves ten basic principles to guide the actions of its members. These principles collectively form the Principles for Forest and Landscape Restoration. While the Principles do not legally bind partnership members, the Principles were articulated as a means of creating a common set of goals to achieve landscape level restoration that both protects the environment while ensuring adequate agricultural production. The principles have been subsequently adopted by the Subsidiary Body on Scientific, Technical and Technological Advice of the Biodiversity Convention as supporting guidance for ecosystem restoration.[133] The ten principles include:[134]

Principle 1: Landscape level restoration works requires continual learning and adaptive management.

Principle 2: Achieving landscape level restoration will require a 'common concern entry point' where stakeholders are willing to make compromises to achieve certain outcomes.

Principle 3: Landscape restoration projects must be aware of the multiple levels at which ecosystems processes function.

Principle 4: Landscapes have multiple uses that are valued in different ways by different stakeholders.

Principle 5: Multiple stakeholders' interests must be recognised and differences in values and approaches may need to be negotiated for the success of a project.

Principle 6: Stakeholders must accept a shared logic in their approach to restoration work and to be aware of risks and uncertainties associated with proceeding on projects.

Principle 7: Stakeholders must clarify their rights and responsibilities in pursuing ecological restoration work. This approach is considered to be an alternative to command and control regulatory approaches to projects.

Principle 8: Monitoring of restoration efforts should be participatory and user-friendly.

Principle 9: Restoration work should be undertaken with the goal of creating a resilient system that is better able to respond to future shocks.

Principle 10: There should be capacity building to ensure that stakeholders can participate fully in ecological restoration efforts.

132 The Global Partnership on Forest Landscape Restoration, Assessing National Potential for Landscape Restoration: A Briefing Note for Decision Makers, <www.forestla ndscaperestoration.org/sites/default/files/topic/assesing_national_potential_for_la ndscape_restoration.pdf>.
133 Biodiversity Convention, Information Note, UNEP/CBD/SBSTTA/20/35 (2016).
134 J. Sayer et al., 'Ten principles for a landscape approach to reconciling agriculture, conservation, and other competing land uses' (2013) 110(21) *PNAS* 8349–8356.

As of 2016, it is unclear how these principles have been adopted by the different members of the Global Partnership. At some point in the future, some of these principles could become the basis for fashioning criteria or indicators to measure the success of various national or regional projects. If these principles were to become indicators, then this would be a fitting contribution to the implementation of Principle 1 encouraging partners to engage in continual learning through the process of restoration work.

A variety of other national and regional landscape restoration targets have also been negotiated with the support of non-governmental organisations such as the World Resources Institute. For example, in Latin America and the Caribbean, countries have pledged by 2020 to restore 20 million hectares of forest.[135] In Africa, states have agreed to restore 100 million hectares of land by 2030.[136]

The Global Partnership and the various parallel commitments such as the New York Declaration on Forests reflect a new governance trend by formally recognising that landscape restoration requires active participation across society.[137] While commitments to pursue restoration have been made by states across a variety of binding international instruments, the domestication of these commitments requires complex negotiations among a variety of actors including both landowners and various other users of land. The domestication of international legal obligations to restore will be discussed further in Chapter 8. The next chapter evaluates the legal obligations of parties to the Convention on Biodiversity (CBD), to participate in ecological restoration efforts. Whereas the obligation to restore is a component of each of the treaties discussed in this chapter, it is not necessarily a central obligation. Under the CBD regime, restoration is one of the core treaty obligations. As a result, states have engaged in an ongoing conversation that includes not just states but also a variety of non-state stakeholders regarding how to implement national and international restoration efforts.

---

135 Initiative 20X20, <www.wri.org/our-work/project/initiative-20x20>.
136 AFR100, <www.wri.org/our-work/project/AFR100/about-afr100>.
137 The role of government actors will be discussed further in Chapter 8 and the role of private actors will be described in Chapter 9.

6 The Convention on Biological Diversity and ecological restoration

The two previous chapters discussing a variety of international treaties and soft-law instruments set the context for this chapter. The Convention on Biological Diversity (CBD, or Biodiversity Convention) is the culmination of many of the previous negotiated international agreements.[1] Like many of the other treaties discussed in the previous chapters that focus on restoration of particular habitats, the CBD adopts an approach towards restoration reflecting an 'obligation of conduct'. Where the CBD differs from the treaties discussed previously is that State Parties to the CBD have subsequently negotiated a voluntary strategy to improve accountability that includes quantitative targets that would, if they were legally binding, qualify as 'obligations of result'. For purposes of thinking about ecological restoration and law, the CBD is of particular interest because of the recent attention in the past few years to mainstreaming restoration as one critical path for extensive ecosystem recovery. More so than any other treaty arrangement, the CBD has prioritised ecological restoration.

This chapter begins with a brief look at the language in the CBD text describing the restoration obligation for states before examining the development of the Aichi Targets in 2010. The chapter continues with a detailed historical look at decisions by CBD Conference of the Parties involving restoration with a special emphasis placed on 2012 documents circulated by the Secretariat of the CBD that focused on defining 'ecosystem restoration' and 2014 documents that describe a number of interim actions to achieve restoration goals.

Restoration obligations in the Biodiversity Convention text

Negotiated in preparation for the 1992 Rio Conference ('Earth Summit'), the CBD was intended to be a 'holistic' convention that would comprehensively address both conservation and the sustainable use of biodiversity.[2] The intent was to fill gaps left by other conventions, not to create an 'umbrella convention that would absorb the existing conservation conventions'.[3] An ad hoc group of

---

1 Convention on Biological Diversity, 5 June 1992 (1992) 31 *ILM* 818 (CBD).
2 Ibid., Article 1.
3 Report of the Ad Hoc Working Group of Experts on Biological Diversity on the Work of its First Session, UNEP/Bio.Div.1/3 (1988) p. 5, para. 16.

experts met in November 1988 to discuss the feasibility of developing a new binding instrument on biological diversity. This group noted that restoration was just one strategy among many on the continuum of conservation and sustainable use of biodiversity.[4] In early discussions about the content of the Convention, experts proposed that one of the general obligations of parties should be that national 'measures to maintain and restore biological diversity should include measures to conserve it'.[5] This suggestion captures the important understanding that restoration is really a long-term conservation strategy. Experts also proposed in early discussions including draft obligations for in-situ conservation that would require parties 'to restore and regenerate plans for animal species and their habitats'.[6] This awkwardly phrased recommendation explicitly recognised the need for states to engage in systematic ecological restoration planning.

The efforts of the ad hoc experts resulted in the formation of a new ad hoc group of legal and technical experts who continued the work of developing a treaty. In their first meeting in November 1990, the experts noted that 'restore' was a term that would need to be defined within the treaty.[7] This proposal was never finalised leaving somewhat uncertain what the parties intended with the inclusion of the term restoration. There are some hints from the early documents about what parties might have been considering but they are not conclusive. For example, in September 1990, at least one delegate suggested that to 'maintain and restore biodiversity' should be interpreted to mean 'the rational and sustainable use of biological diversity'.[8] Ultimately, the Convention included two operative articles committing states to restoration efforts[9] and several other articles that implicitly refer to restoration efforts that are discussed here.

Article 8(f)

The first explicit reference to 'restoration' in the Convention is in Article 8 on 'in-situ conservation'. Parties are expected to 'as far as possible and as appropriate' to undertake a series of affirmative steps, including to 'rehabilitate and

---

4 It can be assumed that the broad term conservation includes restoration because the only language regarding restoration is folded into the section on 'in-situ conservation', which has been defined by the treaty as 'conservation of ecosystems and natural habitats and the maintenance and recovery of viable populations of species in their natural surroundings'. Ibid., Article 2.
5 Report of the Ad Hoc Working Group on the Work of its Third Session in Preparation for a Legal Instrument on Biological Diversity of the Planet, UNEP/Bio.Div.3/ 12 (1990) Annex 1, p. 17, para. 5.
6 Ibid., Annex 1, p. 19, para. 8.
7 United Nations Environment Programme, Ad Hoc Working Group of Legal and Technical Experts on Biological Diversity, Elements for Possible Inclusion in a Global Framework Legal Instrument on Biological Diversity, UNEP/Bio.Div/WG2/1/3 (24 September 1990) p. 6.
8 Ibid., p. 7.
9 CBD, Article 8(f) and Article 14(2).

restore degraded ecosystems and promote the recovery of threatened species ... through the development and implementation of plans or other management strategies'.[10]

The first proposed version of this language provided for the 'adoption of plans for the recovery and rehabilitation of species, habitats and ecosystems'.[11] A number of changes were proposed to this language that are added in brackets in the following text, including the provision for the:

> adoption of [appropriate] plans for the recovery and rehabilitation of species, habitats, and ecosystems [and of plans for restoration and regeneration of animal and vegetal species and for their habitat including rare and fragile ecosystems as well as the habitat of depleted, threatened or endangered species and other forms of marine, terrestrial and aquatic life] [and bearing in mind existing conventions, action plans and programmes at national, regional and global levels].[12]

This proposed language was subsequently examined by a 'lawyers' meeting', with lawyers from a representative number of regions reviewing and revising language. In July 1991, they proposed the following changes to the text: 'States shall ... Adopt [appropriate] plans for the recovery, rehabilitation, management and sustainable use and development of ecosystems, natural habitats and species [as far as possible, and as appropriate for each Contracting Party]'.[13] As of July 1991, the references to restoration had been removed and states focused instead on 'recovery, rehabilitation, management and sustainable use and development' on the basis of common but differentiated responsibilities. This language was again revised by a CBD Working Group who proposed that the language read that states shall:

> as far as possible and as appropriate in accordance with national legislation ... adopt and implement [as far as possible and as appropriate] plans for the recovery, rehabilitation, restoration, management [and sustainable use] of degraded ecosystems and habitats as well as [endangered] [threatened] species [populations and varieties] [representative of those covered in subparagraph (a)] [in order to strengthen biological diversity]. [Each such plan should specify the state of biological diversity selected as its objective] [and, where the Contracting Party is a developing country, the financial and technology requirements to achieve these objectives ...].[14]

10 CBD, Article 8(f).
11 UNEP, Elements for Possible Inclusion in a Global Framework Legal Instrument on Biological Diversity, p. 10, Sec. V(A)(f).
12 Ibid.
13 UNEP, Second Revised Draft Convention on Biological Diversity, UNEP/Bio.Div/INC.4/2 (23 July 1991).
14 UNEP, Intergovernmental Negotiating Committee for a Convention on Biological Diversity, UNEP/Bio.Div/N4-INC.2/5 (2 October 1991) Annex, Draft Articles Adopted by the Plenary as the Basis for Future Negotiations, p. 19.

By February 1992, the language for Article 8(f) appeared in its current treaty form with no explanation from the Working Group as to why it had chosen that language except that the language was reached 'by informal consultation between delegates'.[15]

The ultimate choice of language adopted by states in Rio that became Article 8(f) has the potential to be confusing for implementation purposes because it calls for parties to 'rehabilitate and restore degraded ecosystems and promote the recovery of threatened species'. As used by ecologists, the term ecological restoration can encompass restoration, rehabilitation, remediation and reclamation.[16] On an ecological recovery spectrum, restoration outcomes are often preferable to rehabilitation outcomes. What the drafters meant by the phrase 'rehabilitate and restore', however, is uncertain. The choice of the word 'and' could mean that states must at a minimum rehabilitate degraded ecosystems in preparation for their eventually restoration. This is a potentially expensive proposition in light of the technical and financial challenges of ecologically restoring ecosystem functions to a former industrial site such as an open pit mining site.

If it was not the intention of the drafters to trigger nationwide action to undertake some effort to reverse environmental degradation across the landscape, then the language appears to offer a great deal of latitude for how states might act to fulfil Article 8(f). To implement its obligations, a state could choose an approach that favours rehabilitation outcomes by default over restoration outcomes in order to reduce financial costs. When faced with a degraded ecosystem such as a wetland that could be either rehabilitated or restored, a state may proportionally favour rehabilitation strategies and defend this decision on the basis that the text in Article 8 does not require a party to 'restore to the greatest extent possible and rehabilitate everything else'.

Based on the negotiating history leading up to the language in Article 8(f) and because there has been no clarification of what states must do to satisfy Article 8(f), the choice of how to fulfil one's obligation under this part of the CBD seems intended to be one of practicality. It doesn't seem satisfactory for a state to achieve Article 8(f) through *de minimis* actions. Rather, when a state can achieve ecological restoration, it must 'as far as possible and as appropriate' undertake restoration to satisfy its in-situ conservation obligations. Where it is not possible because of financial constraints or technical impossibilities, then parties can meet the Convention obligations with acts of rehabilitation that are presumably less holistic than would be expected from an ecological restoration project. As with all treaty language, a state is expected to interpret its obligation in good faith to achieve the treaty purposes and objectives. What this means is

15 UNEP, Intergovernmental Negotiating Committee for a Convention on Biological Diversity, UNEP/Bio.Div/N6-INC.4/4 (18 February 1992) p. 10, para. 28.
16 A. Bradshaw, 'Introduction and Philosophy' in M. Perrow (ed) *Handbook of Ecological Restoration* (Cambridge University Press 2002) 4.

that a state must be able to prioritise sites that can be restored from those that are unlikely to be restored as part of its obligation to both 'rehabilitate and restore'.

Article 14(2)

In Article 14 on impact assessment and adverse impacts, the CBD instructs the Conference of the Parties to 'examine, on the basis of studies to be carried out, the issue of liability and redress, including restoration and compensation, for damage to biological diversity, except where such liability is a purely internal matter'. In partial fulfilment of the obligation under this Article, the Conference of the Parties has continued to discuss liability and redress. The CBD Executive Secretary at the request of the parties published a synthesis informational report in 2008 on the topic to assist parties in making future decisions on liability and redress.[17] The report elaborated on the concept of 'damage to biological diversity' and provided that such damage might require a loss of biodiversity 'of an enduring nature' resulting from a significant adverse or negative change.[18] The report also suggested that damage should be assessed on the basis of pre-incident or baseline conditions and that this baseline can serve as the 'reference point for restoration activities'.[19] The synthesis report recognised that the baseline condition can be dynamic and fluctuating.[20]

Regarding the topic of restoration, the report observed that experts from the CBD Group of Legal and Technical Experts preferred primary to compensatory restoration as redress.[21] 'Complementary restoration' at a different site from the damaged site may be appropriate 'where primary restoration to baseline conditions is not possible or practical'.[22] The report contemplates the need for additional rules and suggests that:

> Liability and redress rules could outline a process to facilitate the choice and application of an approach, while defining the respective roles and responsibilities of competent governmental authorities, responsible parties and civil society in the process … [and] could also provide guidance on how to make the determination [of a restoration approach] itself.[23]

---

17 Decision VIII/29 (2006); Executive Secretary of the CBD, Liability and Redress in the Context of Paragraph 2 of Article 14 of the Convention on Biological Diversity, UNEP/CBD/COP/9/20/Add.1. (The Executive Secretary's Report draws heavily on a previous report from the CBD Group of Legal and Technical Experts at UNEP/CBD/COP/8/27/Add.3.)
18 Ibid., pp. 3–4.
19 Ibid., pp. 6–7, para. 37.
20 Ibid., p. 6, para. 38 and p. 7, para. 41.
21 Ibid., p. 10, para. 68.
22 Ibid., p. 9, para. 60.
23 Ibid., p. 9, para. 58.

Criteria that have been proposed for measuring restoration options include:

a   Cost;
b   Extent to which the damaged resource would be returned to its baseline;
c   Likelihood of success;
d   Extent to which future damage would be prevented and the extent to which collateral damage from implementation would be avoided;
e   Extent to which more than one natural resource and/or service is benefited;
f   Effect on public health and safety.[24]

Regarding cost, the Executive Secretary's report indicated that a cost-benefit analysis may be appropriate in deciding whether to restore and how to restore.[25]

The Conference of the Parties has not yet made any decisions on what restoration activities would satisfy redress for biodiversity loss. Many questions remain. Will a party be expected to reintroduce historical structures and functions that correlate with the baseline? In practice this could mean that a state that fails to prevent transboundary harm and damages to a tidal wetland would then be expected to revive ecological function that would have been associated with a tidal wetland, even if habitat conditions may have changed due to sea level rise. Or will it be satisfactory for a state to revive any valuable ecological function? Under this scenario, a state could presumably replace tidal wetland ecosystem functions with any generic wetland ecosystem functions. Liability and redress continue to be subject of attention at the regular Conference of the Parties.

Other CBD articles

In addition to the two explicit references to restoration, there are at least two other articles that imply a commitment by states to undertake restoration activities. For example, Article 8(h) calls on parties 'as far as possible and as appropriate …[to] prevent the introduction of, control or eradicate those alien species which threaten ecosystems, habitats, or species'. The control or eradication of invasive species is ecological restoration, particularly when such species are interfering with the ability of an ecosystem to return to a particular historical trajectory with certain characteristic ecological structure and function. For example, it has become difficult in the San Francisco Bay Area to restore certain coastal hill habitat because of the prevalence of invasive species such as Scotch broom, a hardy perennial that outcompetes other species and thereby prevents the return of native plant communities.

Article 9 also suggests a restoration obligation in relation to ex situ conservation, when the CBD parties agreed to 'adopt measures for the

24 Ibid., p. 10, para. 71.
25 Ibid., p. 11, paras 78–80.

recovery and rehabilitation of threatened species and for their reintroduction into their natural habitats under appropriate conditions'.[26] Reintroduction of depleted species into their historical range would qualify as ecological restoration work.

The text of the Convention provides no further interpretive clues of what is expected from parties to satisfy these obligations but instead relies upon the post-Convention work of the parties. Among the post-Convention efforts of the states to realise the objective to restore are the CBD Strategic Plan, several Conference of the Party resolutions and national biodiversity plans. The CBD Secretariat has supported treaty implementation efforts by issuing research papers from the Executive Secretary for discussion by the parties at meetings such as the Subsidiary Body on Scientific and Technical and Technological Advice or at larger Conference of the Party meetings.

Implementation of CBD restoration obligations

To achieve the collective obligations of conservation and sustainable use of biodiversity embodied in the CBD individual states will need to to domesticate their obligations. International success is defined by domestic effort. This section examines the Aichi Targets, resolutions from a number of the Conference of the Parties and efforts by CBD institutions, as three negotiated approaches by which states have been working together to better define the CBD restoration obligations and to implement the CBD restoration obligations.

Aichi Targets

In 2010, parties to the CBD negotiated as part of the 'CBD Strategic Plan 2011–2020' biodiversity targets that are today referred to as the Aichi Biodiversity Targets, after the Japanese city where the parties agreed to collectively improve domestic implementation of the CBD. Two of these 20 targets refer directly to restoration and operate in a fashion as obligations of result by articulating anticipated outcomes. Target 14 provides that 'By 2020, ecosystems that provide essential services, including services related to water, and contribute to health, livelihoods and well-being, are restored and safeguarded, taking into account the needs of women, indigenous and local communities and the poor and vulnerable.'[27] Target 15 provides that 'By 2020, ecosystem resilience and the contribution of biodiversity to carbon stocks have been enhanced, through conservation and restoration, including restoration of at least 15 per cent of degraded ecosystems, thereby contributing to climate change mitigation and adaptation and to combating desertification.'[28]

---

26 CBD, Article 9(c).
27 Aichi Biodiversity Targets (2011), <www.cbd.int/sp/targets/default.shtml>.
28 Ibid.

Four additional Aichi Biodiversity Targets support restoration efforts even though the term 'restoration' does not appear in the text for any of these targets. First, Target 5 calls for 'the rate of loss of all natural habitats, including forests' to be 'at least halved and where feasible brought close to zero'. To reduce the loss, it is likely that some restoration interventions may be necessary, to reconnect fragmented habitats and provide connectivity through biodiversity corridors. Target 6 calls for better management of fish, invertebrate stocks, and aquatic plants through in part 'recovery plans and measures ... for all depleted species'. The reference to recovery plans is arguably a call for states to engage in either passive or active restoration as part of ongoing management through either no-take zones, hatchery support, habitat restoration or other measures that will increase stocks. Target 9 provides that 'By 2020, invasive alien species and pathways are identified and prioritized, priority species are controlled or eradicated, and measures are in place to manage pathways to prevent their introduction and establishment.' In many projects where restoration ecologists are attempting to revive certain ecosystem structures and function, a substantial part of the project may include the removal of species that out-compete other flora or fauna in the habitat. Often these species will be invasive species but may also include species that natively occur in the area such as the feral pigs on islands off the coast of Southern California.[29] Finally, Target 11 requiring the conservation of 'at least 17 per cent of terrestrial and inland water, and 10 per cent of coastal and marine areas' through 'effectively and equitably managed, ecologically representative and well connected systems of protected areas and other effective area-based conservation measures' is a restoration target. While certain states may have inland and coastal regions that may still be relatively pristine and good candidates for area conservation, there are many states that will require additional restoration interventions in order to achieve in particular 'well connected systems of protected areas'.[30] These interventions may require the removal of certain coastal human infrastructure or active design efforts to restore key habitat connectors such as marshes or wetlands.

While there are no definitions of what the parties were contemplating when they used the word 'restored' or 'restoration' in Targets 14 and 15, both targets appear to be premised on the protection of key ecosystem services. This interpretation is reinforced by a number of the National Biodiversity Strategy and Action Plans that states submitted in response to the adoption of the Aichi Targets to explain how they will domestically achieve the goals of the CBD.

In response to these targets, states have adopted a variety of national targets which have been publicised on the CBD website to encourage national accountability. Table 6.1 provides a list of restoration-based national targets that states

---

29 See, e.g., P.T. Schuyler, D.K. Garcelon and S. Escover, *Eradication of Feral Pigs on Santa Catalina Island*, <www.issg.org/database/species/reference_files/TURTID/Schuyler.pdf> (even though pigs may be native to the islands, the breeding success of the pigs had seriously impacted on the quality of the island habitat).
30 See Chapter 10 for a discussion of restoration and protected areas.

Table 6.1 National targets under the Convention on Biological Diversity

Nation (National Target Number)	National Target
Afghanistan (Preliminary Target 2)	Restore populations of selected taxonomic group.
Antigua and Barbuda (Target 15)	By 2020, restore biodiversity hotspots in Antigua and Barbuda thereby contributing to climate change mitigation and adaptation and to combating desertification.
Australia (Target 5)	By 2015, restore 1,000 km² of fragmented landscapes and aquatic systems to improve ecological connectivity.
Belarus (Target 11)	Ensure the restoration of 15% of degraded and inefficiently used ecological systems.
Belgium (Obj. 3)	Maintain or restore biodiversity and ecosystem services in Belgium to a favourable conservation status (all operational objectives). Ecosystems, their resilience and their services are maintained and enhanced by establishing, inter alia, a green infrastructure and restoring at least 15% of degraded ecosystems.
Bosnia and Herzegovina (National Goal 17)	By 2020, 30 lakes, created in opencast mining pits, restored to marshland habitats.
Botswana (National Goal 14)	By 2025, ecosystem services are identified and restored or maintained in all Botswana's ecoregions, and contribute to livelihood improvement through strategies that enable equitable access by all vulnerable groups, including women, the poor and local communities.

Table 6.1 (continued)

Nation (National Target Number)	National Target
Brazil (National Target 15)	By 2020, ecosystem resilience and the contribution of biodiversity to carbon stocks has been enhanced through conservation and restoration actions, including restoration of at least 15% of degraded ecosystems, prioritising the most degraded biomes, hydrographic regions and ecoregions, thereby contributing to climate change mitigation and adaptation and to combatting desertification.
Cambodia (Target 6) (Target 12)	By 2020, ecosystems and their functioning restored and preserved benefiting local communities particularly women, old people, children and indigenous people. By 2020, the rate of natural habitat loss reduced, and restoration of natural habitat and wildlife corridors improved.
Cameroon (Target 10)	By 2020, wetlands of great significance should be under management plans and at least 10% of degraded fresh water catchment areas and riparian zones restored and protected.
Canada (Target 3)	By 2020, Canada's wetlands conserved or enhanced to sustain their ecosystem services through retention, restoration and management activities.
Dominican Republic (Target 15)	By 2016, ecosystem resilience and the contribution of biodiversity to carbon stocks increased, through conservation and restoration, including the restoration of degraded land, thereby contributing to climate change mitigation and adaptation and to combating desertification.

Table 6.1 (continued)

Nation (National Target Number)	National Target
Eritrea (Target 15)	By 2020, ecosystem resilience and the contribution of biodiversity to carbon stocks enhanced, through conservation and restoration of degraded ecosystems, thereby contributing to climate change mitigation and adaptation and combating desertification.
Ethiopia (Target 10)	By 2020, contribution of biodiversity for ecological services, including climate change adaptation and mitigation improved through increasing forest cover from 12% to 14%; increased designation of wetlands from 4.5% to 9.0% and doubling restoration of degraded areas.
European Union (Target 2)	By 2020, ecosystems and their services maintained and enhanced by establishing green infrastructure and restoring at least 15% of degraded ecosystems.
Finland (Target 14) (Target 15)	By 2020, ecosystems that provide essential services, including services related to water, health, livelihoods and well-being, restored and safeguarded, taking into account socioeconomic and cultural considerations, notably the needs of the indigenous Sámi community. By 2020, ecosystem resilience and the contribution of biodiversity to carbon stocks enhanced through conservation and restoration. Finland participates in global efforts to restore at least 15% of degraded ecosystems, thereby contributing to climate change mitigation and adaptation and to combating desertification.
France (Target 6)	Preserve and restore ecosystems and their functioning.

Table 6.1 (continued)

Nation (National Target Number)	National Target
Gambia (Target 15)	By 2020, ecosystem resilience and the contribution of biodiversity to carbon stocks enhanced, through conservation and restoration, including restoration of at least 50 per cent of degraded ecosystems.
Greece (Specific Target 2.2) (Specific Target 8.2)	Restoration of important species and habitat types. Taking action to restore the impacts of invasive alien species on biodiversity.
Guyana (Strategic Objective 1)	Improve the status of biodiversity by conserving ecosystems, species and genetic diversity and by restoring biodiversity and ecosystem services in degraded areas.
Iraq (Target 8) (Target 9)	By the end of 2020 legislation to address the main pressures on forest ecosystems and native forest species issued, promoting sustainable management, restoration and conservation. By the end of 2020, about 1,000 square km of desertified shrub land and grassland restored.
Ireland (Target 10)	Continued rehabilitation or restoration of biodiversity elements.

Table 6.1 (continued)

Nation (National Target Number)	National Target
Japan (Action Goal B-1–2) (Target B-4) (Key action goal B-4–3) (Key action goal C-2–4) (Target D-1) (Key action goal D-1–4) (Key action goal D-2–2)	Reduce the rate of loss of natural habitats by at least half or bring this close to zero in cases where it is possible to do so by 2020. In addition, carry out the initiatives needed to noticeably reduce the degradation and fragmentation of natural habitats, such as the development of ecological networks and the restoration of wetlands and tidal flats. Identify invasive alien species and … promote a restoration of the habitation status of rare species and restore ecosystems to their original state by controlling or exterminating high priority [invasive] species through such efforts. Control or eradicate high priority invasive alien species, while also restoring the habitation status of rare species and restoring ecosystems to their original state through such efforts. Work to address ex-situ conservation for those species such as the crested ibis and the Tsushima leopard cat that are believed to be at an extremely high risk of extinction … The aim will be to restore ecosystems and revitalise regional communities by promoting the return to wildlife of individual animals that were propagated through artificial breeding through such initiatives. Strengthen the benefits received from biodiversity and ecosystem services in Japan and elsewhere by giving consideration to the needs of women and local communities through the conservation and restoration of ecosystems by 2020. Promote the restoration of coastal forests through the Green Bonds Regeneration Project, which gives forethought to conserving biodiversity. Promote measures for the conservation and restoration of ecosystems, thereby advancing measures that will contribute to climate change mitigation and adaptation.
Kiribati	Develop and initiate actions to protect and restore at least two threatened species in each of the Gilberts, Line and Phoenix Groups.
Malawi (Target 6)	By 2020, at least 50% of the degraded terrestrial habitats restored and protected.

Table 6.1 (continued)

Nation (National Target Number)	National Target
Malta (Target 13)	By 2020, vulnerable ecosystems that provide essential services safeguarded, with at least 15% of degraded ecosystems restored, while 20% of the habitats of European Community Importance in the Maltese territory have a favourable or improved conservation status.
Namibia (Target 4.1)	By 2022, ecosystems that provide essential services and contribute to health, livelihoods and well-being safeguarded, and restoration programmes initiated for degraded ecosystems covering at least 15% of the priority areas.
Nepal	By 2020, at least 10,000 hectares degraded mountain ecosystems restored through implementation of ecosystem based adaptation programmes. By 2020, at least 5% of the forested ecosystems restored through implementation of REDD+ programme.
Niue (Objective 9)	Following invasive species management, the best methods determined and implemented to facilitate effective restoration of native biodiversity or recovery of other values.
Oman (Target D.1)	By 2020, watershed areas, forest lands, outstanding scenic terrestrial, coastal and marine views, oases, and Aflaj (traditional water channels) restored and safeguarded.
Moldova	Ensure the conservation of rare, vulnerable and endangered species by restoring the habitats of five endangered plant and five endangered animal species.
Saint Lucia (3.5)	Agriculture, fisheries including aquaculture and forestry biological resources conserved, restored and sustainably managed and GMOs/Living Modified Organisms (LMOs) effectively managed to minimise genetic erosion and safeguard genetic diversity.
Serbia (Objective 1.1)	Restore biological diversity in degraded areas.
Seychelles (Project 11)	Conserve and restore lowland and highland wetlands.

have agreed to in order to implement the Aichi Targets. States could have simply adopted the Aichi Targets as their national targets. In many instances, the states adopted a version of the Aichi Targets and either changed language as in the case of Botswana taking Aichi Target 14 and extending the target deadline or Gambia increasing its goals of restoring 50 per cent of the degraded land. Other states have removed language as in the case of Eritrea taking out the quantitative target. These targets then reflect states interpreting for themselves what it means to achieve Article 8 CBD objectives for restoration.

While there are clear similarities among many of these national targets as modelled on the Aichi Targets, it is noteworthy that a number of states, particularly states in the Global South, have extended the deadline by which they intend to meet the target. These extensions suggest that some of these States may be applying the common but differentiated responsibility principle. Other national targets are very detailed such as Bosnia and Herzegovina who intends to restore 30 mining pits or marshlands or Iraq who intends to restore 1000 square kilometres of shrub and grass land (approximately the area of Los Angeles). These targets provide easy metrics against which to assess state accountability. Some national targets such as those set by Finland take on a global twist with the intent of Finland to not just restore its own land but also to contribute to restoring 15 per cent of global lands. Even though the goals vary greatly, the very fact that states have publicly articulated these national goals suggests that states feel a compliance pull to achieve the Aichi Targets in order to meet CBD Article 8(f) obligations.

Some states have voluntarily created national biodiversity strategies to implement the Aichi Biodiversity Targets which vary greatly depending on how broadly the concept of ecological restoration is understood by each state.[31] Elaborating on the national targets Table 6.2 provides a partial list of national approaches in National Biodiversity Strategy and Action Plans (NBSAPs): that involve restoration activities 64 out of 196 parties have provided NBSAPs to the CBD Secretariat since the Aichi Targets were adopted. For purposes of this chapter, all of the NBSAPs on file with the CBD were reviewed but only the national strategies where the state describes more than just a general commitment to implement the Aichi Targets are included in the table.[32]

Why should one care about these highly specific targets and strategies in a book on restoration and law? These targets and strategies are important because they reflect the domestication of a party's international obligations under the CBD and in particular Article 8(f). The domestication of the obligations reinforces the universal value of the international legal obligations. The development

---

31 As of 2014, approximately 23 states of the 194 parties had submitted revised National Biodiversity Strategies to reflect the Aichi Targets.
32 Copies of NBSAPs are available at <www.cbd.int/nbsap/about/latest/default.shtml>.

Table 6.2 Approaches to restoration under National Biodiversity Strategies

Nation	Partial strategies provided in National Biodiversity Strategy Documents illustrating diverse approaches to restoration [33]
Belgium[34]	* Restore areas capable of supporting ecological networks. * Restore both abiotic and biotic components. * Regards restoration as a 'process'.
United Kingdom[35]	* Restore ecological networks capable of delivering ecosystem services. * Restore as a contribution to climate change and adaptation.
Ireland[36]	* Restore at the landscape level rural agricultural landscapes.
Spain[37]	* Restore areas threatened with disappearing. * Restore ecological function.
France[38]	* Restore ecosystem functions.
Malta[39]	* Restore integrity, structure and functioning of important ecosystems including ecosystems acting as carbon sinks. * Adopt guidelines on management and restoration.
Finland[40]	* Restore degraded ecosystems cost-effectively or where that is not possible leave degraded ecosystems to 'revert to their natural states through natural processes'. * Identify legal obstacles to the restoration of habitats and reform economic incentives in order to promote a restoration of 'original natural values' in agricultural environments.

33 The partial strategies are grouped in the table according to geographical regions.
34 Belgian National Focal Point to the Convention on Biological Diversity (ed) Biodiversity 2020 – Update of Belgium's National Biodiversity Strategy (Royal Belgian Institute of Natural Sciences 2013) 10, <www.cbd.int/doc/world/be/be-nbsap-v2-en.pdf>.
35 Biodiversity 2020: A Strategy for England's Wildlife and Ecosystem Services (Department for Environment, Food and Rural Affairs 2011) 12, <www.cbd.int/doc/world/gb/gb-nbsap-v3-en.pdf>.
36 Actions for Biodiversity 2011–2016, Ireland's National Biodiversity Plan (Department of Arts, Heritage and the Gaeltacht) 33, <www.cbd.int/doc/world/ie/ie-nbsap-v2-en.pdf>.
37 Government of Spain, Ministry of the Environment, Rural Affairs, and Marine Environment, 'Plan estratégico del patrimonio natural y de la biodiversidad 2011–2017', Boletín Official del Estado (30 September 2011) 103074 and 103207–103209, <www.cbd.int/doc/world/es/es-nbsap-v3-es.pdf>.
38 Government of France, National Biodiversity Strategy 2011–2020 (2010) 18, <www.cbd.int/doc/world/fr/fr-nbsap-v2-en.pdf>.
39 Malta's National Biodiversity Strategy and Action Plan (2012–2020) 11, <www.cbd.int/doc/world/mt/mt-nbsap-01-en.pdf>.
40 Government Resolution on the Strategy for the Conservation and Sustainable Use of Biodiversity in Finland for the years 2012–2020 'Saving Nature for People', 2, <www.cbd.int/doc/world/fi/fi-nbsap-v3-p1-en.pdf>.

Table 6.2 (continued)

Nation	Partial strategies provided in National Biodiversity Strategy Documents illustrating diverse approaches to restoration
Belarus[41]	* Restore rare and endangered species of wild animals and plants, their populations and genetic diversity, sustaining their number which is necessary for their sustainable existence. * Restore ecosystems in demilitarised zones. * Restore and use of swamps, peat bogs that are not under industrial use anymore, renaturalising of wetlands (swamps). * Restore 'viable populations of at least five species of wild animals and plants, included into [sic] the Red Book of the Republic of Belarus, or that used to live in the territory of the Republic of Belarus, but disappeared'.
Serbia[42]	* Develop and implement restoration guidelines and standards for using native species, evaluating projects, and working with private landowners and regional institutions to restore areas of national concern.
Switzerland[43]	* Promote 'renaturation',[44] 'revitalisation'[45] and 'regeneration'.[46]
Japan[47]	* Ecosystems destroyed in last 100 years shall be restored in next 100 years. * Restore coastlines to 'restore the connection between people and the sea'.
Myanmar[48]	* Restore forest in critical watersheds and stop fishing until fish populations are restored.

41 Belarus Strategy on Conservation and Sustainable Utilization of Biological Diversity for 2011–2020, 9, <www.cbd.int/doc/world/by/by-nbsap-v2-p1-en.pdf>.
42 Biodiversity Strategy of the Republic of Serbia 2011–2018, 71 and 89–90, <www.cbd.int/doc/world/cs/cs-nbsap-01-en.pdf>.
43 Swiss Biodiversity Strategy (2012), <www.cbd.int/doc/world/ch/ch-nbsap-v2-en.pdf>.
44 Ibid., 38 (renaturation is a 'return of an anthropogenically altered habitat to a near natural state' through 'structural measures' such as improving 'small flowing watercourses').
45 Ibid., 85 (revitalisation is 'A form of renaturation' and 'measures ... for the restoration of the dynamic water and sediment processes in impaired alluvial zones').
46 Ibid. (regeneration is 'a form of renaturation' specifically 'measures ... for the re-establishment of wetland forming and long-term self-regulating hydrological processes in raised bogs and fenlands').
47 Japanese Ministry of Environment, Overview of the national Biodiversity Strategy of Japan 2012–2020 (2012) 8, <www.cbd.int/doc/world/jp/jp-nbsap-v5-en.pdf>.
48 The Republic of the Union of Myanmar, National Biodiversity Strategy and Action Plan (2011) 92, <www.cbd.int/doc/world/mm/mm-nbsap-01-en.pdf>.

Nation	Partial strategies provided in National Biodiversity Strategy Documents illustrating diverse approaches to restoration
Australia[49]	* Establish four continent-wide programmes to improve ecological connectivity. * Require all jurisdictions to align laws and policies to support biodiversity strategy.
Timor-Leste[50]	* By 2015 to plant at least one million trees especially mangroves in critical watersheds and on degraded lands while also removing invasive species as a means of supporting sustainable livelihoods for local communities through ecosystem restoration. * Restore in the medium term 100% habitat connectivity for terrestrial protected areas and 50% habitat connectivity for marine protected areas.
El Salvador[51]	* Prioritise restoration of critical ecosystems including mangroves and beach ecosystems, rivers, wetlands, reefs and 'gallery forests' (riparian forests).
Guatemala[52]	* Develop a National Plan for Conservation and Restoration of Ecosystems to improve resilience. * Create an incentive programme for restoration of biodiversity and ecosystem services.
Dominican Republic[53]	* By 2016 improve the connectivity of protected ecosystems while taking into account the needs of local groups. * Compile information about how ecosystems can contribute to carbon storage and the role of biodiversity to promoting resilience.
Surinam[54]	* Environmentally rehabilitate mining areas. * Limit spread of invasive species. * Restore forest areas where there has been clear-cutting.

49 Australia's Biodiversity Conservation Strategy, Natural Resource Management Ministerial Council 2010–2030 (2011), <www.cbd.int/doc/world/au/au-nbsap-v2-en.pdf>.
50 The National Biodiversity Strategy and Action Plan of Timor-Leste (2011–2020), Ministry of Economy and Development (2011) 23, <www.cbd.int/doc/world/tl/tl-nbsap-01-en.pdf>.
51 El Salvador, La Estrategia Nacional del Medio Ambiente, Ministerio de Medio Ambiente y Recursos Naturales (2013), <www.cbd.int/doc/world/sv/sv-nbsap-v2-es.pdf>.
52 Guatemala, Política Nacional de Diversidad Biológica (Acuerdo Gubernativo 220–2011) Estrategia Nacional de Diversidad Biológica y su Plan de Acción (Resolución 01-16-2012), <www.cbd.int/doc/world/gt/gt-nbsap-v2-es.pdf>; Ibid., p. 76 (restoration is regarded as a 'recovery' strategy for ecosystems and populations).
53 Dominican Republic, Ministerio de Medio Ambiente y Recursos Naturales, Estrategia Nacional de Conservación y Uso Sostenible de la Biodiversidad, Plan de Acción 2011–2020, 63, <www.cbd.int/doc/world/do/do-nbsap-01-es.pdf>.
54 Surinam National Biodiversity Action Plan 2012–2016, Ministry of Labor, Technological Development, and the Environment (2013) 27, <www.cbd.int/doc/world/sr/sr-nbsap-v2-en.pdf>.

Table 6.2 (continued)

Nation	Partial strategies provided in National Biodiversity Strategy Documents illustrating diverse approaches to restoration
Columbia[55]	* Restore or rehabilitate, by 2014, 280,000 hectares for uses such as biological connectivity corridors and 'the prevention of deforestation'. * Restore areas affected by forest fire. * Promote the formulation of programmes for the development of technologies for the restoration, recuperation and rehabilitation of ecosystems and the sustainable use of biodiversity, in coordination with the National System of Science and Technology.
Senegal[56]	* Restore for ecosystem resilience. * Restore wetlands. * Develop a payment for ecosystem service to restore degraded environments. * Coordinate with neighbouring states on coastal mangrove restoration efforts.
Burundi[57]	* Restore productivity of soil and ecosystem services for this generation and future generations. * Restore '*services essentiels*' including water, fisheries and forests. * Improve knowledge of restoration techniques for endangered resources.
Cameroon[58]	* Restore and protect at least 10% of degraded freshwater catchment areas. * Intensify programmes of mangrove restoration.
Congo[59]	* Create biological corridors between protected areas to assist restoration efforts. * Mainstream the best practices and methods of ecosystem restoration.

55 Republic of Columbia, National Policy for the Integral Management of Biodiversity and its Ecosystemic Services (PNGIBSE) (2012), <www.cbd.int/doc/world/co/co-nbsap-v2-en.pdf>.
56 Republique du Senegal, Strategie Nationale & Plan National D'actions Pour La Biodiversite (August 2015), <www.cbd.int/doc/world/sn/sn-nbsap-v2-fr.pdf>.
57 Burundi Ministère de l'Eau, de l'Environnement, de l'Aménagement du Territoire et de l'Urbanisme, Stratégie Nationale et Plan d'Action sur la Biodiversité (2013), <www.cbd.int/doc/world/bi/bi-nbsap-v2-p1-fr.pdf>.
58 Republic of Cameroon, National Biodiversity Strategy and Action Plan Version II (2012), <www.cbd.int/doc/world/cm/cm-nbsap-v2-en.pdf>.
59 Republic of Congo, Stratégie Nationale et Plan D'Actions Sur la Diversité Biologique (2015), <www.cbd.int/doc/world/cg/cg-nbsap-v2-fr.pdf>.

Nation	Partial strategies provided in National Biodiversity Strategy Documents illustrating diverse approaches to restoration
Afghanistan[60]	* Restore water, forests and rangeland through incentives and technical support. * Restore at least 15% of degraded forests (195,000 hectares) and rangelands (4.5 million hectares). * Restore populations of species of selected taxonomic groups. * Restore degraded wetlands. * Restore degraded lands: stabilise sand dunes and soils; reseed highly degraded rangeland; reduce grazing and dry land cultivation in vulnerable areas; map areas vulnerable to desertification.

of these national targets and NBSAPs by many state actors reflect state practice and may be contributing to the formation of customary obligations. Based on these national strategies, one might argue that restoration is becoming part of the law of nations where states have a shared understanding that they have a minimum legal duty to incorporate restoration efforts into national biodiversity strategies and identify degraded ecosystems. Going even further, it might be possible to review both the Aichi Targets and the content of some of the National Biodiversity Strategies and conclude that there may be an emerging customary obligation for states to actively participate in the restoration of certain essential supporting ecosystem services such as soil fertility to protect the needs of citizens.[61] What this means in practice is that states may need to reconsider their national priorities to ensure that ecological restoration is a substantive priority and that states are investing in restoration work.

Crystallising a narrow but important customary obligation based on practice and *opinio juris* may clarify the vague CBD treaty obligation that calls upon states 'as far as possible and as appropriate' to 'rehabilitate and restore degraded ecosystems and promote the recovery of threatened species ... through the development and implementation of plans or other management strategies'.[62] The recurring question for CBD parties has been what it means to restore degraded ecosystems 'as appropriate'. The Aichi Targets have now defined a minimum international standard for what states deem to be an 'appropriate' restoration of degraded

60 Islamic Republic of Afghanistan, National Biodiversity Strategy and Action Plan: Framework for Implementation 2014–2017 (2014), <www.cbd.int/doc/world/af/af-nbsap-01-en.pdf>.

61 A.G. Power, 'Ecosystem services and agriculture: tradeoffs and synergies' (2010) 365 *Phil. Trans. R. Soc. Bm* 2959–2971, DOI: 10.1098/rstb.2010.0143 (describing soil formation and structure, soil fertility, nutrient cycling and the provision of water as supporting ecosystem services).

62 CBD, Article 8(f).

habitats. This international standard is further reinforced by the objectives articulated within many of the NBSAPs.

CBD decisions on ecosystem restoration, 1998–2010

After the Convention went into force, much of the work to implement the Convention is done by the Conference of the Parties who provide decisions that encourage specific action by State Parties or by the CBD Secretariat. The content of these decisions can be informative about how states perceive their obligations to restore.

The CBD has an active Conference of the Parties which is expected to review the Convention and its implementation and, when necessary, to agree by consensus to make changes to improve treaty implementation and achievement of the treaty objectives.[63] The Convention's Conference of the Parties has adopted a number of decisions that call for direct action by states to achieve restoration. These decisions are not legally binding but reflect political commitments and normative development in the field.[64] States could in theory rely upon these decisions in formulating their own domestic interpretations of international obligations.[65]

Restoration has been an interest of the CBD parties since its inception and has been a theme at nearly every conference. At the Third Conference of the Parties in 1996 parties were requested to take action to 'achieve the restoration of habitats, including their biological diversity component'.[66] This decision reflected the intent for restoration to be holistic and complex enough to accommodate for 'diversity within species, between species, and of ecosystems'.[67] At the next several Conference of the Parties, states focused on specific restoration problems for particular habitats and ecosystems. Table 6.3 provides a sample of decisions made at Conference of the Parties from 1998 to 2010 that incorporate direct references to restoration.

---

63 Decisions are made on the basis of 2/3rd majority; ibid., Article 23(3).
64 R. Churchill and G. Ulfstein, 'Autonomous institutional arrangements in multilateral environmental agreements: a little-noticed phenomenon in international law' (2000) *American Journal of International Law* 636–642.
65 See, e.g., J. Verschuuren, *Ramsar Soft Law is Not Soft at All*, <http://archive.ramsar. org/pdf/wurc/wurc_verschuuren_bonaire.pdf> (This explains how a Dutch Court relied upon the Ramsar Convention text and guidelines adopted in a subsequent Ramsar Convention COP decision to decide that the Governor of Bonaire Islands, an island in the Netherlands Antilles, had correctly annulled development permits on the basis of violations of Ramsar commitments and that an EIA was necessary. In making this determination, the Court observed that because the Ramsar Convention text is very brief in describing State obligations that the Court was entitled to review subsequent resolutions, recommendations, and guidelines and to give these documents legal weight because they directly assisted with the interpretation of the Convention text and were unanimously adopted by the Ramsar Convention Parties).
66 Decision III/9 (1996) para. 6.
67 CBD, Article 2.

Table 6.3 Convention on Biodiversity Diversity parties' decisions on restoration, 1998–2010

Habitat or Ecosystem Service	CBD Decision	Language
Inland Water resources	Decision IV/4 (1998) Annex 1, para. 9 (a)(ii)	States should develop "watershed, catchment and river basin management strategies to maintain, restore or improve the quality and supply of inland water resources and the economic, social, hydrological, biological diversity and other functions and values of inland water ecosystems.'
Agroecosystems	CBD Decision X/34 (2000) para. 5(j)	Encouraging Parties to promote 'opportunities for sustainable increases in agricultural productivity including through maintaining and/or restoring the functioning of agro-ecosystems, the biodiversity within them and the services they deliver...').
Mountain areas	Decision VII/27 (2004) Action 1.2.1.	'Develop and implement programmes to restore degraded mountain ecosystems and protect natural dynamic processes and maintain biological diversity in order to enhance the capacity of mountain ecosystems to resist and adapt to climate change, or recover from its negative impacts...'.
Forest restoration	Decision IX/5 (2008) para. 2(g)	Invites Parties to 'Promote forest restoration, including reforestation and afforestation, in line with sustainable forest management through, inter alia, the Global Partnership on Forest Landscape Restoration and other regional cooperation mechanisms, paying particular attention to genetic diversity.'
Inland water resources	Decision X/28 (2010) para. 44	'Encourages Parties and other Governments to conserve, sustainably use and, where necessary, restore ecosystems so that freshwater flows and water resources sustain biodiversity and thus contribute to human well-being.'

Table 6.3 (continued)

Habitat or Ecosystem Service	CBD Decision	Language
Coastal habitat	Decision X/29 (2010) para. 72	Urges Parties to stop the degradation and loss of ecologically important ecosystems and habitats, such as estuaries, coastal sand dunes, mangroves forests, salt marshes, seagrass beds, and biogenic reefs, due to coastal development and other factors in coastal area, to facilitate their recovery through the management of human impacts and restoration, where appropriate.'

In addition to the numerous COP decisions that invoke expectations of habitat restoration by parties, the parties to the CBD have also adopted decisions linking the private business community to biodiversity conservation and restoration. In 2006 in Decision VIII/17, CBD parties adopted a decision on private-sector engagement in conserving and restoring biodiversity.[68] The decision recognized that 'The private sector is arguably the least engaged of all stakeholders in the implementation of the Convention, yet the daily activities of business and industry have major impacts on biodiversity.'[69] States were encouraged to work with the private sector including small and medium-sized businesses to 'adopt practices that support the implementation of national biodiversity strategies and action plans and the objectives of the Convention' and encourage measures for enhancing business participation.[70] Additional discussion of private sector involvement in restoration can be found in Chapter 9.

Hyderabad Conference of the Parties' decisions on ecosystem restoration (2012)

2012 was the year that restoration became a priority issue for the CBD parties. In October 2012, the governments of India, South Korea and South Africa in cooperation with the Secretariats of the Biodiversity Convention, the United Nations Convention to Combat Desertification, the United Nations Framework Convention on Climate Change and the Ramsar Convention on Wetlands, plus a

---

68 Decision VIII/17 (2006).
69 Ibid., Preamble.
70 Ibid., p. 3, para. 1 and p. 2 (listing a series of potential measures to enhance business participation including voluntary or mandatory reporting and performance standards, certification schemes, international standards on business activities, biodiversity policies to define and operationalise companies' biodiversity commitments, biodiversity benchmarks, guidelines for environmental impact assessments).

number of intergovernmental and non-governmental organisations,[71] agreed to a 'Call for a Concerted Effort on Ecosystem Restoration'. In this call, the governments, intergovernmental organisations, multilateral environmental agreement secretariats and NGOs agreed to 'make concerted and coordinated long-term efforts to mobilize resources and facilitate the implementation of ecosystem restoration activities on the ground for sustaining and improving the health and well-being of humans and all other species with whom we share the planet'.[72] While the document captures only a voluntary political commitment, it reflects a crucial recognition that substantial resources will need to be invested in restoration efforts if restoration is to enable environmental recovery.

The Hyderabad document captures a number of key themes that have been reflected in the current trend to legalise restoration as a sustainability strategy. First, the document notes the lack of investment in the 'productivity, health, and sustainability' of ecosystems and the need for the 'restoration and rehabilitation of degraded lands, ecosystems, and landscapes (including seascapes)'.[73] Second, restoration is considered to be a high political priority because it not only underpins goods and services but may also alleviate other global pressures such as food and water reductions, climate change and land degradation. Finally, restoration has the possibility of addressing socio-economic equity concerns such as chronic poverty by reducing 'environmental risks and scarcities'.[74]

Recognising the need to improve efforts to restore under Article 8(f) of the Convention, states adopted a decision on 'Ecosystem Restoration' in December 2012.[75] Prior to the negotiations, the Secretariat provided CBD members with three information notes regarding ecosystem restoration offering a variety of definitions of ecosystem restoration.[76] At the Conference of the Parties, states agreed

---

71 The groups included the Global Environmental Facility, United Nations Development Programme, United Nations Forum on Forests, the Food and Agriculture Organization, the International Union for Conservation of Nature and the Society for Ecological Restoration.

72 Hyderabad Call for a Concerted Effort on Ecosystem Restoration, <www.cbd.int/doc/restoration/Hyderabad-call-restoration-en.pdf>.

73 Ibid.

74 Ibid.

75 Decision XI/16. Ecosystem Restoration, UNEP/CBD/COP/DEC/XI/1 (2012).

76 Available Guidance and Guidelines on Ecosystem Restoration, UNEP/CBD/COP/11/INF17 (2012) (providing an annotated bibliography of resources on restoration), <www.cbd.int/doc/meetings/cop/cop-11/information/cop-11-inf-17-en.pdf>; Available Tools and Technologies on Ecosystem Restoration, UNEP/CBD/COP/11/INF18 (2012), <www.cbd.int/doc/meetings/cop/cop-11/information/cop-11-inf-18-en.pdf>; Most Used Definitions/Descriptions of Key Terms Related to Ecosystem Restoration, UNEP/CBD/COP/11/INF19 (2012), <www.cbd.int/doc/meetings/cop/cop-11/information/cop-11-inf-19-en.pdf> (providing four definitions of the term 'ecosystem restoration' including 1) a 1993 definition by John Cairns, 'Human intervention ... designed to accelerate the recovery of damaged habitats, or to bring ecosystems back to as close an approximation as possible of their pre-disturbance states' ('Ecological restoration: replenishing our national and global ecological capital' in D. Saunders, R.J. Hobbs and P. Ehrlich (eds), *Nature*

to focus greater resources on achieving ecosystem restoration. Decision XI/16 also captures some additional shared perspectives on the strategy of promoting ecosystem restoration. While recognising that restoration may be an important economic development driver by creating public work opportunities, the decision also observes that 'restoration is not a substitute for conservation',[77] because it is instead a conservation strategy of the last resort if harm cannot be avoided.

As part of the decision, states 'depending on national circumstances' agreed to undertake additional commitments including:

- implementing restoration provisions of previous Conference of the Parties' decisions;
- preventing and reducing activities that degrade ecosystems;
- identifying ecosystems that may be eligible for restoration;
- using best practices for ecosystem restoration;
- protecting indigenous and local communities from negative social impacts of restoration projects;
- linking poverty eradication efforts with ecosystem restoration efforts;
- providing for 'full and effective participation' of indigenous and local communities; and
- promoting ecosystem restoration that will 'restore critical ecosystem functions and the delivery of benefits to people'.[78]

States are particularly encouraged to cooperate on capacity building with civil society groups and intergovernmental organisations including the 'Society for Ecological Restoration, the International Union for Conservation of Nature, the World Resources Institute, the Global Partnership on Forest Landscape Restoration, the International Tropical Timber Organization and other relevant organizations and initiatives such as the Sub-Global Assessment Network'.[79] Countries that are capable of supporting ecosystem restoration efforts should contribute financial and technical assistance.[80] The Executive Secretary of the CBD through his

Conservation: Reconstruction of Fragmented Ecosystems, Surrey Beatty & Sons); 2) a 1996 definition by M.K. Briggs, 'the process of returning an ecosystem to a natural pre-disturbance structure and function' (*Riparian Ecosystem Recovery in Arid Lands: Strategies and References*, University of Arizona Press); 3) a 2003 UNEP definition, 'The process of intentionally altering a site to establish a defined, indigenous ecosystem. The goal of this process is to emulate the structure, function, diversity and dynamics of the specified ecosystem' (Freshwater Management Series No. 7 (2003) Phytotechnologies: A Technical Approach in Environmental Management); and 4) a 2004 Society for Ecological Restoration definition, 'The process of assisting the recovery of an ecosystem that has been degraded, damaged, or destroyed' (*The SER Primer on Ecological Restoration*, Society for Ecological Restoration International, Science and Policy Working Group).

77 Decision XI/16. Ecosystem Restoration, UNEP/CBD/COP/DEC/XI/1 (2012).
78 Ibid., Section 1.
79 Ibid., Section 2.
80 Ibid., Section 3.

Secretariat staff is charged with a number of responsibilities including facilitating the development of practical guidance on ecosystem restoration, creating a clearinghouse for best restoration practices and collaborating with other treaty secretariats 'to enhance and harmonize efforts in ecosystem restoration and avoid duplication'.[81] Effective ecological restoration is appropriately conceived of as a multi-stakeholder project.

For the purposes of creating effective law of ecosystem restoration, the COP decision makes two important requests of the Executive Secretary. First, it calls for the Executive Secretary to identify 'implementation tools for ecosystem restoration'.[82] Most of the documents that have been provided to the Executive Secretary by interested stakeholders on ecosystem restoration focus on the technical fieldwork aspects of restoration as a scientific challenge. While addressing challenges of how to set realistic implementable standards for restoration is critical, there has been surprisingly little discussion of the role of law as an 'implementation tool'.

The second request of the Executive Secretary by the CBD parties is related to the need for additional 'implementation tools'. The Secretary is expected to 'develop clear terms and definitions of ecosystem rehabilitation and restoration and clarify the desired outcomes of implementation of restoration activities'.[83] If national laws are to be developed or revised to take into consideration ecosystem restoration practices, states seem to recognise a need for some uniform set of restoration practices to set a baseline for restoration work.

Pyeongchang Conference of the Parties' decisions on ecosystem restoration (2014)

The Conference of the Parties in 2014 continued to discuss strategies for enhancing restoration efforts. States recognised that more research was needed to improve implementation efforts of the Strategic Plan for Biodiversity 2011–2020 and the Aichi Targets. Specifically, the states recognised a 'need for better understanding of ecosystem processes and functions and their implications for ecosystem conservation and restoration, ecological limits, tipping points, socio-ecological resilience and ecosystem services; and improved methodologies and indicators for monitoring ecosystem resilience and recovery, in particular for vulnerable ecosystems'.[84]

Decision XII/5 highlighted the social aspects of restoration work. This decision included the Chennai Guidance for the Integration of Biodiversity and Poverty Reduction, as a voluntary set of standards for addressing biodiversity concerns in implementing international development priorities. The Guidance proceeds from the assumption that the poor depend on access to biodiverse resources to meet their needs. The Guidance urges states to ensure that communities have access to 'mechanisms of redress, at the national and local level including restoration and

---

81 Ibid., Section 5.
82 Ibid., Section 5(h).
83 Ibid., Section 5(i).
84 Decision XII/1 (2014) Annex 1.

compensation for damages caused to biodiversity and the poor'.[85] While the connection between ecological restoration and poverty alleviation has been recognised in the field by international development practitioners, this reference in the Chennai Guidance is one of the first soft-law acknowledgments of the dependence of economically vulnerable communities on restoration as a means of securing their livelihoods.

Decision XII/19 on 'conservation and restoration' noted 'with concern that not enough progress has been made' towards achieving the Aichi Biodiversity Targets 14 and 15, the two targets that explicitly require restoration.[86] The decision 'invites' states and other stakeholders to take a number of steps towards increasing the salience of restoration work. In particular the decision encourages states to:

- 'develop spatial planning approaches at the landscape and seascape level ... to promote ecosystem restoration';
- 'promote cross-sectoral approaches, including with the public sector, private sector and civil society, to develop a coherent framework for ecosystem conservation and restoration';
- 'promote, where appropriate, holistic and integrated planning for ecosystem ... restoration in indigenous and local community conserved areas';
- 'promote ecosystem restoration activities, in particular large-scale restoration activities' after 'taking into consideration that priority should be given to avoiding or reducing ecosystem losses';
- 'provide appropriate incentives to promote ... sustainable management and best practices in the conservation and restoration of ecosystems at the national and subnational levels, in the public and private sectors';
- 'develop and strengthen monitoring of ecosystem degradation and restoration' and to 'give due attention to both native species and genetic diversity in ... restoration activities, while avoiding the introduction of invasive alien species'.[87]

While this decision does not create any legal binding obligations for states, it is noteworthy because it signals the need for states to cooperate on restoration across boundaries in order to achieve large-scale restoration objectives and emphasises the need for initiatives that can network across an array of public and private actors to achieve restoration goals. While these two points are not developed in this decision, they suggest that more resources must be invested in building multi-stakeholder governance networks and potentially in recognising legal pluralism as a tool for furthering restoration goals across the landscape. Private actors do play a prominent role in ecological restoration projects as explained in Chapter 9, but for the most part they are not invested with the

85 Decision XII/5 (2014) Annex, Section 2(iv).
86 Decision XII/19 (2014).
87 Ibid. para. 4(a)–(h).

responsibility of creating and implementing rules for restoration. As this chapter and the preceding chapter illustrate, international law alone is insufficient to achieve restoration goals, in spite of obligations under the Ramsar Convention, the CBD and many other treaties. The obligations to restore landscapes are only beginning to be mainstreamed into practice at the community, municipal or business level.

Involvement by the Subsidiary Body on Scientific, Technical and Technological Advice (SBSTTA)

States created the SBSTTA in the framework of the CBD to provide multi-disciplinary advice on the implementation of the CBD. One of the jobs that the CBD assigns the SBSTTA is to 'identify innovative, efficient and state-of-the-art technologies and know-how relating to the conservation and sustainable use of biological diversity and advise on the ways and means of promoting development and/or transferring such technologies'.[88] In recent years the SBSTTA has become increasingly involved in developing a shared understanding of the practice of ecological restoration. In 2016, the SBSTTA requested the Executive Secretary of the CBD in cooperation with a variety of other organisations including the secretariats of other conventions to develop 'a short-term action plan on ecosystem restoration' designed to deliver working restoration plans before 2020.[89]

The short-term plan to speed up the implementation of Aichi Targets 14 and 15 has been designed to achieve the 'restoration of natural and semi-natural eco-systems' in order to stop the loss of biodiversity, improve ecosystem resilience, improve ecosystem services, mitigate and adapt to the effects of climate change, combat desertification, improve human well-being, and reduce environmental risks and scarcities.[90] More specifically, parties intend to increase public communication about the benefits of ecosystem restoration, 'accelerate action in the planning, implementation and monitoring of ecosystem restoration activities at all levels' and 'identify and formalize regional, national and local targets, policies and actions for ecosystem restoration'.[91] Parties understand the plan as contributing not just to the achievement of party obligations under the CBD but also to the achievement of party objectives under the UN Framework Convention on Climate Change, the UN Convention to Combat Desertification, the Ramsar Convention on Wetlands, the Convention on the Conservation of Migratory Species of Wild Animals, the

---

88 CBD, Article 25(2)(c).
89 Protected Areas and Ecosystem Restoration, Note by the Executive Secretary, UNEP/CBD/SBSTTA/20/12 (2016) paras 2–3 (indicating the active participation of the Secretariat of the UNCCD, the Society for Ecological Restoration and the Global Partnership for Forest and Landscape Restoration in providing technical advice).
90 Ibid., Annex 1, para. 1.
91 Ibid., Annex 1, para. 3.

UN Sustainable Development Agenda and the Sendai Framework for Disaster Risk Reduction 2015–2030.[92]

For purposes of the short-term plan, ecological restoration is understood as 'the process of actively managing or assisting the recovery of an ecosystem that has been degraded, damaged or destroyed as a means of sustaining ecosystem resilience and conserving biodiversity'.[93] The CBD Subsidiary Body emphasised that restoration is not a substitute for conservation but where it is appropriate there needs to be created 'an enabling institutional framework' to ensure that restoration work could take place across a number of different habitats by a variety of actors for a range of purposes.[94] Restoration must be 'planned and implemented using the best available science and local knowledge' as well as the best practices associated with ensuring community participation including free, prior and informed consent.[95]

The Subsidiary Body offers four specific recommendations. First, states must create a priority list of areas requiring ecological restoration. This formal assessment should be conducted in a manner that incorporates the need to obtain free, prior and informed consent from indigenous groups and local communities. In addition, the plan should implement safeguards to protect indigenous groups and local communities from habitat loss and degradation that might be inadvertently triggered by a restoration project.[96] In preparing an assessment, planners should take into consideration ecological baseline information, costs of taking actions, costs of failing to act, drivers of biological diversity loss (e.g. lack of land tenure), and gaps in law and policy that might be contributing to ecological degradation.[97]

Second, states in the short term should improve the enabling environment for restoration work. This recommendation is probably the most significant recommendation in the context of this book on restoration and law. Specifically, states are expected to 'improve or establish a legal and policy framework for the protection and restoration of ecosystems' including developing 'as appropriate, laws, regulations, policies and other requirements for protecting, and restoring vulnerable habitats'.[98] Here states are expected to create appropriate governance systems that may include land tenure reforms or establishing national targets for ecosystem restoration in national biodiversity strategies, national sustainable development plans or national climate mitigation/adaptation plans.[99] States arc

---

92 Obligations to restore under the UNFCCC are discussed in Chapter 11. Obligations to restore under the remainder of the treaties referenced by the Subsidiary Body are discussed in Chapter 5. Expectations to restore under the Sustainable Development Goals and the Sendai Framework are discussed in Chapter 4.

93 Protected Areas and Ecosystem Restoration, Annex 1, para. 4.

94 Ibid., Annex 1, paras 5 and 8.

95 Ibid., Annex 1, para. 10.

96 Ibid., Annex 1, paras 13(1) and 13(7).

97 Ibid., Annex 1, paras 13.A(3), 13.A(4), 13.A(5).

98 Ibid., Annex 1, para. 14.B(1).

99 Ibid., Annex 1, para. 14.B(4).

expected to offer some national accounting that values intact ecosystems and not just extraction activities.[100] Finally, as part of the enabling environment, states should consider the role of economic incentives in behavioural change and mobilise financial resources in order to support better restoration outcomes.[101] Achieving this framework has the potential to be transformative for the practice of ecological restoration by making restoration a state action priority and ensuring that there is both political authority behind restoration decisions (e.g. laws) and financing.

The third step in designing a short-term plan requires states to engage in ecological restoration planning for the priority sites that they identified in step one and within the enabling environment created under step two. Here states are expected to mainstream restoration work into landscape planning and explore various alternatives for achieving restoration goals based on factors such as cost and 'ecological appropriateness' (e.g. taking into consideration impacts of climate change).[102] States may want to consider evaluating options to achieve national restoration goals as part of an environmental impact assessment process.[103] Significantly, the SBSTTA recommends that state restoration plans should avoid vague objectives but instead include objectives that are measurable such as specific ecological or socioeconomic attributes like an increase in species diversity or an increase in delivery of ecosystem services.[104]

Finally, once locations have been identified and states have planned as enabled by the new or existing governance framework, states must invest in monitoring, evaluation, feedback and dissemination. Restoration is recognised as an iterative process in which parties learn through doing and observing others doing similar projects.[105] The implementation success of the proposed short-term ecosystem restoration plan will depend largely on the political will of states at the highest levels of decision-making to invest in an enabling framework that shifts ecological restoration from the jurisdiction of a few government offices to a government-wide mandate. If ecological restoration is to take place at the landscape scale that ecologists are urging, it will be the enabling frameworks that provide the necessary means for achieving socio-ecological ends that value environmental assets for more than simply commodity value.

Taken in total, what do all of these national plans, Conference of the Parties' decisions and initiatives by CBD bodies tell us about the practice of restoration? They suggest the potential for soft law in the form of Conference of the Parties' decisions and subsidiary body recommendations to shape the implementation of pre-existing duties. Here, states have explicit legal duties under the CBD. To the extent that these duties remain vague, the subsequent Conference of the Parties'

100 Ibid., Annex 1, para. 14.B(5).
101 Ibid., Annex 1, paras 14.B(6) and 14.B(7).
102 Ibid., Annex 1, paras 15.C(1) and 15.C(2).
103 Ibid., Annex 1, para. 15.C(2).
104 Ibid., Annex 1, para. 15.C(3).
105 Ibid., Annex 1, para. 16.

decisions and subsidiary body recommendations build on the Aichi Biodiversity Strategy and provide an evolving clarification of states' understandings of their original obligations of conduct.

A CBD Protocol on ecological restoration?

One means of implementing a plan to promote effective governance across various levels in the context of restoration might be through a binding Protocol on Ecological Restoration that commits states to seeking to develop restoration strategies in collaboration with other social sectors. A potential protocol should build on recommendations of the CBD Subsidiary Body for States to engage in short-term planning for ecosystem restoration. Under the Biodiversity Convention, the Conference of the Parties has the authority to adopt protocols by a two-thirds majority.[106] Any protocol must, of course, be ratified by parties under national treaty ratification practices. The idea of negotiating a protocol under the CBD is not new. The CBD parties have already negotiated and adopted the Cartagena Protocol on Biosafety to elaborate on those portions of the CBD involving transfer of species and the Nagoya Protocol on Access and Benefit Sharing to provide a legal approach to the equitable use and sharing of resources.[107]

Parties might consider negotiating a Protocol on Ecological Restoration of Habitats and Species that would establish a set of minimum international standards for national restoration planning. This type of negotiating exercise may be especially fruitful in light of the species and habitat restoration duties already dispersed across international legal regimes. A protocol could provide a common language to discuss socioecological restoration objectives and serve as a focal point for technical and financial transfers. Requirements for stakeholder participation, ongoing monitoring, advance planning and coordination across international boundaries might be subjects covered in such a protocol. A protocol might also provide a legal structure for technical and financial exchanges to support national restoration efforts. The political feasibility of this depends, of course, on the political will of states. Given the specific restoration goals set in the Aichi Targets and the 2030 Sustainable Development Goals, the time may now be ripe for a legal agreement regarding minimum national standards for restoration to assist states in meeting their obligation to restore. This could fit into the proposed work plan of the Conference of the Parties to the CBD as a 'related means of implementation' of the Strategic Plan for Biodiversity 2011–2020.[108]

---

106 CBD, Article 23(4)(c).
107 Cartagena Protocol on Biosafety to the Convention on Biological Diversity, 29 January 2000, 2226 *UNTS* 208; (2000) 39 *ILM* 1027; Nagoya Protocol on Access to Genetic Resources and the Fair and Equitable Sharing of Benefits Arising from their Utilization to the Convention on Biological Diversity, 29 October 2010, UNEP/ CBD/COP/DEC/X/1.
108 Decision XII/31 (2014) (describing a work plan for the CBD Conference of the Parties meetings in 2016, 2018 and 2020).

An international duty to restore?

In the last three chapters, this book has reviewed a series of soft law documents (e.g. Stockholm Declaration, Rio Declaration, Sustainable Development Goals) that reflect formal political commitments and a series of treaty documents (e.g. Law of the Sea Convention, Convention on Migratory Species, Ramsar Convention and Convention on Biological Diversity) that reflect legal commitments. Taken together this body of documents suggests that the restoration of important species including endangered species and some commercial species and the restoration of critical biodiverse habitats has matured into a customary obligation. In other words, there may not simply be a desire to restore but a duty to restore by states.

If one accepts that there is a legal duty to restore, what more can we conclude from the international legal documents about the content of this duty? The answer is unfortunately very little since restoration is rarely defined. Appearing largely as an 'obligation of conduct' in legal texts the duty to restore is left to each state to define for itself. At best, we can identify a shared objective for restoration, a trigger for the duty and at least one condition for restoration. Reviewing the commitments under the various multilateral environmental agreements, it becomes clear that restoration is regarded as a last-resort strategy. When avoidance of harm is no longer possible, then states must logically restore for what harm has occurred. As far as when the restoration duty is triggered, states have acknowledged that there is a duty to engage in restoration either when a given habitat is in a degraded state or a species is threatened, endangered, or below the carrying capacity necessary for achieving a maximum sustainable yield in the case of commercial species. This duty to restore is not a total duty but a duty based on the existence of certain conditions. We know from the treaty commitments that restoration is only expected when it is feasible or appropriate. What constitutes feasibility is currently defined state by state. It is fair to assume that some habitats are simply too damaged and degraded to be restored. It is probably also fair to assume that some habitats such as some coastal freshwater wetlands may be in too much flux given climate shifts to be capable of being restored as freshwater wetlands. If one is replacing a freshwater wetland with a saltwater wetland, should this be considered restoration for purposes of the law since one distinctive habitat is replacing another?

A review of the international legal frameworks including the Conference of the Parties' decisions and recommendations yields remarkably little guidance regarding how to measure compliance with the general restoration obligations in the text. As of 2016, there are no internationally agreed standards from which to measure whether a state is making progress towards satisfying its restoration obligations, although there are some promising national developments that could form the basis for a protocol.[109] If restoration projects are to have meaningful socio-ecological outcomes, the international community must invest negotiating

---

109 See Chapter 12 for a discussion of the 2016 National Standards for the Practice of Ecological Restoration in Australia.

resources in the development and implementation of an internationally agreed set of standards for restoration projects. A protocol such as this chapter proposes may be one means of continuing the dialogue among states about what constitutes the content of a state duty to restore.

The need for some set of minimal standards is highlighted by the current global political commitments including the Aichi Targets, the Sustainable Development Goals and the Bonn Challenge where a wide spectrum of states have agreed to undertake substantial restoration efforts. If these projects are to make a difference ecologically, in terms of recovering valuable habitats and species and be socially sustainable, in terms of being supported by local communities, there must be some consistency in how they are designed and implemented.

Since international law has only announced a general obligation to restore, the specific content of national and regional laws implementing this obligation is essential to achieving restoration outcomes. Chapter 7 explores the regional approach taken by the European Union to mainstreaming restoration obligations. Chapter 8 explores a number of novel domestic legal approaches that states have adopted to achieve restoration that run the gamut from creating prescriptive formulas for how to perform restoration to assigning a 'right to restoration' to nature.

7 Regional approaches: law and policy on ecological restoration in the European Union

Although several regional environmental treaties can form the legal basis for ecological restoration (see Chapter 5 for some examples), the European Union's policy and legal framework on restoration are particularly worth mentioning. Not only has the European Union (EU) – as a supranational organisation – law-making and enforcement powers, but it also has concrete policy targets and legislation on restoration. Furthermore, recent case law from the European Court of Justice provides us with some very relevant insights into the application of restoration obligations in the EU's Member States. After a short introduction to the EU's environmental law and policy, and the need for restoration in the EU, this chapter will discuss the legal and policy frameworks relevant for restoration in the EU.

EU environmental law and policy

As a supranational organisation, the EU has important competences with regards to environmental policy and law.[1] According to the Treaty on the European Union, the EU shall work for the sustainable development of Europe, aiming at 'a high level of protection and improvement of the quality of the environment'.[2] While the environmental competences of the EU are generally shared with the Member States,[3] Article 191 of the EU Treaty mandates the EU to contribute to the 'preserving, protecting and improving the quality of the environment'.[4] Although Article 191 does not explicitly refer to restoration, it is clear that 'improving' the quality of the environment can include ecological restoration measures. Article

---

1 On EU environmental law, see: J. Jans and H. Vedder, *European Environmental Law* (Europa Law Publishing 2012); L. Kramer, *EU Environmental Law* (Sweet & Maxwell 2012); E. Morgera, 'European environmental law' in S. Alam, J. Bhuiyan, T. Chowdhury and E. Techera (eds), *Routledge Handbook Of International Environmental Law* (Routledge 2013) 427–442.
2 See Article 3(3), Treaty on European Union (hereafter TEU), consolidated version *OJ* C 326, 26 October 2012.
3 Article 4(2)(e), Treaty on the Functioning of the European Union (hereafter TFEU), consolidated version *OJ* C 326, 26 October 2012; but see Article 3(1)(d), TFEU (assigning conservation of marine biological resources under the Common Fisheries Policy to the exclusive competence of the EU).
4 Article 191(1), TFEU.

191 authorises the adoption of measures that result directly or indirectly in an improvement of the environment, such as conservation and restoration.[5] Since the development of environmental policy within the EU a vast body of policy instruments and legislation has been developed, including action plans and binding legislation (directives and regulations).

The European Commission (and more specifically the Environment Directorate General) plays an important role as it proposes environmental policies and legislation. An important feature of the EU is the enforcement framework, with the European Commission as the 'guardian' of EU law ensuring the transposition and implementation of EU law in the Member States. The European Court of Justice interprets EU law and settles disputes between the EU Commission and Member States.

Restoration in Europe

Most landscapes in Europe have been heavily influenced by human activities, such that the natural ecosystems have long since been replaced by semi-natural or even artificial or created systems. These 'cultural landscapes' depend on continued human intervention for maintenance of their structural and ecological diversity. Since the middle of the twentieth century, intensified agriculture, transport infrastructure and an increasing urbanisation have led to an ecological impoverishment in Europe.[6] Biodiversity loss occurs throughout Europe. Major threats include habitat degradation and destruction, landscape fragmentation, invasive alien species, pollution, resource over-exploitation and climate change.[7]

Because of the continuous decline in biodiversity, restoration in Europe becomes increasingly prominent in environmental policies.[8] Because of the long-time human settlement in large parts of Europe and the presence of culture landscapes, the focus of restoration is different than in other parts of the world. European restorationists often mention the need to ensure that restored ecosystems have cultural value and that threatened cultural practices, such as grazing and hay-making that created particular landscapes, are maintained.[9] In contrast, restoration practitioners from North America and Australia rather see restoration as a desire to return a damaged ecosystem to its original historical characteristics,

5   Morgera, 'European environmental law' 435.
6   F. Madgwick and T. Jones, 'Europe' in M. Perrow and A. Davy (eds), *Handbook of Ecological Restoration. Volume 2: Restoration in Practice* (Cambridge University Press 2002) 32–33.
7   On landscape fragmentation, see: European Environment Agency, *Landscape Fragmentation in Europe* (Joint EEA-FOEN report, No 2/2011, European Environment Agency 2011); on biodiversity decline in Europe, see: European Environment Agency, *Assessing biodiversity in Europe – the 2010 report* (No 5/2010, European Environment Agency 2010); European Environment Agency, *The European Environment – State and Outlook 2015: Synthesis Report* (European Environment Agency 2015) 56.
8   On restoration in Europe, see Madgwick and Jones, 'Europe' 32–56.
9   See, for example, the restoration of a semi-natural ecosystem, as depicted on the cover of this book.

or at least to return an ecosystem to its historical trajectory, allowing for continuing dynamic ecological and evolutionary change within the ecosystem.[10] The need for restoration becomes increasingly accepted as part of environmental law and policy and nature conservation management. Restoration targets and obligations can be found in EU biodiversity policy and law. The next sections will discuss the legal provisions in EU law that are relevant for restoration. First, the core nature conservation legal instruments within the EU, namely the Birds Directive and the Habitats Directive, will be discussed. Second, other relevant EU instruments will briefly be discussed. Finally, the EU Biodiversity Strategy will be discussed, which implements the Convention on Biological Diversity (CBD) Aichi Biodiversity Targets within the EU and holds some specific targets for restoration.[11]

EU biodiversity law on restoration: the Birds and Habitats Directives

The core legislation for nature conservation and restoration in the EU are two directives: the Birds Directive from 1979 (consolidated in 2009)[12] and the Habitats Directive from 1992.[13] The overall goal of the Nature Directives is to contribute towards ensuring biodiversity through the conservation of natural habitats and of wild fauna and flora in the European territory of the Member States.[14] This section first discusses the general obligations in the Directives that are relevant for maintaining or restoring a favourable conservation status for both habitats and species. Restoration obligations can also be found explicitly or implicitly in the articles on area protection and species protection. The restoration obligations within the EU-protected areas that form the Natura 2000 network will be dealt with in Chapter 10. This section examines very briefly restoration in the wider landscape and restoration of species.

10 S. Allison, *Ecological Restoration and Environmental Change: Renewing Damaged Ecosystems* (Routledge 2012) 123.
11 See also A. Cliquet, K. Decleer and H. Schoukens, 'Restoring nature in the EU: the only way is up?' in C.-H. Born, A. Cliquet, H. Schoukens, D. Misonne and G. Van Hoorick (eds), *The Habitats Directive in its EU Environmental Law Context: European Nature's Best Hope?* (Routledge 2015) 267–272.
12 Directive 2009/147/EC of the European Parliament and of the Council of 30 November 2009 on the conservation of wild birds, *OJ L* 20, 26 January 2010, replacing the original Birds Directive, Directive 79/409/EEC of 2 April 1979 on the Conservation of Wild Birds, *OJ L* 103, 25 April 1979 (further referred to as Birds Directive).
13 Directive 92/43/EEC of 21 May 1992 on the conservation of natural habitats and of wild fauna and flora, *OJ L* 206, 22 July 1992 (further referred to as Habitats Directive).
14 Article 2(1), Habitats Directive; the Birds and Habitats Directives are not applicable to the overseas territories; however their application includes the Macaronesian biogeographical region (including the Azores, Madeira and the Canary Islands).

In addition to the EU legislation supporting restoration, the case law on EU's nature conservation legislation is of tremendous importance in the interpretation and enforcement of the EU Nature Directives.[15] The cases in the European Court of Justice on the Birds and Habitats Directives have been numerous and constitute a large part of environmental cases.[16] Most of the cases for the Court of Justice are infringement cases brought forward by the European Commission,[17] including cases on incomplete or incorrect transposition (non-conformity) and bad application of the Birds and Habitats Directives. There are relatively few though interesting judgments on preliminary reference.[18]

General obligations on restoration in the Birds and Habitats Directives

In the Birds and Habitats Directives there are explicit references to restoration. The Preamble of the Birds Directive states: 'The preservation, maintenance or *restoration* of a sufficient diversity and area of habitats is essential to the conservation of all species of birds.'[19] According to Article 3(1), Member States 'shall take the requisite measures to preserve, maintain or *re-establish* a sufficient diversity and area of habitats for all the species of birds referred to in Article 1'. 'The preservation, maintenance and re-establishment of biotopes and habitats shall include … [the] *re-establishment* of destroyed biotopes and the *creation* of biotopes.'[20] Article 3 of the Birds Directive has a wide scope as it applies to all wild bird species covered by the Birds Directive.[21] According to the Court of Justice it is clear from the Birds Directive preamble 'that the preservation, maintenance or restoration of a sufficient diversity and area of habitats is essential to the conservation of all species of birds. The Member States are therefore required to

---

15 On enforcement of EU environmental law, see: L. Krämer, 'Regional economic integration organizations: the European Union as an example' in D. Bodansky, J. Brunnée and E. Hey (eds), *The Oxford Handbook of International Environmental Law* (Oxford University Press 2007) 853–876, 868–870; see specifically on enforcement of the Habitats Directive: L. Krämer, 'Implementation and enforcement of the Habitats Directive' in C.-H. Born, A. Cliquet, H. Schoukens, D. Misonne and G. Van Hoorick (eds), *The Habitats Directive in its EU Environmental Law Context: European Nature's Best Hope?* (Routledge 2015) 229–244.

16 See L. Krämer, *Environmental Judgments by the Court of Justice and their Duration* (Research Papers in Law No 4/2008, College of Europe 2008) 4, Table 2.

17 Based on Article 258 TFEU.

18 Based on Article 267 TFEU; the reference for a preliminary ruling is a procedure exercised before the European Court of Justice. This procedure enables national courts to question the Court of Justice on the interpretation or validity of European Union law. The reference for a preliminary ruling therefore offers a means to guarantee legal certainty by uniform application of EU law.

19 Preamble, para. 8, Birds Directive (emphasis added).

20 Article 3(2)(c–d), Birds Directive (emphasis added).

21 Cf. Article 1, Birds Directive: 'This Directive relates to the conservation of all species of naturally occurring birds in the wild state in the European territory of the Member States to which the Treaty applies.'

adopt the measures necessary for the conservation of those species'.[22] This can include, where appropriate, restoration measures.

The Habitats Directive also refers explicitly to restoration. In the Directive 'conservation' is defined as 'a series of measures required to maintain or *restore* the natural habitats and the populations of species of wild fauna and flora at a favourable status'.[23] Other definitions in the Directive also refer to restoration. A Site of Community Importance (SCI) is defined as a site that contributes significantly to the maintenance or *restoration* at a favourable conservation status.[24] A Special Area of Conservation (SAC) is described as a site where the necessary conservation measures are applied for 'the maintenance or *restoration*, at a favourable conservation status'.[25] While neither Directive contain a specific definition on restoration or include specific targets on restoration, both of the Directives contemplate restoration as a core conservation strategy. The general aim of the Habitats Directive, as put forward in Article 2(1) is to contribute towards ensuring biodiversity through the conservation of natural habitats and of wild fauna and flora in the European territory of the Member States to which the Treaty applies. This is considered a result obligation.[26] Measures taken pursuant to this Directive shall be designed to maintain or restore at favourable conservation status, natural habitats and species of wild fauna and flora of Community interest.[27] The favourable conservation status of a natural habitat is defined as follows:

- its natural range and areas it covers within that range are stable or increasing, and
- the specific structure and functions which are necessary for its long-term maintenance exist and are likely to continue to exist for the foreseeable future, and
- the conservation status of its typical species is favourable ...[28]

The favourable conservation status of species is defined as:

- population dynamics data on the species concerned indicate that it is maintaining itself on a long-term basis as a viable component of its natural habitats, and
- the natural range of the species is neither being reduced nor is likely to be reduced for the foreseeable future, and

---

22 Case C-235/04 *Commission v Spain* [2007] ECR I-0000, para. 23.
23 Article 1(a), Habitats Directive (emphasis added).
24 Article 1(k), Habitats Directive (emphasis added).
25 Article 1(l), Habitats Directive (emphasis added).
26 European Commission, *Managing Natura 2000 sites: The provisions of Article 6 of the 'Habitats' Directive 92/43/EEC* (Office for Official Publications of the European Communities 2000) 18.
27 Article 2(2), Habitats Directive.
28 Article 1(e), Habitats Directive.

- there is, and will probably continue to be, a sufficiently large habitat to maintain its populations on a long-term basis.[29]

In light of the overall objective of the Habitats Directive and the unfavourable conservation status for many habitats and species, restoration measures in order to reach a favourable conservation status can be considered as a core obligation under the Nature Directives.[30]

Restoration in the wider landscape

The restoration obligations of the Habitats Directive are not limited to the habitats and species within the designated sites of the Natura 2000 network.[31] According to Article 2 of the Habitats Directive the aim of the Directive shall be to contribute towards ensuring biodiversity through the conservation of natural habitats and of wild flora and fauna in the European territory of Member States.[32] Measures shall be designed to maintain or restore, at favourable conservation status, natural habitats and species of wild fauna and flora of Community interest.[33] The designation and management of the Natura 2000 network is an important means to obtain this overall goal, but is not the only way to obtain a favourable conservation status. This is also clear from the monitoring and reporting obligation under Articles 11 and 17 of the Habitats Directive. If the monitoring shows that there is an unfavourable conservation status of the habitats and species, then additional restoration measures will be required. This could entail the designation, conservation and restoration of additional protected sites under the Natura 2000 network in order to reach a favourable conservation status of habitats or habitats of species. This is in line with the case law of the Court, which sees the designation of Nature 2000 sites as a continuous process.[34]

But measures outside Natura 2000 sites can also contribute to restoring to a favourable conservation status. Several provisions form a legal basis for taking restoration measures outside protected areas. For example Article 4(4) of the Birds Directive provides that 'Outside these protection areas, Member States shall also strive to avoid pollution or deterioration of habitats.' In a ruling against Ireland[35] the European Court of Justice stated that, although the second sentence of Article 4(4) of the Birds Directive does not require that certain results be achieved, the

---

29 Article 1(i), Habitats Directive.
30 See also J. Verschuuren, 'Climate change: rethinking restoration in the European Union's Birds and Habitats Directives' (2010) 28(4) *Ecological Restoration* 431–439, 432.
31 The Natura 2000 network consists of the protected areas designated under the Birds and Habitats Directives; see Chapter 10.
32 Article 2(1), Habitats Directive.
33 Article 2(2), Habitats Directive.
34 Case C-209/04 *Commission v Austria* [2006] ECR I-2755.
35 The case concerned the inadequate transposition and application of several articles of the Birds and Habitats Directives.

Member States must nevertheless make a serious attempt at protecting those habitats which lie outside the Special Protection Areas (SPAs).[36] Although the Court itself does not specifically refer to restoration obligations in this case, the Advocate General[37] does when she writes that the obligation under Article 4(4) is a duty to use best endeavours. Best endeavours encompass the taking of all reasonable measures. The framework for determining what is reasonable is set out in Article 2 of the Birds Directive. Consequently, the Advocate General states:

> the measures taken in connection with endeavours made pursuant to the second paragraph of Article 4(4) of the Birds Directive must be arranged – on an ornithological basis – in such a way that they – in conjunction with other measures required under the directive – restore or maintain the level of the relevant species required under Article 2.[38]

In the Habitats Directive, Article 6(2) includes the obligation to prevent deterioration of habitats and disturbance of species. According to the Commission, this provision applies also to activities outside the designated sites, if external events may have an impact on the species and the habitats inside the site.[39] This was also confirmed by the Court of Justice in a ruling against the United Kingdom[40] where the Court stated that, in implementing Article 6(2), it may be necessary to adopt measures intended to avoid external human-caused impairment and disturbance that may cause the conservation status of species and habitats in SACs to deteriorate.[41]

Furthermore, the Habitats Directive includes obligations for restoration in the wider landscape through its obligations on connectivity.[42] These articles taken together will be an important element for connecting the EU-protected areas.[43] Lastly, species protection measures apply on the whole of a Member State's territory, regardless of it being designated a protected area.

Restoration of species and habitats of species

Articles 12 and 13 of the Habitats Directive contain the obligation for Member States to take the requisite measures to establish a system of strict protection for

---

36 Case C-418/04 *Commission v Ireland* [2007] ECR I-10947, para. 179.
37 The Advocate General presents a legal opinion to the Court. The opinion is advisory and does not bind the Court, however opinions are considered to be very influential.
38 Opinion AG Kokott, Case 418/04 *Commission v Ireland* [2007] ECR I-10947, paras 109–112.
39 European Commission, *Managing Natura 2000 sites: The provisions of Article 6 of the 'Habitats' Directive 92/43/EEC* (Office for Official Publications of the European Communities 2000) 24; see also Case C-6/04 *Commission v United Kingdom* [2005] ECR I-9017, para. 34.
40 This case concerned the incorrect transposition of the Habitats Directive.
41 Case C-6/04 *Commission v United Kingdom* [2005] ECR I-9017, para. 34.
42 Articles 3 and 10, Habitats Directive.
43 See Chapter 10 on connectivity.

the species listed in Annex IV in their natural range including prohibiting a number of activities such as killing and disturbing animal species and collecting and destruction of plants. Article 12(1)(d) of the Habitats Directive prohibits the deterioration or destruction of breeding sites or resting places of animal species listed in Annex IV. While the species protection regime seems to be an example of classic prohibition rules, the question is to what extent these provisions also allow for restoration of species and their habitats.

Resolving this question is particularly important as the species protection measures are not limited to the sites of the Natura 2000 network, but apply across all EU territory, and can thus provide for wider possibilities for restoration. According to the Commission guidelines on species protection the strict protection measures adopted under Article 12 must contribute to fulfilling the main objective of the Directive, namely maintaining or restoring a favourable conservation status.[44] Although the Commission guidelines are not legally binding, they provide valuable guidance on the interpretation of the Directives. The Court ruled in a case against Greece on the protection of the Milos Viper that the system of strict protection presupposes the adoption of coherent and coordinated measures of a preventive nature.[45] Such a system of strict protection must therefore enable the effective avoidance of deterioration or destruction of breeding sites or resting places of the animal species listed in Annex IV. However, also according to the Commission species guidelines:

> it is important to recognise that proactive habitat management measures (such as restoration of habitats/populations, improvement of habitats) are not an obligation under Article 12, even though they might well be under Article 6. For example, if proactive biotope restoration is needed for a butterfly species listed only in Annex IV(a) because its habitat has nearly disappeared and only a larger habitat would ensure long-term survival, such a measure would not be covered by Article 12. Such situations could be avoided or corrected in the medium to long term by revision of the annexes or the Directive itself.[46]

Although this interpretation seems in line with the literal wording of the provision of the Habitats Directive, it might seem at odds with the overall objective of the Directive to maintain or restore to a favourable conservation status. If a species is in a bad conservation status, or on the brink of disappearing, habitat restoration measures seem essential to fulfil the overall objective of the Directive.

44 European Commission, *Guidance document on the strict protection of animal species of Community interest under the Habitats Directive 92/43/EEC* (European Commission 2007).

45 Case C-518/04 *Commission v Greece* [2006] ECR I-42, para. 16; also judgment of 11 January 2007 in Case C-183/05 *Commission v Ireland* [2007] ECR I-137, para. 30.

46 European Commission, *Guidance document on the strict protection of animal species of Community interest under the Habitats Directive 92/43/EEC* (European Commission 2007) 20.

In a case before the European Court of Justice, the so-called 'European hamster' case (Alsace, France),[47] both the Advocate General and the Court shed additional light on the restoration obligations under Article 12 of the Directive. In this case the Commission took France to court for breaching its obligations under Article 12(1)(d) of the Habitats Directive, by failing to establish a programme of measures to ensure strict protection of the European hamster.

The French government relied on the Commission guidance document in order to justify its apparent lack of effective proactive habitat measures. However, Advocate General Kokott did not follow this line of argumentation. In her opinion the Commission's position is based essentially on the fact that proactive habitat measures are associated mainly with territorial protection under Articles 4, 5, and 6 of the Habitats Directive. According to her opinion, the fact that states rely on these Articles to drive restoration work 'does not altogether preclude proactive measures from also being included in the protection of species under Article 12(1). This applies particularly to species such as the European hamster for which no such protected areas are provided.'[48] According to the Advocate General 'prohibitions are of a defensive nature and therefore aim primarily to prevent the deterioration of an existing condition. However, prohibitions can also help to restore or improve habitats in so far as they enable positive natural developments to take place'.[49] The Advocate General further explains that prohibitions for the protection of species can of course also influence habitat management, such as a ban on deep ploughing in agriculture because that is likely to destroy the hamster burrows.[50]

The territorial scope of any potential restoration measures will also depend on the conservation status of the species:

> If its conservation status is good, it may be sufficient to make general provision for the prohibitions laid down in Article 12(1) and to monitor the species. An unfavourable conservation status gives rise to more far-reaching obligations for the Member States, however, because the system of protection is intended to help to restore a favourable conservation status. The protection of breeding sites and resting places of a species with a very unfavourable conservation status, as in the case of the European hamster in Alsace, therefore requires a generous delimitation of territory in order to prevent the species from disappearing, and thus the functionality of the sites from being lost. The

47 Case C-383/09 *Commission v France* [2011] ECR I-4869; for an extensive review of this case see H. Schoukens, 'Going beyond the status quo: towards a duty for species restoration under EU law?' in V. Sancin and M.K. Dine (eds), *International Environmental Law: Contemporary Concerns and Challenges* (IUS Software, d.o.o., GV Zalozba 2014) 343–358.
48 Opinion A.G. Kokott, Case C-383/09 *Commission v France* [2011] ECR I-4869, para. 44.
49 Ibid., para. 45.
50 Ibid., para. 46.

protection measures must, so far as possible, be adjusted specifically to the circumstances giving rise to the unfavourable conservation status.[51]

The Advocate General also elaborated on the question of which measures are required to prevent deterioration and destruction of the breeding sites or resting places. She is of the opinion that behaviour which impairs or eliminates the ecological functionality of breeding sites or resting places must be regarded as deterioration and destruction. But, on the other hand, the Advocate General finds that the obligations are geographically limited. As she opined,

> [M]easures in areas where there are no hamster burrows are not necessary. Measures of that kind are certainly sensible for the future repopulation of those habitats by the European hamster and, therefore, presumably also necessary for the restoration of a favourable conservation status for the species in Alsace generally. However, the measures required by Article 12(1)(d) of the Habitats Directive relate only to the breeding sites and resting places of existing populations. The Commission has not asserted, and it appears unlikely, that a favourable conservation status for those specific populations would require a particular form of management of land outside the vicinity of their burrows.[52]

From an ecological and conservation point of view, the Advocate General's analysis is regrettable as restoration measures outside of current breeding sites or resting places could be important for relict populations. At least in this opinion, it seems that the Advocate General was not prepared to let nature conservation prevail over the literal content of Article 12(1).[53]

Advocate General Kokott did not uphold the claim of the European Commission that there is an obligation to restore hamster populations that previously existed, on the grounds that France may not have given sufficient protection to the European hamster in the past. However, this was due to procedural reasons. The Advocate General acknowledged that a system of strict protection had to be introduced for the European hamster, and it is possible that past omissions may give rise to an obligation on the part of Member States to provide for restoration. However, the Commission did not make a claim in respect of restoration in the pre-litigation procedure or in the application, but only indirectly in the reply, which is an impermissible extension of the subject-matter of the proceedings.[54]

In the ruling of the Court itself, the Court did not explicitly elaborate on the Member States' duty to restore the population of an endangered species. However, it did require France to come up with sufficient measures that are able to reverse

---

51 Ibid., para. 37.
52 Ibid., para. 50.
53 Schoukens, 'Going beyond the status quo' 353.
54 Opinion AG Kokott, Case C-383/09 *Commission v France* [2011] ECR I-4869, para. 51.

the negative trend. Such measures could include the establishment of repopulation areas covering a large part of the hamster's historical range and the stricter application of rules on the development of maize crops and urbanisation projects. By scrutinising the French repopulation measures, the Court underscored its willingness to consider a duty to restore a species that finds itself in an unfavourable conservation status, at least in such cases where a decline of a protected species can be ascribed to a faulty protection policy of a Member State.[55] The Court, as well as the Advocate General did not, however, interpret Article 12 in such a way as also to include a restoration duty that goes beyond the maintenance or improvement of the actual sites of the European hamster.[56]

Other legal obligations in the EU on restoration

In addition to the core nature legislation, the EU has some other instruments that are relevant for restoration. We will briefly discuss the Water Framework Directive, the Floods Directive, the Marine Strategy Directive, the Rural Development Regulation and the Environmental Liability Directive.

Restoration in the 'Water' Directives

The Water Framework Directive[57] aims to establish a framework for the protection of inland surface waters, transitional waters, coastal waters and groundwater, which prevents further deterioration and protects and enhances the status of aquatic ecosystems, promotes sustainable water use, aims at the reduction of pollution and contributes to mitigating the effects of floods and droughts.[58] The Directive opts for coordination of administrative arrangements within river basin districts.[59] For each river basin district a programme of measures shall be established,[60] as well as a river basin management plan,[61] which shall be reviewed and updated every six years. Restoration of wetlands is mentioned as one of the supplementary measures which Member States within each river basin district may choose to adopt as part of the programme of measures required under Article 11.[62]

The ultimate environmental objective of the Directive is to achieve a good water status for both surface and groundwater 15 years after the entry into force of the Directive (which was 22 December 2015). A 'good' water status includes

55 Cliquet, Decleer and Schoukens, 'Restoring nature in the EU' 283.
56 Schoukens, 'Going beyond the status quo' 357.
57 Directive 2000/60/EC of the European Parliament and of the Council of 23 October 2000 establishing a framework for Community action in the field of water policy, *OJ L 327*, 22 December 2000 (hereafter referred to as Water Framework Directive).
58 Cf. Article 1, Water Framework Directive.
59 Cf. Article 3, Water Framework Directive.
60 Article 11, Water Framework Directive.
61 Article 13, Water Framework Directive.
62 Annex VI. Lists of measures to be included within the programmes of measures, Part B, Water Framework Directive.

the objectives of good ecological and chemical status for surface waters and good quantitative and chemical status for groundwater.[63] The deadline to achieve a 'good status' can be extended for the purposes of phased achievement of the objectives but requires taking into account certain restrictions including no extension beyond 2027.[64]

States not only have to implement the necessary measures to prevent deterioration of the status of all bodies of surface and groundwater, but they also have to protect, enhance and restore all bodies of surface water and groundwater.[65] Good water status has to be reached everywhere, and is not limited to protected areas. For protected areas, the Directive states that Member States shall achieve compliance with all standards and objectives at the latest 15 years after the date of entry into force of this Directive, unless otherwise specified in the Community legislation under which the individual protected areas have been established.[66] There can obviously be a large overlap between the water bodies that are covered by the Water Framework Directive and the protected areas under the Birds and Habitats Directives.[67]

The Water Framework Directive has the potential to be an important driver for ecological restoration. The monitoring obligations in the Directive[68] should be able to reveal if restoration is needed, what type of restoration is needed and if restoration was successful.[69] However, the overall objective of the Directive has not yet been met. According to the State of the Environment report of 2015, an assessment showed that only 43 per cent of surface water bodies were in good or high ecological status in 2009. The objective of reaching a good ecological status by 2015 was only likely to be met by 53 per cent of surface water bodies. This constitutes a modest improvement and is far from meeting policy objectives. The 2015 report shows that only half of surface water bodies met the objective to achieve a good ecological status by the end of 2015.[70] The modest improvement can be explained, in part, because river basin management is a new approach for

63 European Commission, *Report from the Commission to the European Parliament and the Council on the Implementation of the Water Framework Directive (2000 /60/EC). River Basin Management Plans* (COM(2012) 670 final, European Commission 2012) 3.

64 Article 4(4), Water Framework Directive.

65 Article 4(1)(a)(ii) and (b)(ii), Water Framework Directive.

66 Article 4(1)(c), Water Framework Directive.

67 On the relationship between the Water Framework Directive and the Habitats Directive, see P. De Smedt and M. van Rijswick, 'Nature conservation and water management' in C.-H. Born, A. Cliquet, H. Schoukens, D. Misonne and G. Van Hoorick (eds), *The Habitats Directive in its EU Environmental Law Context: European Nature's Best Hope?* (Routledge 2015) 417–433.

68 Cf. Article 8 and Annex V, Water Framework Directive.

69 D. Hering et al., 'The European Water Framework Directive at the age of 10: a critical review of the achievements with recommendations for the future' (2010) 408 *Science of the Total Environment* 4007–4019.

70 European Environment Agency, *The European environment – state and outlook 2015: synthesis report* (EEA 2015) 62–63; see also European Environment Agency, *European waters – current status and future challenges. Synthesis* (EEA Report No 9/2012, European Environment Agency 2012).

some Member States, but the minimal improvement is also a reflection of the difficulties in restoring aquatic systems.[71]

Two 'sister' Directives of the Water Framework Directive also include obligations on restoration. The purpose of the Floods Directive[72] is 'to establish a framework for the assessment and management of flood risks, aiming at the reduction of the adverse consequences for human health, the environment, cultural heritage and economic activity associated with floods in the Community'.[73] The Directive requires Member States to carry out a preliminary assessment by 2011 to identify the river basins and associated coastal areas at risk of flooding.[74] For these areas flood risk maps have to be completed by 2013 and flood risk management plans must be established by 2015.[75] Flood risk management plans should focus on prevention, protection and preparedness. With a view to giving rivers more space, the plans should consider where possible the maintenance and/or restoration of floodplains.[76] In a note by the European Commission, the Commission stressed the importance of building up green infrastructure. This offers a potential triple-win for the EU through its contribution to the protection and restoration of floodplain and coastal ecosystems; its mitigation of climate change impacts; and its provision of cost-effective protection against some of the threats that result from climate change such as increased floods.[77]

For the marine environment the Marine Strategy Directive[78] includes obligations to restore the marine environment. This Directive establishes a framework within which Member States shall take the necessary measures to achieve or maintain good environmental status in the marine environment by the year 2020 at the latest.[79] States are expected to develop and implement marine strategies in order to protect and preserve the marine environment, prevent its deterioration or, where practicable, restore marine ecosystems in areas where they have been adversely affected.[80] The Directive establishes four European marine regions: the Baltic Sea, the North-east Atlantic Ocean, the Mediterranean Sea and the Black

---

71 G. Phillips, 'Progress towards the implementation of the European Water Framework Directive (2000–2012)' (2014) 17(4) *Aquatic Ecosystem Health and Management* 424–436.

72 Directive 2007/60/EC of the European Parliament and of the Council of 23 October 2007 on the assessment and management of flood risks, *OJ* L 288, 6 November 2007 (hereafter referred to as Floods Directive).

73 Article 1, Floods Directive.

74 Articles 4–5, Floods Directive.

75 Articles 6–8, Floods Directive.

76 Preamble, para. 14, Floods Directive.

77 DG Environment, *Towards Better Environmental Options for Flood Risk Management* (DG ENV D.1 (2011) 236452, 2011); see also the Annex to this Note, in which some examples of restoration of floodplains are given.

78 Directive 2008/56/EC of the European Parliament and of the Council of 17 June 2008 establishing a framework for community action in the field of marine environmental policy, *OJ* L 164, 25 June 2008 (hereafter referred to as Marine Strategy Framework Directive).

79 Article 1(1), Marine Strategy Framework Directive.

80 Article 1(2)(a), Marine Strategy Framework Directive.

Sea.[81] Each Member State, in respect of each marine region or sub-region concerned, is required to develop a strategy for its own marine waters. The strategy includes: an initial assessment of the current environmental status of the marine waters of the Member States and the human impact on the environment; a determination of what a good environmental status means; the establishment of environmental targets and associated indicators; the establishment and implementation of a monitoring programme by 15 July 2014; and the development of a programme of measures designed to achieve or maintain good environmental status by 2015.[82]

Monitoring programmes need to include activities to identify the cause of the change and hence the possible corrective measures that would need to be taken to restore the good environmental status.[83] The development of a programme of measures to maintain or improve environmental status of the water is expected to take into consideration the measures listed in Annex VI including restoring damaged components of marine ecosystems.[84] Where the Member States consider that the management of a human activity at the Community or international level is likely to have a significant impact on the marine environment, particularly in a marine protected area, they shall, individually or jointly, address the competent authority or international organisation concerned with a view to the consideration and possible adoption of measures that may be necessary to achieve the objectives of this Directive. These measures may include rules to enable the maintenance or, where appropriate, the restoration of the integrity, structure and functioning of a given marine ecosystem.[85] Although legal instruments for the protection of the marine environment are in place, the results are insufficient. An assessment by the European Environment Agency shows that Europe's seas cannot be considered healthy or clean.[86] An assessment by the European Commission of the implementation of the Directive shows that more efforts are urgently needed if the EU is to reach its goal.[87]

Restoration in the Common Agricultural Policy

Some limited support for restoration can also be found in the rural development policy and regulations. The rural development policy of the EU is the so-called 'second pillar' of the EU Common Agricultural Policy (CAP). The rural

---

81 Article 4, Marine Strategy Framework Directive.
82 Article 5, Marine Strategy Framework Directive.
83 Annex V, para. 4, Marine Strategy Framework Directive.
84 Annex VI, para. 7, Marine Strategy Framework Directive.
85 Article 13(5), Marine Strategy Framework Directive.
86 European Environment Agency, *State of Europe's seas* (EEA Report No 2/2015, European Environment Agency 2015).
87 European Commission, *Report from the Commission to the Council and the European Parliament, The first phase of implementation of the Marine Strategy Framework Directive (2008/56/EC). The European Commission's assessment and guidance* (COM (2014) 97 final, European Commission 2014).

development policy aims to help the rural areas of the EU to address a wide range of economic, environmental and social challenges and opportunities. The CAP was reformed in 2013 for the period 2014–2020,[88] with the aim of 'greening' the CAP. In spite of this objective the reform of the CAP is deemed insufficient for a real greening of the CAP.[89] One of the main legal instruments to support this policy is Regulation 1305/2013 from 2013.[90] Financed by the European Agricultural Fund for Rural Development, this Regulation lays down general rules governing EU support for rural development including ecological restoration activities in rural areas.[91] One of the EU priorities in this policy is 'restoring, preserving and enhancing biodiversity, including in Natura 2000 areas, and in areas facing natural or other specific constraints, and high nature value farming, as well as the state of European landscapes'.[92] Presently, the scope of restoration activities financed under the Regulation is limited to various restoration measures for forests including the restoration of damage to forests from forest fires, natural disasters, catastrophic events (including pest and disease outbreaks) and climate-related threats.[93]

Restoration in the Environmental Liability Directive

Specific obligations with regards to restoration can be found in the Environmental Liability Directive.[94] The purpose of this Directive is to establish a framework of

88 See on the rural development policy, the EU website: <http://ec.europa.eu/agri culture/rural-development-2014-2020/index_en.htm>.
89 I. Doussan and H. Schoukens, 'Biodiversity and agriculture: greening the CAP beyond the status quo?' in C.-H. Born, A. Cliquet, H. Schoukens, D. Misonne and G. Van Hoorick (eds), *The Habitats Directive in its EU Environmental Law Context: European Nature's Best Hope?* (Routledge 2015) 437–451.
90 Regulation (EU) No 1305/2013 of the European Parliament and of the Council of 17 December 2013 on support for rural development by the European Agricultural Fund for Rural Development (EAFRD) and repealing Council Regulation (EC) No 1698/2005, *OJ L* 347, 20 December 2013 (hereafter referred to as Rural Development Regulation).
91 For an assessment of the rural development policy and some examples of restoration projects, see J. Poláková, G. Tucker, K. Hart, J. Dwyer and M. Rayment, *Addressing biodiversity and habitat preservation through Measures applied under the Common Agricultural Policy* (Report Prepared for DG Agriculture and Rural Development, Contract No. 30-CE-0388497/00–44, Institute for European Environmental Policy 2011); some more examples of restoration projects can be found on the website of the European Network for Rural Development (ENRD): <http://enrd.ec.europa.eu/ en/policy-in-action/cap-towards-2020/rdp-programming-2014-2020/rural-devel opment-priorities/ecosystems#twoTab>.
92 Article 5(4)(a), Rural Development Regulation.
93 Article 6 and 21(1)(a), (c) and (d), Rural Development Regulation.
94 Directive 2004/35/CE of the European Parliament and of the Council of 21 April 2004 on environmental liability with regard to the prevention and remedying of environmental damage, *OJ L* 143, 30 April 2004 (hereafter referred to as (Environmental) Liability Directive); see for an extensive analysis on the Environmental Liability Directive: V. Fogleman, 'The threshold for liability for ecological damage in the EU' in C.-H. Born, A. Cliquet, H. Schoukens, D. Misonne and G. Van Hoorick (eds), *The*

environmental liability based on the polluter pays principle to prevent and remedy environmental damage.[95] This section will only focus on those provisions which aim to restore environmental damage.

Environmental damage

'Environmental damage' includes damage to protected species and natural habitats under the Birds and Habitats Directives, water damage and land damage.[96] For the application of the Environmental Liability Directive, the species and habitats only include Annex I species listed in the Birds Directive, as well as regularly occurring migratory species, and their habitats. Thus, the protection under the Environmental Liability Directive is more limited than the Birds Directive itself, which protects all wild bird species in Europe. The regime of the Environmental Liability Directive is also applicable to the species and habitats listed in the Habitats Directive, including Annex II and IV species, the habitats of Annex I, the habitats of species of Annex II and breeding sites and resting places of Annex IV species. The scope of the Environmental Liability Directive is not limited to the species and habitats of the Birds and Habitats Directives which occur within the Natura 2000 sites. The Environmental Liability Directive also applies to damage to species and habitats outside the Natura 2000 sites including habitats which are eligible for designation but where Member States have failed to timely designate the sites and habitats that do not meet the criteria for designation as a protected site.[97] In the original Directive water damage was limited to waters covered by the Water Framework Directive. This was extended to marine waters in 2013.[98] Furthermore, a Member State can extend the Liability Directive to any habitat or species not listed in the Birds and Habitats Directives.[99]

At first glance, it may seem that the Environmental Liability Directive has a broad scope. However, the application of the Directive is limited as not all damage will lead to remedial action. Only damage that fulfils the threshold criteria as provided for in the Directive will necessitate restoration measures. Environmental damage to protected species and natural habitats is defined in the Directive as 'any damage that has significant adverse effects on reaching or maintaining the favourable conservation status of such habitats or species. The significance of such effects is to be assessed with reference to the baseline condition, taking account of the

Habitats Directive in its EU Environmental Law Context: European Nature's Best Hope? (Routledge 2015) 181–214.

95 Article 1, Environmental Liability Directive.
96 Article 2(1), (3) and (5), Environmental Liability Directive.
97 G.M. van den Broek, 'Environmental liability and nature protected areas: will the EU Environmental Liability Directive actually lead to the restoration of damaged natural resources? (2009) 5(1) *Utrecht Law Review* 118–119.
98 Directive 2013/30/EU of the European Parliament and of the Council of 12 June 2013 on safety of offshore oil and gas operations and amending Directive 2004/35/ EC, *OJ L* 178, 28 June 2013. This Directive extended the scope of damage to marine waters.
99 Cf. Article 2(3)(c), Environmental Liability Directive.

criteria set out in Annex I'.[100] The 'baseline condition' is defined as 'the condition at the time of the damage of the natural resources and services that would have existed had the environmental damage not occurred, estimated on the basis of the best information available'.[101] 'Damage' in turn is defined as 'a measurable adverse change in a natural resource or measurable impairment of a natural resource service which may occur directly or indirectly'.[102] This reduces the protection of the Environmental Liability Directive to a measurable adverse change in a natural resource or measurable impairment of a natural resource service, which has a significant adverse effect on reaching or maintaining the favourable conservation status, compared with the baseline condition.[103]

Legal literature criticises the use of the concept of favourable conservation status, which is the overall aim of the Habitats Directive, as a threshold under the Environmental Liability Directive. This concept may not be appropriate for setting liability standards.[104] Furthermore, there is legal uncertainty about the 'significance level'.[105] Because the legal standard is quite high, it can be presumed that the threshold for damage to habitats and species is not likely to be exceeded in many cases.[106] In cases where accurate measurable data is not available, it might be difficult to assess the significance of the damage.[107]

Restoration obligations

This Directive contains two liability regimes. It shall apply first to environmental damage caused by any of the activities listed in Annex III,[108] and to any imminent threat of such damage occurring by reason of any of those activities. This liability is regardless of fault or negligence. Second, the Directive applies to damage to protected species and natural habitats caused by any activities other than those listed in Annex III, and to any imminent threat of such damage occurring by reason of any of those activities, whenever the operator has been at fault or negligent.[109]

Where environmental damage has occurred the operator shall, without delay, inform the competent authority of all relevant aspects of the situation and take the necessary remedial measures.[110] The competent authority may require the

---

100 Article 2(1)(a), Environmental Liability Directive.
101 Article 2(14), Environmental Liability Directive.
102 Article 2(2), Environmental Liability Directive.
103 Van den Broek, 'Environmental liability and nature protected areas' 119.
104 Fogleman, 'The threshold for liability for ecological damage in the EU' 200.
105 Ibid., 207–211.
106 This appears also from a Dutch study, see reference in van den Broek, 'Environmental liability and nature protected areas' 122.
107 Van den Broek, 'Environmental liability and nature protected areas' 123–124.
108 Annex III contains a variety of activities, which have been regulated in other EU Directives, such as the deliberate introduction of GMOs into the environment, the transport of dangerous goods, etc.
109 Article 3(1), Environmental Liability Directive.
110 Article 6(1)(b), Environmental Liability Directive.

operator to take the necessary remedial measures or the authority may take the necessary remedial measures.[111] 'Remedial measures' are defined as 'any action, or combination of actions, including mitigating or interim measures to restore, rehabilitate or replace damaged natural resources and/or impaired services, or to provide an equivalent alternative to those resources or services as foreseen in Annex II'.[112]

Annex II of the Directive sets out a common framework to be followed in order to choose the most appropriate measures to ensure the remedying of environmental damage. Remedying of environmental damage, in relation to water or protected species or natural habitats, is achieved through the restoration of the environment to its baseline condition by way of primary, complementary and compensatory remediation. 'Primary remediation' is any remedial measure that returns the damaged natural resources and/or impaired services to, or towards, baseline condition.[113]

Where the damaged natural resources and/or services do not return to their baseline condition, then 'complementary remediation' will be undertaken. The purpose of complementary remediation is to provide a similar level of natural resources and/or services as would have been provided if the damaged site had been returned to its baseline condition. Complementary remediation may, as appropriate, be provided at an alternative site that should be geographically linked to the damaged site, taking into account the interests of the affected population.[114]

In addition, 'compensatory remediation' will be undertaken to compensate for interim losses. Compensatory remediation shall be undertaken to compensate for the interim loss of natural resources and services pending recovery. This compensation consists of additional improvements to protected natural habitats and species or water at either the damaged site or at an alternative site. It does not consist of financial compensation to members of the public.[115]

However, the application of the Directive is limited, first in scope and second in practical relevance. So far, there has been little implementation of this Directive. A report from the Commission on the implementation of the Directive from 2010 estimated that around 50 cases had been treated under the Liability Directive. It appears that most cases relate to damage to water and land and only a limited number to protected species and natural habitats. In most cases primary remediation measures were applied immediately (excavation and soil replacement as well as clean-up of water, aiming to restore the site's baseline condition). However, none of the cases reported included information about the other two types of remediation (complementary and compensatory). The activities involved were almost exclusively listed in Annex III of the Directive.

111 Article 6(2)(c) and (e), Environmental Liability Directive.
112 Article 2(11), Environmental Liability Directive.
113 Annex II, 1(a), Environmental Liability Directive.
114 Annex II, 1.1.2., Environmental Liability Directive.
115 Annex II, 1.1.3., Environmental Liability Directive.

The Commission's report gives some possible reasons for this low number of Liability Directive cases, including limited knowledge by operators. But it may also reflect the preventive effect that the Directive is already having. Another reason for the limited number of cases relying upon the Directive may be that some Member States maintained their existing laws for soil or water remediation which included more stringent measures than the Liability Directive. Finally, the exceptions and defences of the Directive including insolvency and non-identification of the responsible operators may have led to fewer Directive cases. The Commission's report concludes that there is insufficient data to draw reliable conclusions on the effectiveness of the Directive in terms of actual remediation of environmental damage.[116]

EU biodiversity policy on restoration

In addition to the legal framework discussed in the previous sections, restoration has also become the focus of attention in the EU biodiversity policy. In the first Biodiversity Strategy of 1998,[117] 'restoration' was mentioned albeit without any concrete targets. Under the theme on conservation and sustainable use of biodiversity, the Strategy mentions that 'the Community should seek the conservation and, where relevant, restoration of ecosystems and populations of species in their natural surroundings'. Specifically under in-situ conservation, the Strategy puts forward that the Community should seek to support recovery plans for the most threatened species.[118] The Strategy also sums up different objectives within several policy areas. Under the conservation of natural resources, one of the objectives is to protect wetlands within the EU and restore the ecological character of degraded wetlands.[119] Regarding forests, the EU's objective is to promote restoration and regeneration of areas that have suffered deforestation.[120]

The European Commission, in its new Biodiversity Strategy to 2020,[121] adopted in 2011, sets several targets that are important for ecological restoration. As a party

116 European Commission, *Report from the Commission to the Council, the European Parliament, the European Economic and Social Committee and the Committee of the Regions under Article 14(2) of Directive 2004/35/CE on the environmental liability with regard to the prevention and remedying of environmental damage* (COM(2010) 581 final, European Commission 2010).

117 European Commission, *Communication from the Commission to the Council and the European Parliament on a European Community biodiversity strategy* (COM(1998) 42 final, European Commission 1998); this Strategy was developed in response to the Biodiversity Convention, to which the EU is a party.

118 Biodiversity Strategy 1998, II, para. 3.

119 Biodiversity Strategy 1998, III, para. 3.

120 Biodiversity Strategy 1998, III, para. 29.

121 European Commission, *Communication from the Commission to the European Parliament, the Council, the Economic and Social Committee and the Committee of the Regions, Our life insurance, our natural capital: an EU biodiversity strategy to 2020* (COM (2011) 244 final, European Commission 2011) (further referred to as EU Biodiversity Strategy); the Biodiversity Strategy was endorsed by the Council of the European

to the Biodiversity Convention, the European Union is legally bound by the Convention. Although the decisions by the Conference of the Parties (COP decisions) of the Biodiversity Convention are usually considered to be legally non-binding,[122] they are at least politically binding for the parties to the Convention, including the EU. As a response to the Aichi Targets the EU concluded the EU Biodiversity Strategy 2011. The EU Strategy includes a 2050 vision and a 2020 headline target, both of them referring to restoration. The 2050 vision is:

> By 2050, European Union biodiversity and the ecosystem services it provides – its natural capital – are protected, valued and appropriately restored for biodiversity's intrinsic value and for their essential contribution to human wellbeing and economic prosperity, and so that catastrophic changes caused by the loss of biodiversity are avoided.

The 2020 headline target is:

> Halting the loss of biodiversity and the degradation of ecosystem services in the EU by 2020, and restoring them in so far as feasible, while stepping up the EU contribution to averting global biodiversity loss.

The Strategy includes six mutually supportive and inter-dependent targets that respond to the objectives of the 2020 headline target. Each target addresses a specific operational issue: protecting and restoring biodiversity and associated ecosystem services (Targets 1 and 2), enhancing the positive contribution of agriculture and forestry and reducing key pressures on EU biodiversity (Targets 3, 4 and 5), and stepping up the EU's contribution to global biodiversity (Target 6). Each target has a set of actions, listed in the Annex to the Communication of the Commission. The most relevant targets relating to restoration are Targets 1 and 2, which will be further discussed in detail below. Implemented together, the targets should halt biodiversity loss and the degradation of ecosystem services.

Target 1

Target 1 states:

> To halt the deterioration in the status of all species and habitats covered by EU nature legislation and achieve a significant and measurable improvement in their status so that, by 2020, compared to current assessments: (i) 100%

Union in its Decision of 21 June 2011 (EU Biodiversity Strategy to 2020 – Council conclusions, 11978/11).

122 See, for instance, J. Brunnée, 'COPing with consent: law-making under multilateral environmental agreements' (2002) 15 *Leiden Journal of International Law* 1–52; A. Wiersema, 'The new international law-makers? Conferences of the parties to multilateral environmental agreements' (2009) 31 *Michigan Journal of International Law* 231–287.

more habitat assessments and 50% more species assessments under the Habitats Directive show an improved conservation status; and (ii) 50% more species assessments under the Birds Directive show a secure or improved status.[123]

The fact that the EU Target 1 mentions 'improvement' of the status is important with regard to restoration, as the improvements represent the restoration efforts that are needed to achieve the overall 2020 biodiversity headline target.[124] As this target demands a serious improvement in the conservation status of EU protected habitats and species,[125] it is important to know the current conservation status. According to a 2009 assessment, based on the period of 2001–2006, only 17 per cent of the habitat assessments were favourable. Also for species, other than birds, 17 per cent of the species assessments had favourable conservation status.[126] For bird species 52 per cent of the assessed bird species had a favourable conservation status.[127] However, although the birds species protected under the Birds Directive seem to do relatively better, compared with other species, there has been a decline in 'common' bird species since 1990. According to the European Environment Agency common bird populations have decreased by around 12 per cent in 27 European countries. The decline of common farmland birds was even more pronounced at 30 per cent, whereas common forest birds declined by 8 per cent.[128] These assessment trends are confirmed by a recent scientific study.[129]

123 Target 1, EU Biodiversity Strategy.
124 European Commission, *Commission Staff Working Paper. Impact Assessment. Accompanying the document Communication from the Commission to the European Parliament, the Council, the Economic and Social Committee and the Committee of the Regions, Our life insurance, our natural capital: an EU biodiversity strategy to 2020* (SEC (2011) 540 final, European Commission 2011) 26.
125 For the reporting period 2001 to 2006, Member States provided detailed assessments on the conservation status of each of the habitat types (216) and species (nearly 1182) listed in the Directive and found within their territory (European Commission, *Report from the Commission to the Council and the European Parliament Composite – Report on the Conservation Status of Habitat Types and Species as required under Article 17 of the Habitats Directive* (COM/2009/0358 final, European Commission 2009) 3).
126 European Commission, *Report from the Commission to the Council and the European Parliament Composite – Report on the Conservation Status of Habitat Types and Species as required under Article 17 of the Habitats Directive* (COM/2009/0358 final, European Commission 2009); in total, 701 habitat assessments and 2240 species assessments were made at bio-geographic level (ibid., 6).
127 European Environment Agency, *State of nature in the EU. Results from reporting under the nature directives 2007–2012* (EEA Technical report No 2/2015, European Environment Agency 2015) 23 (based on the assessment of 447 bird species).
128 European Environment Agency, *Abundance and distribution of selected species (SEBI 001)* (European Environment Agency 2015), <www.eea.europa.eu/data-and-maps/indicators/abundance-and-distribution-of-selected-species/abundance-a nd-distribution-of-selected-2>.
129 R. Inger, R. Gregory, J.P. Duffy, I. Stott, P. Voříšek and K.J. Gaston, 'Common European birds are declining rapidly while less abundant species' numbers are rising' (2015) 18(1) *Ecology Letters* 28–36.

As the goal in the EU Biodiversity Strategy is to have an improved status by either 100 per cent for habitats or 50 per cent for species, this leads to the following concrete goals: by 2020, 34 per cent of habitats assessments (plus 100 per cent of the 2009 assessment), more than 25 per cent of species assessments (plus 50 per cent of the 2009 assessment) have to show an improved conservation status and nearly 80 per cent of birds' species assessments (plus 50 per cent of the 2009 assessment) are in a secure or improved status.[130]

A 2015 assessment from the European Environment Agency confirms the poor conservation status of most species and habitats in the European Union and even shows a further decline for habitats.[131] The assessment under the Habitats Directive for the period 2007–2012 shows that only 23 per cent of plant and animal species assessments and 16 per cent of habitats assessments were considered to be in a favourable conservation status (compared to 17 per cent for species and 17 per cent habitats in the previous 2009 assessment), which means an improvement in status for species but a slight decline for habitats. However, looking at the number of species and habitats in an unfavourable conservation status, 60 per cent of the species and 77 per cent of the habitats are in an unfavourable conservation status (compared with 52 per cent for species and 65 per cent for habitats in the previous assessment period). The main change with regards to the previous assessment period (2001–2006) was the number of assessments where conservation status was now known.[132] It is impossible to say whether the decrease in unfavourable conservation status for species and habitats is caused by a deterioration in condition of species or habitats, or reflects the improvements in the knowledge base.[133]

Several actions are mentioned under Target 1. Under action 1c Member States will ensure that management plans or equivalent instruments, which set out conservation and restoration measures, are developed and implemented in a timely manner for all Natura 2000 sites. Measures under this target should focus on speeding up the completion of the Natura 2000 network and on making the network fully operational through the effective management and restoration of Natura 2000 sites.[134]

130 K. Decleer, 'The new European Biodiversity Strategy: a challenge to the restoration community' (2012) 30(2) *Ecological Restoration* 93–94.
131 In total, 804 EU regional habitat assessments and 2665 EU regional species assessments were produced.
132 For species assessment, there was a decline in unknown status from 31 per cent to 17 per cent, for habitat assessment, there was a decrease in unknown status from 18 per cent to 7 per cent.
133 European Environment Agency, *The European environment – state and outlook 2015: synthesis report* (European Environment Agency 2015) 57; for a thorough analysis and a comparison between the assessment period 2001–2006 and 2007–2012, see European Environment Agency, *State of nature in the EU. Results from reporting under the nature directives 2007–2012* (EEA Technical report No 2/2015, European Environment Agency 2015) 144–150.
134 European Commission, *Commission Staff Working Paper. Impact Assessment. Accompanying the document Communication from the Commission to the European Parliament, the Council, the Economic and Social Committee and the Committee of the*

In the 2015 State of Nature report, an overview is given of the number of management plans for Natura 2000 sites as well as the type of management measures that are taken. Notably, several of these management measures include restoration measures, such as restoration of the hydrological regime, restoration of forest habitats or restoration of water quality.[135] It is clear that Member States are conducting restoration activities within Natura 2000 sites, but the figures on the unfavourable conservation status also show the pressing need to increase the restoration efforts within the EU.

Target 2

Target 2 is the most explicit of the targets in the EU Biodiversity Strategy on the need for the EU to undertake restoration. According to the target, 'by 2020, ecosystems and their services are maintained and enhanced by establishing green infrastructure and restoring at least 15% of degraded ecosystems'. According to the Strategy the rationale behind this target is that in the EU, many ecosystems and their services have been degraded, largely as a result of land fragmentation. Nearly 30 per cent of the EU territory is moderately to very highly fragmented. Target 2 focuses on maintaining and enhancing ecosystem services and restoring degraded ecosystems by incorporating green infrastructure in spatial planning. This will contribute to the EU's sustainable growth objectives and to mitigating and adapting to climate change, while promoting economic, territorial and social cohesion and safeguarding the EU's cultural heritage. It will also ensure better functional connectivity between ecosystems within and between Natura 2000 areas and in the wider countryside. Target 2 incorporates the Aichi global target agreed by EU Member States and the EU to restore 15 per cent of degraded ecosystems by 2020.[136]

Although Target 2 includes two ways of enhancing ecosystems and their services – through green infrastructure on the one hand and restoration on the other – both are strongly linked, and this is, as the rationale behind the target mentions, related to the high level of fragmentation in Europe. When this target was concluded, the Commission considered three options: 1) establishment of green infrastructure; 2) restoration of degraded ecosystems; or 3) establishment of green infrastructure and restoration of degraded ecosystems. The Commission chose the third option. With option 1, the focus would have been on spatial integration of ecosystem services. This had the appropriate qualitative focus but risked not being concrete enough in terms of desired output for restoration, especially as the concept of green infrastructure is still being developed. Option 2 would have focused on ecosystems only, as a proxy for ecosystem services. This had the merit of being

Regions, Our life insurance, our natural capital: an EU biodiversity strategy to 2020 (SEC (2011) 540 final European Commission 2011) 38.

135 European Environment Agency, *State of nature in the EU. Results from reporting under the nature directives 2007–2012* (EEA Technical report No 2/2015, European Environment Agency 2015) 130–135.

136 EU Biodiversity Strategy, para. 3.2.

more easily quantified than ecosystem services. However, there was a risk that the focus on restoration would be mainly on restoration areas, without due consideration to the type and level of services restored. Elements such as connectivity and integration in the landscape might not be taken into consideration, and opportunities for benefits from multiple services might be lost. Option 3 was the preferred option as it combined the two elements above and had the merit of combining a concrete focus on outcome by linking the qualitative elements of ecosystem services to a network of green infrastructure.[137]

Specific actions under Target 2 of the EU Biodiversity Strategy include that Member States, with the assistance of the Commission, by 2014 will develop a strategic framework to set priorities for ecosystem restoration at the sub-national, national and EU level (action 6a). This prioritisation framework has to define the scale of the restoration target and the criteria on which prioritisation should be based.[138] Another action under Target 2 was for the Commission to develop a Green Infrastructure Strategy by 2012 to promote the deployment of green infrastructure in the EU in urban and rural areas (action 6b). By 2013, the Commission had developed a Green Infrastructure Strategy.[139] Another action under Target 2 aims to ensure no net loss of biodiversity and ecosystem services.

By 2015, the Commission will introduce an initiative to ensure there is no net loss of ecosystems and their services (e.g. through compensation or offsetting schemes) (action 7b).[140] The no net-loss approach could be necessary to ensure no further loss or degradation of ecosystems and their services overall. There are legal requirements for compensation under the Habitats Directive and the Environmental Liability Directive. However, in EU law there is no requirement for systematic compensation which leads to net losses outside the Natura 2000 framework. An option might be to have an EU legal framework for no net loss of ecosystems.[141] The no net-loss target could especially be important for restoration outside Natura 2000 areas, although there is still debate on the precise scope and

---

137 European Commission, *Commission Staff Working Paper. Impact Assessment. Accompanying the document Communication from the Commission to the European Parliament, the Council, the Economic and Social Committee and the Committee of the Regions, Our life insurance, our natural capital: an EU biodiversity strategy to 2020* (SEC (2011) 540 final European Commission 2011) 27.

138 Ibid., 41.

139 See European Commission, *Communication from the Commission to the European Parliament, the Council, the European Economic and Social Committee and the Committee of the Regions, Green Infrastructure (GI) – Enhancing Europe's Natural Capital* (COM(2013) 249 final, European Commission 2013).

140 On the no net-loss policy, see the Commission website: <http://ec.europa.eu/envir onment/nature/biodiversity/nnl/index_en.htm>.

141 European Commission, *Commission Staff Working Paper. Impact Assessment. Accompanying the document Communication from the Commission to the European Parliament, the Council, the Economic and Social Committee and the Committee of the Regions, Our life insurance, our natural capital: an EU biodiversity strategy to 2020* (SEC (2011) 540 final) 42.

implementation of the no net-loss initiative.[142] In the Council Conclusions of 23 June 2011 the Council stressed the importance 'of further work to operationalise the "no net loss" objective of the Strategy for areas and species not covered by existing EU nature legislation and of ensuring no further loss or degradation of ecosystems and their services'. The no net-loss concept entails 'that conservation losses in one geographically or otherwise defined area are balanced by a gain elsewhere provided that this principle does not entail any impairment of existing biodiversity as protected by EU nature legislation'.[143] It is mandatory that development projects strictly follow the mitigation hierarchy in order to achieve no net loss of – or net positive impact on – biodiversity. Although the no net-loss policy might include interesting possibilities for restoration as a way of offsetting outside the scope of the Natura 2000 obligations on compensation, no further steps have been taken by the Commission to develop further initiatives in this regard.

As was shown in previous chapters, the restoration target under the CBD Aichi Targets leaves much uncertainty over what is meant by restoration, which makes it difficult to measure the progress on this target. It is interesting to explore whether in the EU there is more clarity regarding the global and regional restoration targets because subsequent Commission documents and studies have shed light on the definition of 'restoration', the meaning of a '15 per cent' restoration target, and where and how restoration should take place. A first question is: what is understood by restoration in EU policy? The accompanying document from the Commission to the Biodiversity Strategy defines 'restoration' as follows:

> The restoration of ecosystems and their services is understood as actively assisting the recovery of an ecosystem that has been degraded, damaged, or destroyed, although natural regeneration may suffice in cases of low degradation. The objective should be the return of an ecosystem to its original community structure, natural complement of species, and natural functions to ensure the continued provision of services in the long term, although in cases of extreme degradation, the focus on specific services may be justified.[144]

According to a 2013 study on the costs of implementing Target 2 of the Biodiversity Strategy, this level of restoration would be prohibitively expensive in most

---

142 Working Group on no net loss of ecosystems and their services. Sub-Group on the scope and objectives of the no net loss initiative, *Scope and objectives of the no net loss initiative* (12/7/2013); see Commission website <http://ec.europa.eu/environment/nature/biodiversity/nnl/pdf/Subgroup_NNL_Scope_Objectives.pdf>.

143 EU Biodiversity Strategy to 2020 – Council conclusions, 11978/11 (21 June 2011) 5, footnote 1.

144 European Commission, *Commission Staff Working Paper. Impact Assessment. Accompanying the document Communication from the Commission to the European Parliament, the Council, the Economic and Social Committee and the Committee of the Regions, Our life insurance, our natural capital: an EU biodiversity strategy to 2020* (SEC (2011) 540 final European Commission 2011) 21.

situations and often impossible.[145] In the study the costs for restoration have been estimated on the basis of restoration of key species, properties and processes of ecosystems and their functions.[146]

Another question is what the '15 per cent' refers to in Target 2: will the EU restore 15 per cent of all degraded ecosystems? As all ecosystems in the EU are degraded to some extent, would this mean that 15 per cent of the EU should be restored? In the study estimating the costs for the implementation of Target 2, it is assumed that the target means 15 per cent of those areas of each ecosystem type that are degraded. If, for example, 10 per cent of the area of an ecosystem is degraded, then restoration is required for 1.5 per cent of the ecosystem's total area.[147] However, for the species and habitats covered by the Birds and Habitats Directives, which have to reach a favourable conservation status, this might require more than 15 per cent, depending on the ecological requirements of the site.

Is 15 per cent enough restoration? The European Commission stated in the accompanying document to the EU Biodiversity Strategy that a 15 per cent restoration target is a minimum, but a higher level (for example 30 per cent) could be considered for a number of reasons. First, the EU has adopted a more ambitious headline target in the EU Biodiversity Strategy than the global one and the EU's 2050 vision requires that ecosystem services are appropriately restored. Second, the EU is the most fragmented continent in the world, and a significant amount of restoration is expected to take place under existing legislation.[148] The Commission thus implicitly recognises the legal obligations on restoration in the Nature Directives. Third, according to the Commission, a higher percentage of restoration is likely to be cost-beneficial given the climate change mitigation and adaptation benefits of many ecosystems. The Commission, however, states that as there is not sufficient evidence of how much restoration would take place under existing EU policy and whether additional efforts would be needed to reach 30 per cent, the EU's chosen target level of 15 per cent reflects minimum compliance with international commitments.[149]

In a reaction to the EU Biodiversity Strategy, the European Parliament adopted a resolution.[150] Although resolutions by the European Parliament are not binding, they reflect a political will to act in a certain way. The European Parliament

---

145 G. Tucker, E. Underwood, A. Farmer, R. Scalera, I. Dickie, A. McConville and W. van Vliet, *Estimation of the financing needs to implement Target 2 of the EU Biodiversity Strategy. Report to the European Commission* (Institute for European Environmental Policy 2013) 31.

146 Ibid., 60.

147 Ibid., 58.

148 European Commission, *Commission Staff Working Paper. Impact Assessment. Accompanying the document Communication from the Commission to the European Parliament, the Council, the Economic and Social Committee and the Committee of the Regions, Our life insurance, our natural capital: an EU biodiversity strategy to 2020* (SEC (2011) 540 final, European Commission 2011) 27–28.

149 Ibid.

150 European Parliament resolution of 20 April 2012 on our life insurance, our natural capital: an EU biodiversity strategy to 2020 (2011/2307(INI)).

welcomed and supported the EU Biodiversity Strategy to 2020, but also took the view that some actions may have to be strengthened and specified more clearly, and that more concrete measures should be deployed in order to ensure effective implementation of the strategy. For Target 1 the European Parliament is of the opinion that at least 40 per cent of habitats and species should be in a favourable conservation status by 2020 and by 2050 100 per cent (or almost 100 per cent) of habitats and species must have a favourable conservation status. Also, the Parliament wishes the EU to set a considerably higher restoration target reflecting its own more ambitious headline target and its 2050 vision. The Parliament urges the Commission to define clearly what is meant by 'degraded ecosystems' and to set a baseline against which progress can be measured. Furthermore, the creation of natural environments should not be limited to designated areas alone, but should be encouraged in different places, such as industrial sites, in order to develop a truly green infrastructure.

It is also important to know where to restore and how to prioritise. The Commission published a study in 2013 to support Member States in prioritising the restoration of degraded ecosystems, as was set forward by action 6a under Target 2.[151] The study contains interesting insights on how to prioritise restoration efforts within the EU. It introduces a four-level model for ecosystem restoration: the model divides the continuum of ecosystem condition from poor to excellent into four distinct levels, in which level 1 is the highest level and includes wilderness areas and Natura 2000 sites in a favourable conservation status. For each level there are sets of ecosystem descriptors[152] and associated threshold values that are regarded as typical for that level.[153] Restoration means moving from a lower level to a higher level in the four-level model. Restoration levels need to be described for each ecosystem type by means of a well-defined set of descriptors and well-defined threshold values between the restoration levels. An EU-wide common understanding on how to determine the levels for ecosystem condition is very important; that is, an agreed list of applied descriptors as well as a shared understanding on the transitions between levels (threshold values).

The study published by the Commission also clarified what should be included in Target 2's 15 per cent restoration target. First, the target applies to each Member State. Second, the 15 per cent restoration target should apply separately to both the marine as well as the terrestrial area, meaning that each state should

---

151 J. Lammerant, R. Peters, M. Snethlage, B. Delbaere, I. Dickie and G. Whiteley, *Implementation of 2020 EU Biodiversity Strategy: Priorities for the restoration of ecosystems and their services in the EU. Report to the European Commission* (ARCADIS, in cooperation with ECNC and Eftec 2013).

152 A descriptor characterises ecosystem condition. A descriptor consists of one or more indicators and distinguishes ecosystem condition levels by means of threshold values between levels; for each descriptor an indicator and indicator unit (e.g. ha, %) needs to be defined. These indicators allow measuring the state of the descriptors. As an example the indicator for the descriptor 'connectivity' is the level of fragmentation (Lammerant et al., *Implementation of 2020 EU Biodiversity Strategy* 18).

153 Ibid., 15–18.

provide for 15 per cent restoration targets within its marine environment and 15 per cent restoration targets within its terrestrial environment.[154] However, the study on restoration prioritisation status is only an information document. The Commission has not transformed its findings into policy guidelines, let alone binding legislation at the EU level that would legally concretise the restoration obligations for EU Member States. According to the Commission, from the mid-term review it appears that only two Member States, the Netherlands and Germany, have provided the Commission with 'Restoration Priority Framework' documentation concerning their priorities for the restoration of degraded ecosystems. However, according to the Commission, ecosystem restoration work is under way in many Member States.[155] According to an October 2015 mid-term review of the EU Biodiversity Strategy, there are few comprehensive restoration strategies at national and sub-national levels. Some restoration is taking place in Member States, often in response to legislation such as the Water Framework Directive, the Marine Strategy Framework Directive and the Birds and Habitats Directives. The mid-term review also states that increased efforts will be needed to complete and implement national restoration prioritisation frameworks. A lot remains to be done in relation to halting the loss of ordinary biodiversity in the 80 per cent of the EU territory that is not protected under the Natura 2000 network. This will require consideration of the most suitable approach to ensure no net loss of biodiversity and ecosystem services.[156] The European Commission has been criticised for not taking a more active role on restoration since it has left the methodology to set the targets and develop the frameworks to the Member States. As a consequence, it seems most Member States have not even begun the development of the frameworks, and the work that has been undertaken seems to be of low quality. This fact hinders severely the effectiveness of the Biodiversity Strategy.[157]

Links between Target 1 and 2

While Target 1 of the EU Biodiversity Strategy is focused on those habitats and species that are covered by the Birds and Habitats Directives, Target 2 deals with the restoration of ecosystems and ecosystem services and is not limited to the habitat types or species from the Habitats and Birds Directives. This means that Target 1 and its associated actions focus on measures within the Natura 2000 network and that the 15 per cent restoration target in Target 2 is not limited to

154 Ibid., 19–21.
155 European Commission, *Commission Staff Working Document. EU Assessment of Progress in Implementing the EU Biodiversity Strategy to 2020. Accompanying the document Report from the Commission to the European Parliament and the Council. The Mid-Term Review of the EU Biodiversity Strategy to 2020* (SWD(2015) 187 final, Part 2/3, European Commission 2015) 10–11.
156 European Commission, *Report from the Commission to the European Parliament and the Council. The Mid-term review of the EU Biodiversity Strategy to 2020*, Brussels, 2015, COM(2015) 478 final.
157 W. Langhout, *Why the EU Will Fail to Deliver on Ecosystem Restoration* (Birdlife International 2014), <www.birdlife.org/europe-and-central-asia/news/why-eu-will-fail-deliver-ecosystem-restoration>.

restoration within the Natura 2000 network and can be realised outside the Natura 2000 network. Arguably, the 15 per cent target should predominantly be realised outside the Natura 2000 network because there is already a separate target on restoration for the Natura 2000 network in Target 1. The two targets are however mutually dependent: Target 1 measures will contribute to the achievement of Target 2, and Target 2 is aimed at restoration in the wider environment, which is important for the coherence of the Natura 2000 network and connectivity measures.[158]

There is, however, no clear and formal view on the link between both targets. The study on the prioritisation for restoration in the EU states that the 15 per cent restoration target includes Natura 2000 targets (i.e. the achieved progress on Target 1 of the Biodiversity Strategy contributes to the achievement of Target 2). In their view all areas that are not in a favourable conservation status, including Natura 2000 sites, can be considered as a 'restorable area' and the restoration of these areas can contribute to the realisation of Target 2.[159]

Conclusion

The EU has accepted in its Biodiversity Strategy up to 2020 several restoration targets, including the 15 per cent restoration target that has been set at the international level. These targets are meant as an implementation of the Aichi Targets. However, as is the case for the Aichi Targets, there are still uncertainties about the restoration targets in the EU. Specific actions have been agreed upon, including a prioritisation framework by the Member States. So far, however, most Member States have not come up with a prioritisation framework, and although some restoration initiatives have been taken, much more needs to be done, especially in light of the unfavourable conservation status of most protected habitats and species in the EU.

Even though there are explicit provisions on restoration in EU law, most notably in the Birds and Habitats Directives, the EU may fall short of reaching its policy targets on time. In spite of these provisions on restoration, either explicit or implicit, Member States have not taken their restoration duties seriously. Apparently, more concrete and clear restoration obligations and guidance is needed to persuade Member States to undertake restoration efforts. Some recent case law of the European Court of Justice and the opinions of the Advocate General have clarified certain restoration obligations of Member States.

While restoration obligations are present for the Natura 2000 network, many of these obligations lack clarity. Given the amount of degraded land in the EU, the implementation of restoration obligations outside the Natura 2000 network is clearly insufficient. Although the Birds and Habitats Directives have certain legal

---

158 See also G. Tucker, E. Underwood, A. Farmer, R. Scalera, I. Dickie, A. McConville and W. van Vliet, *Estimation of the financing needs to implement Target 2 of the EU Biodiversity Strategy. Report to the European Commission* (Institute for European Environmental Policy 2013) 29.

159 Lammerant et al., *Implementation of 2020 EU Biodiversity Strategy* 20–21.

provisions that apply outside the ecological network, such as the connectivity provision and restoration duties for species and their habitats, these provisions are not sufficiently implemented by Member States. While other instruments such as the Water Framework Directive also include restoration obligations, biodiversity is still declining, especially in areas outside Natura 2000. In order to overcome these shortcomings more guidance should be provided by the European Commission and there may be a need for additional legislation aimed at conservation and restoration for biodiversity outside the Natura 2000 sites.[160]

160 See, for instance, G. Van Hoorick, 'Biodiversity outside protected areas: an outlaw waiting to be saved?' in C.-H. Born, A. Cliquet, H. Schoukens, D. Misonne and G. Van Hoorick (eds), *The Habitats Directive in its EU Environmental Law Context: European Nature's Best Hope?* (Routledge 2015) 467.

8 National approaches to ecological restoration

All states have ecological restoration challenges. Even with its global affluence and capability to take action, the United States had lost by 1990 nearly 53 per cent of its mainland wetlands and had identified 4.3 million acres of degraded lakes and 3.2 million miles of degraded river and streams.[1] International law alone is not sufficient to achieve national results. This chapter will explore a number of different national legal approaches that have been implemented to promote ecological restoration efforts. While there are some similarities between states' approaches to restoration, there is also a surprising amount of variation in the legal standards that have been designed to address ecosystem restoration. In addition to the laws that are discussed below, most of the states in this section also have a statutory mechanism for the state to recover civil or criminal costs for damage to the environment that can be applied to restoring resources.[2]

While each of these national systems offers the possibility of reviving ecological function for damaged ecosystems, very few of the systems, with the exception of the Ecuadorian described below, approach ecological restoration as a key state obligation and value. Instead restoration is approached in a much more functional fashion as a programmatic response for a particular degraded biome or a particular geographical area. When mentioned in national or regional laws without reference to a particular locale, restoration is generally identified as either an alternative to conservation or as a technical remedy to enhance ecological attributes for a particular location. It is rarely identified as a mandatory environmental protection measure to be realised at the landscape level across society.

---

1 T.E. Dahl, *Wetlands Losses in the United States 1780's to 1980's* (US Department of the Interior, Fish and Wildlife Service, Washington, DC 1990); National Research Council, *Restoration of Aquatic Ecosystems: Science, Technology, and Public Policy* (National Academy Press 1992) 6 and 10 (recommending the restoration of 2 million acres of the 4.3 million acres of degraded lakes and 400,000 miles of river-riparian ecosystems, 12 per cent of the degraded rivers and streams).
2 See, e.g., United States Clean Water Act 33 USC 1321(f)(5) (providing trustee powers to the president or an authorised representative of a state to recover costs for restoring natural resources damaged by oil or hazardous substances and to apply these recovered sums to 'restore, rehabilitate, or acquire the equivalent of such natural resources').

Most of the state efforts in restoration are part of an ad hoc practice that is typically triggered by an undesirable situation that impacts on human development goals. For the most part, national restoration efforts tend not be coordinated across the landscape and, with the exception of a few large ecosystem restoration projects such as the Florida Everglades, there is relatively little direct government engagement in planning restoration efforts and financing. Most of the national interventions in restoration have been either to seek implementation of restoration orders for individual damage to ecosystems or to provide short-term financing for sub-national, civil society or private groups to undertake restoration work. It is possible, as Chapter 6 suggests, that the level of government involvement in ecological restoration work will rapidly change in the years to come as states attempting to achieve the Aichi Targets on restoration are encouraged to implement short-term ecosystem restoration plans that should include the creation or the expansion of an enabling framework for effective governance of ecosystem restoration projects.

Even though the focus of this book is primarily on international law, this book includes this chapter on national approaches in recognition that most international environmental obligations are implemented through domestic mechanisms. This is particularly true in the case of ecological restoration work since restoration must be place-based. Even where states conceive of restoration as the creation of global ecological networks, the implementation of these networks depends on distinct national locations. Because almost all restoration projects to date take place on sovereign territory governed by domestic legal systems, national approaches to fulfilling international legal duties to restore are highly relevant.[3]

United States – Ecological restoration as a post-conservation recovery strategy relying on partnerships

The concept of ecological restoration has been long embedded in United States statutory law. The National Environmental Protection Act (NEPA), considered by some to be the Magna Carta of US environmental law, recognised the 'critical importance of restoring and maintaining environmental quality to the overall welfare and development of man'.[4] Specifically, NEPA required all agencies of the federal government to 'make available to States, counties, municipalities, institutions and individuals, advice and information useful in restoring, maintaining, and enhancing the quality of the environment'.[5] Government agencies are also expected to 'use all practicable means, consistent with the requirements of the Act and other essential considerations of national policy, to restore and enhance the quality of the human environment'.[6] These policies focused especially on the 'quality of

---

3 There are proposals to initiate restoration on the high seas. C.L. van Dover et al., 'Ecological restoration in the deep sea: desiderata' (2014) 44 *Marine Policy* 98–106.
4 Public Law 91–190 (1970) Section 101(a).
5 Ibid., Section 102(2)(F).
6 40 CFR 1500.2(f).

the human environment' may explain the enthusiasm with which ecosystem service concepts have been adopted by US agencies. For example, the Federal Emergency Management Agency (FEMA) in its regulations implementing NEPA is expected to 'take action to restore and preserve the natural and beneficial values served by floodplains'.[7]

A number of US executive agencies now include specific environmental restoration programmes. For example, the National Oceanic and Atmospheric Administration (NOAA), an agency within the Department of Commerce, created in 1991 a Restoration Center to coordinate policy on habitat restoration to improve fisheries by opening rivers, reconnecting coastal wetlands, restoring coral reefs and rebuilding shellfish populations.[8] Most of NOAA's work is done in partnership with non-governmental organisations such as The Nature Conservancy, American Rivers and Restore America's Estuaries.[9] To the extent that it is possible to discern general approaches within a legal system to a topic such as restoration, the US views ecological restoration as the appropriate legal response to environmental degradation particularly where degradation has been triggered by pollution or by habitat destruction.

Clean Water Act

A variety of environmental laws and regulations, mostly water-quality laws, include some reference to an obligation to restore. The Clean Water Act provides that the 'The objective of this Act is to restore and maintain the chemical, physical, and biological integrity of the Nation's waters.'[10] To achieve this objective, the US government intended to achieve restoration of waters in part by eliminating discharges of pollutants into navigable waters, setting interim water-quality standards based on 'fishable' and 'swimmable' standards and developing waste treatment management programmes.[11] The term restore is not defined in the Clean Water Act but the use of the term in the text implies some interest in the ecosystem functions delivered by clean water. While there is no restoration reference point provided in the congressional text, practice by states under the Act indicates that the baseline for restoration depends on the designation of use of a given water body.[12]

---

7 44 CFR 9.2(b)(7).
8 NOAA Habitat Conservation, Restoration Center.
9 E. Schrack et al., *Restoration Works: Highlights from a Decade of Partnership between The Nature Conservancy and the National Oceanic and Atmospheric Administration* (The Nature Conservancy 2012) xi.
10 Clean Water Act, 33 USC 1251.
11 Ibid.
12 Water Quality Standards, 40 CFR 131.10 ('Each State must specify appropriate water uses to be achieved and protected. The classification of the waters of the State must take into consideration the use and value of water for public water supplies, protection and propagation of fish, shellfish and wildlife, recreation in and on the water, agricultural, industrial, and other purposes including navigation. In no case shall a State adopt waste transport or waste assimilation as a designated use for any waters of the United States').

Common designations for use include 'fishable', 'swimmable', agricultural, industrial or navigational use.

Operating under the Clean Water Act, there are a number of geographically specific restoration efforts provided for by statutes. For example, on the East Coast of the United States, the Chesapeake Bay Restoration Act of 2000 provided for federal funding to support water quality improvement projects to bolster the efforts of Maryland, Virginia, Delaware and the District of Columbia in reviving the Chesapeake Bay and its watersheds.[13] The Lake Pontchartain Basin Restoration Act was designed to 'restore the ecological health of the Basin' in Louisiana.[14] The Long Island Sound Restoration Act provided authority for wetland restoration work off the coast of New York state.[15] Other water-quality restoration programmes cover Lake Champlain and the Great Lakes.[16] Most of these statutorily created restoration programmes are focused on providing financial assistance to partners such as non-governmental organisations, states, municipalities and universities to help achieve specific management strategies.

The text of the Chesapeake Bay Restoration Act provides a discrete list of potential restoration management objectives including 'to restore living resources' and to restore habitat for wetlands and riparian forests and other associated Chesapeake Bay habitats.[17] In spite of these laws, there remain substantial gaps between the law and the management of the Bay for a variety of ecological values. In 2009, the President observed in an executive order, calling for increased public and private investment in Chesapeake Bay restoration, that 'at the current level and scope of pollution control within the Chesapeake Bay's watershed, restoration of the Chesapeake Bay is not expected for many years'.[18] In response to these critiques, the EPA has undertaken additional efforts to regulate non-point sources for nitrogen, phosphorous and sediment, using the Clean Water Act. The industry has challenged this authority.[19]

In addition to the restoration programmes that cross sub-national boundaries within the US, there are also water-quality restoration efforts crossing national boundaries. Within the Clean Water Act, Section 1268 defines US obligations to improve water quality in the Great Lakes which is the United States' largest source of freshwater. These obligations are based on the 1978 Great Lakes Water Quality Agreement adopted by Canada and the US in order 'to restore and maintain the chemical, physical and biological integrity of the waters of the Great Lakes Basin

---

13 Clean Water Act 33 USC 1267.
14 Clean Water Act 33 USC 1273.
15 Clean Water Act 33 USC 1269.
16 Clean Water Act 33 USC 1270; Clean Water Act 33 USC 1268.
17 Clean Water Act 33 USC 1267(g)(1).
18 Executive Ord. No. 13508, (President Barack Obama) Chesapeake Bay Protection and Restoration, 74 F.R. 23099.
19 American Farm Bureau Federation v. USEPA (3rd Cir. 2013) Case 13–4079 (appealing a judgment finding that EPA's introduction of Total Maximum Daily Load caps on nitrogen, phosphorus and sediment loadings were legal for waters throughout the entire 64,000-square-mile Chesapeake Bay watershed).

Ecosystem'.[20] The US designed a Lakewide Management Plan which was intended to 'provide a systematic and a comprehensive ecosystem approach to restoring and protecting the beneficial uses of the open waters of each of the Great Lakes'[21] and a Remedial Action Plan which is expected to serve the same function as the Lakewide Management Plan but applies to the 'beneficial uses of areas of concern'.[22] Restoration of aquatic habitat is expected to proceed at the same time as remediation of contaminated sediment.[23] As of 2004, there were approximately 140 US-funded programmes operating in the Great Lakes area, focused on environmental restoration and management objectives that operate today in the context of the Great Lakes Interagency Task Force that brings together 11 cabinet members and heads of agency to coordinate restoration.[24]

Coastal Zone Management Act, Estuary Restoration Act, and Coastal Wetland Planning, Protection, and Restoration Act

The Coastal Zone Management Act (CZMA) defines additional restoration obligations focused on water quality. Some of these obligations are similar to the open-ended Clean Water Act obligations and provide that the federal government is expected to encourage and assist states with the management of coastal development in order to improve, safeguard and restore the quality of coastal waters.[25] The CZMA continues to build on the flexible Clean Water Act's non-point source pollution efforts by requiring the development of management measures 'for nonpoint source pollution to restore and protect coastal waters'.[26] Recognising the land and water nexus in the coastal zone, the Act includes a land conservation programme where the federal government supports state, regional and municipal efforts to protect lands that could be 'restored to effectively conserve, enhance, or restore ecological function'.[27]

In response to declining habitat, there is also legislation focused on specific habitat restoration. One important legal tool for post-conservation recovery of ecosystem function is the Estuary Restoration Act.[28] This law was negotiated to

---

20 1978 Great Lakes Water Quality Agreement, <http://epa.gov/greatlakes/glwqa/1978/articles.html#AGREEMENT%20BETWEEN%20CANADA>.
21 33 USC 1268(3)(I).
22 33 USC 1268(3)(J).
23 33 USC 1268(12)(B).
24 Exec. Order 13340 (George W. Bush 2004) Establishment of a Great Lakes Interagency Taskforce and Promotion of a Regional Collaboration of National Significance for the Great Lakes (69 F.R. 29043); Agencies that cooperate in the Taskforce include Environmental Protection Agency, Department of State, Department of Interior, Department of Agriculture, Department of Commerce, Department of Housing and Urban Development, Department of Transportation, Department of Homeland Security, Department of the Army, Department of Health and Human Services and Council on Environmental Quality.
25 16 USC 1452(1)(C).
26 16 USC 1455b.
27 16 USC 1456-1.
28 Public Law-106-457.

respond to a decline in healthy estuary habitat across the United States and set a target of restoring 1,000,000 acres of estuary habitat by 2010.[29] With the passage of the law, the federal government agreed to adopt common monitoring standards for estuary habitat, a common system for tracking restoration acreage and a national estuary habitat restoration strategy that would establish 'effective estuary habitat restoration partnerships among public agencies at all levels of government and … new partnerships between the public and private sectors'.[30] Habitat restoration partnerships are expected to restore habitats for the purposes of supporting wildlife including migratory birds, fish and shellfish; improving surface and ground water; ensuring flood control and providing outdoor recreation.[31]

Unlike many other US laws that simply reference restoration obligations, the Estuary Restoration Act provides a significant definition for restoration that reflects certain key understandings of ecological science that are largely overlooked by other statutes. The term 'estuary habitat restoration activity' means 'an activity that results in improving degraded estuaries or estuary habitat or creating estuary habitat (including both physical and functional restoration), with the goal of attaining a self-sustaining system integrated into the surrounding landscape'.[32] This definition explicitly recognises that successful restoration results in the return of sufficient ecological functions and structures to allow for a system to survive without constant management. This definition also implicitly identifies the value of managing systems at a landscape level where there are significant concerns such as building in connectivity, compatibility and, in some cases, redundancies.

What is less clear from the Estuary Restoration Act's definition is what limits might be placed on what qualifies as a successful restoration. Should activities that create new estuary habitat qualify as 'restoration activity'? In its plain meaning context, the term restoration suggests that the party sponsoring a restoration activity starts with some baseline upon which improvements will be made to revive ecological values rather than creating potentially new ecological values. As drafted, the language in the Estuary Act might encourage the generation of novel ecosystems that may or may not maintain the historical trajectory of ecological functions for a given site. For example, would it be acceptable for purposes of the Act to generate a mangrove forest where a salt-water estuary might have formerly existed and then consider the mangroves to be restored 'estuary habitat'?

There is no requirement for purposes of the Act that the restoration work be holistic in nature. Government agencies or partnerships between public and private actors can select appropriate activities that might include re-establishing chemical, physical, hydrologic and biological estuary features and components; cleaning up pollution; controlling invasive species; reintroducing endemic species; constructing reefs and other activities 'that improve estuary habitat'.[33] To qualify for funding

29 Ibid., Section 106(b).
30 Ibid., Section 102(2).
31 Ibid., Section 106(d)(3).
32 Ibid., Section 103(4).
33 Ibid., Section 103(4)(B).

under the Act, the federal or regional programme is expected to be 'developed with the substantial participation of appropriate public and private stakeholders'.[34]

In addition to the Estuary Restoration Act and Coastal Zone Management Act, the US also has a Coastal Wetland Planning, Protection, and Restoration Act focused on Louisiana, the single state with the largest amount of coastal wetlands.[35] Under this Act, a 'coastal wetlands restoration project' refers to 'any technically feasible activity to create, restore, protect, or enhance coastal wetlands through sediment and freshwater diversion, water management, or other measures' that contribute to 'the long-term restoration or protection of the physical, chemical and biological integrity of coastal wetlands in the State of Louisiana'.[36] This definition presents some of the same interpretive challenges as the broad definition of 'estuary habitat restoration activities' because it extends the concept of what constitutes restoration. For example, the use of the terms 'protect' and 'create' do not seem to belong in an ecological 'restoration' definition. Protection is a different project than restoration; it is a predecessor effort. While installing barrier islands in front of a coastal wetland may protect existing or restored wetlands, is it really a restoration effort to install islands in a place where they did not historically exist? The difference matters because restoration is largely a conservative process rather than a fully fledged creative process. While restoration involves interventions, it is about maintaining all of the parts of a system including those that we do not fully understand. It is not a project about re-sculpting the landscape away from reference ecosystems. Key decisions may need to be made in the near future that we need to create certain landscape features but these are not restoration projects.

Comprehensive Everglades Restoration Plan

Of all of the US large restoration projects, the Comprehensive Everglades Restoration Plan has received the most public attention. Covering 16 counties, the project is focused on restoring historical aspects of South Florida's hydrology so that the state can better sustain ecosystems in the Everglades and Florida Bay. The restoration project is a multi-stakeholder project including the US Army Corps of Engineers representing the federal government, the state of Florida and local municipalities.

The $11.9 billion project is a reaction to a former federal infrastructure act. The 1936 Flood Control Act provided that the federal government should 'improve or participate in the improvement of navigable waters or their tributaries for flood control purposes, if the benefits to whomsoever they may accrue, are in excess of the estimated cost, and if the lives and social security of the people are otherwise adversely affected'.[37] In 1948, the US Congress authorised

---

34 Ibid., Section 103(6)(A).
35 R. Dean, *New Orleans and the Wetlands of Southern Louisiana* (The Bridge, National Academy of Engineering 2006) 36.
36 16 USC 3951 (6).
37 Public Law 74–738 (1936).

the first phase of the Central and Southern Florida project that would radically change the hydrology in Southern Florida through installing 30 pumping stations, 212 control and diversion structures, 990 miles of levees, 978 miles of canals and 25 navigation locks.[38]

In 1989, US Congress passed the Everglades National Park Expansion Act providing the Secretary of the Army, as part of his responsibilities for the Army Corps of Engineers, with the authority to construct modifications to the Central and Southern Florida project in order to improve water deliveries that would benefit the park.[39] According to the legislation, the Secretary 'shall, to the extent practicable, take steps to restore the natural hydrological conditions within the park'.[40] The Secretary's restoration efforts are 'justified by the environmental benefits…and shall not require further economic justifications'.[41]

In the 1990s, the US Congress recognised that long-term environmental damage was occurring beyond Park boundaries and authorised under the Water Resource Development Act of 1996 the creation of the Comprehensive Everglades Restoration Plan (CERP).[42] As part of the Water Resource Development Act of 2000, the CERP was eventually approved by Congress 'to ensure the protection of water quality in, the reduction of the loss of fresh water from, and the improvement of the environment of the South Florida ecosystem and to achieve and maintain the benefits to the natural system and human environment described in the Plan'.[43] The restoration work is expected to take 30 years and cost between \$9.5 and 11.9 billion.[44] While the CERP focuses on 'getting the water right', the project is intended to also jumpstart ecosystem restoration and community sustainability efforts.[45] To achieve these goals, CERP planning is currently operating across agencies to achieve restoration defined as 'recovery and sustainability of the defining characteristics of the greater Everglades ecosystems'.[46]

38 Development of the Central and South Florida Project, <www.evergladesplan.org/about/restudy_csf_devel.aspx>.
39 Public Law 101–229, 13 December 1989; Section 104(a)(1).
40 Ibid.
41 Ibid., Section 104(a)(3).
42 Public Law 104–303, 12 October 1996.
43 Public Law 106–541, 11 December 2000, Section 601(b)(1)(A); additional amendments have been made to the plan in 2007 Public Law 110–114, 8 November 2007 particularly regarding financing including approximately \$1.3 trillion for the restoration, water supply, flood control and protection of water quality for the Indian River Lagoon in Florida and \$375 million for Picayune Strand restoration.
44 USACE South Florida Restoration Office, US Army Corps of Engineers Jacksonville District and South Florida Water Management District; About CERP: Brief Overview, <http://www.evergladesplan.org/about/about_cerp_brief.aspx>.
45 CERP, The Plan in Depth, <http://www.evergladesplan.org/about/rest_plan_pt_02.aspx>.
46 Recover: REstoration, COordination, VERification (January 2012), <http://www.evergladesplan.org/docs/fs_recover_jan_2012.pdf>.

US Forest Service's ecological restoration policy

In addition to the various agency commitments to restore wetlands, coastal areas and estuaries, the US Forest Service within the Department of Agriculture has introduced an 'ecological restoration policy' to govern its activities in restoring forests and grasslands. The Forest Service as the steward for the US timberlands has been attempting to respond to the changes wrought by climate change arising in the form of bark beetle infestations and increasing wildfire arising from decades of fire suppression policies. In 2013, the Forest Service proposed amending its Forest Service Manual by adding a chapter on 'ecological restoration' intended to 'provide a clear, comprehensive, and science-based restoration policy to guide achievement of sustainable management and ecological integrity under changing environmental conditions, such as those driven by a changing climate and increasing human uses'.[47] This directive was finalised in 2014 and the incorporation of restoration into the Forest Service Manual should influence a variety of agency activities including efforts 'to restore watershed condition and function, control invasive species, re-create natural stream channel complexity, improve or re-establish habitat for threatened and endangered species, and restore natural fire regimes'.[48]

What is particularly interesting about the Forest Service's 2014 directive is that the Forest Service indicates that it wants to distinguish between the more generic term of 'restoration' which is already part of existing Forest Service programmes and 'ecological restoration'. Under the authority of current laws, the Forest Service has historically pursued as 'restoration' efforts various projects such as revegetation to improve range values, reforestation, fuel reduction, and rehabilitation of wildlife, fish and game species.[49] The directive offers a definition of restoration that closely parallels definitions offered by the Society for Ecological Restoration. 'Ecological restoration' is:

> the process of assisting the recovery of an ecosystem that has been degraded, damaged or destroyed. Ecological restoration focuses on reestablishing the composition, structure, pattern, and ecological processes necessary to facilitate terrestrial and aquatic ecosystem sustainability, resilience, and health under current and future conditions.[50]

This definition in the directive is perhaps the most comprehensive and definitely the most science-based definition of 'ecological restoration' in current US national law. The new directive offers a different level of agency engagement by requiring

---

47 Federal Register Vol. 78, No. 177 (12 September 2013), Department of Agriculture, Ecological Restoration Policy, p. 56202.
48 Ibid.
49 Anderson-Mansfield Reforestation and Revegetation Joint Resolution Act of 1949, 16 USC 581; Granger-Thye Act (16 USC 580 g–h); Sikes Act (16 USC 670g); Healthy Forests Restoration Act of 2003 (16 USC 6501–6591).
50 Federal Register, p. 56208 (FSM Sec. 2020.5).

all resource management programmes to pursue 'ecological restoration' and all planning to include 'ecological restoration goals'. The goals should be established

> within the framework defined by laws; Indian treaties and Tribal values and desires; regulations; public values and desires; natural range of variation; current and likely future ecological capabilities; a range of climate and other environmental change projections; the best available scientific information; and technical and economic feasibility to achieve desired conditions for National Forest System lands.[51]

While the definition for 'ecological restoration' in the directive is somewhat ambiguous in terms of what approach managers should take to 'facilitate terrestrial and aquatic ecosystem sustainability, resilience, and health under current and future conditions', the directive's other definitions suggest that managers should plan their restoration work in the context of a 'natural range of variation' (NRV). The concept of the NRV identifies a reference period that incorporates a historic range of ecosystems in a given area to assist managers in assessing the quality of ecological integrity for a given area and to 'help identify key structural, functional, compositional, and connectivity characteristics, for which plan components may be important for either maintenance or restoration of such ecological conditions'.[52] Except for environments that have been 'irreversibly altered', the directive requires that 'ecological restoration activities' be 'planned, authorized, implemented, monitored, and evaluated within the context of the NRV, current and desired conditions, and the potential for future changes in environmental conditions due to climate change and human uses'.[53] The US Forest Service 'ecological restoration' approach is one based largely on following a historical trajectory in terms of restoring composition, structure, function and connectivity within a given ecosystem. Potential amendments to restoration plans may be necessary to adapt to stressors such as climate change.

The US Forest Service, like its sister agencies working on wetland restoration and other water quality restoration efforts, also relies upon building partnerships. Region 5 of the Forest Service covering the State of California has embarked on 'An All Lands Approach to Ecological Restoration: Working Across Boundaries with our Partners' because the 'increasing demand for ecosystem services is growing faster than our current treatments and restoration work, and the scale of work is not adequate to influence the trend of growing impact'.[54] The Region has published a relatively detailed ecological restoration plan including numerous proposed projects across each of the national forests.[55] What is particularly significant is the amount of thought that the plan reflects in terms of planning for

51 Ibid., 56207 (FSM Sec. 2020.3(2)).
52 Ibid., 56208 (FSM Sec. 2020.5).
53 Ibid., 56207 (FSM Sec. 2020.3(5)).
54 Pacific Southwest Region, National Forest Service, Ecological Restoration, <www.fs. usda.gov/detail/r5/landmanagement/?cid=STELPRDB5308848>.
55 Region 5, Ecological Restoration Implementation Plan (2011), <www.fs.usda.gov/ Internet/FSE_DOCUMENTS/stelprdb5411390.pdf>.

long-term restoration goals. The following Region 5 National Forest Restoration goals resonate more broadly with most ecological restoration efforts:

- Restore for 'ecosystem resilience' so that key ecological processes and functions persist in the face of catastrophic events, disturbance processes, and intense public use;[56]
- Ensure that restoration planning does not stop at ownership boundaries;[57]
- Apply best science to restoration efforts;[58]
- Prevent and control invasive species infestations;[59]
- Treat landscapes in a holistic fashion;[60] and
- Provide a broad range of ecosystem services.[61]

Incorporating concepts of ecological restoration into US statutes, regulations and agency practices has been an increasing priority over the past few decades. While one of the drivers of this trend may be a recognition that core natural resources such as fish (particularly salmon) and timber are threatened by the status quo, there also appears to be an increasing emphasis on planning for long-term 'ecological restoration'. The Forest Service directive and the Estuary Restoration Act reflect a new political trajectory where restoration is an active and holistic strategy of national renewal rather than a piecemeal effort.

The United States approach to restoration has evolved over the past couple of decades from one focused on reversing declining environmental trends to an approach that emphasises landscape planning and ecological recovery. With increasing amount of financing for restoration projects becoming available through the various laws detailed above, restoration has become a US federal agency priority particularly for institutions charged with protecting public lands and potentially assisting public lands to respond more robustly to climate change impacts.[62] Restoration funding was a part of the US economic recovery plan during the economic recession of 2009. The American Recovery and Reinvestment Act whose objective was in part to 'to preserve and create jobs and promote economic recovery' as well as 'to invest in … environmental protection' allocated millions in funding to support restoration of Bureau of Land Management lands, restoration projects under the California Bay-Delta Restoration Act and US fish and wildlife habitat restoration work.[63]

---

56 Ibid., 32.
57 Ibid., 39.
58 Ibid., 42.
59 Ibid., 60.
60 Ibid., 75.
61 Ibid., 89.
62 See, e.g., United States Department of Interior, National Park Service Draft Environmental Assessment, Ecological Restoration Plan on Department of Interior in Western Pima County, Arizona (2014) (proposing to restore desert areas by removing invasive species and removing undesignated vehicle routes).
63 American Recovery and Reinvestment Act of 2009, Public Law 111–5 (17 February 2009).

China: restoration as a national security measure and a poverty reduction programme

With China's rapid economic development, the Chinese landscape is undergoing continuous transformation from sleepy rural posts to urban islands. One area that has attracted great attention has been the Loess plateau (Huangtu Plateau, 黄土高原) bordering China's Yellow River and including portions of Shanxi, Shaanxi, Gansu and Inner Mongolia. At its historical apex during the Han Dynasty, this region had highly fertile soils capable of feeding a growing population. By the 1980s, by some calculations, 23 per cent of the land area in China was ecologically degraded even though 35 per cent of the Chinese population depended on that land.[64] Until quite recently, the area was a wasteland of loess soil blowing in the wind and home to some of the poorest residents of China. In 1998, erosion-prone areas including some in the Loess Plateau region turned deadly: 22 million acres of farmland were submerged in flood waters and 4000 citizens died when multiple rivers breached their banks during several flood events.[65]

In 1999, in order to address in part the unexpected flood events tied to landscape erosion and the uncontrolled loss of sediment, the Chinese central government initiated the Grain to Green programme (Conversion of Sloping Farmland to Forest programme) with an investment of $40 billion dollars up until 2050.[66] This programme is characterised in China as an ecosystem restoration project with a focus on restoring lost ecosystem functions such as erosion control. This programme was structured as a payment for ecosystem restoration project to encourage rural households to retire land from intensive agriculture use and convert the land to habitats less prone to erosion. Participants are paid either in the form of Chinese currency or kilos of rice both to retire land and to plant trees on the land. Over its lifetime, the programme has operated in 25 of China's 34 provinces and has replaced 14.4 million acres of vegetation on previously cultivated slopes and 17.3 million acres on land that was abandoned as agricultural land.[67] Notably, the programme's payments for ecosystem services are linked to the success of tree plantings to provide some incentive to participants to ensure that they monitor their plantings.[68]

---

64 Y. Lü et al., 'A policy-driven large scale ecological restoration: quantifying ecosystem services changes in the Loess Plateau of China' (2012) *PlosOne* 1, <www.plosone.org/article/info%3Adoi%2F10.1371%2Fjournal.pone.0031782>.

65 The Economics and Ecosystems of Biodiversity, National-Level Soil Erosion Control Policies in China (2010), <www.teebweb.org/wp-content/uploads/2013/01/National-level-Soil-Erosion-Control-Policies-in-China.pdf>.

66 X. Feng et al., 'How ecological restoration alters ecosystem services: an analysis of carbon sequestration in China's Loess Plateau' (2013) 3 (2846) *Scientific Reports* 1–5, <www.nature.com/srep/2013/131003/srep02846/full/srep02846.html>.

67 Ibid.; The Economics and Ecosystems of Biodiversity, National-Level Soil Erosion Control Policies in China (2010) 3.

68 Ibid.

The replanting restoration programme is more than simply a conservation-oriented programme. It is also considered to be a national security programme because Grain to Green reduces the impacts of soil erosion upstream of China's large hydroelectric dam.[69] Chinese researchers also regard the restoration of vegetation capable of absorbing substantial carbon as holding definite potential for terrestrial carbon sequestration.[70] Creating opportunities for sequestration through restoration activities mirrors the goals of the Aichi Biodiversity Targets.

The primary challenge for the programme has been creating sufficient economic incentives to result in behavioural change so that farmers do not resume farming on areas deemed to be overly vulnerable to erosion or flooding. A related challenge is whether the changes that have occurred will survive the sunset of the central government subsidies. Some field questionnaires suggest that some users will revert to former agriculturally intensive uses of land without payments for restoration efforts.[71] A third challenge for the programme is planning what vegetation will be planted so that it can survive changing climatic trends without too much ongoing human intervention such as irrigation.[72]

China is also pursuing programmes to combat desertification that are characterised as restoration programmes, but may be better considered environmental improvement and poverty reduction programmes. Both the Grain to Green programme and the desertification programmes are run by central government agencies and require extensive resources. Scholars suggest that the Chinese government may want to be more strategic about how it is pursuing restoration efforts. As two scholars commented recently, to 'fundamentally improve the ecosystem functions and services, it is essential for China to have a more balanced and comprehensive approach to ecological restoration; adopt better planning and management practices; strengthen the governance of program implementation; emphasize local people's active engagement; establish an independent, competent monitoring network; and conduct time and high-quality assessments of the program effectiveness and impacts'.[73]

Law may play an increasingly important role in terms of how China pursues ecosystem restoration programmes since it may set more clearly articulated standards for what China is attempting to achieve through its restoration projects. At the present time, it is not clear that the projects promote 'ecological' restoration so much as environmental restoration work that serves national development purposes. The Chinese policy offers a couple of interesting insights in the role that

---

69 Ibid.
70 Feng et al., 'How ecological restoration alters ecosystem services' 3.
71 R. Yin et al., 'Assessing China's ecological restoration: what's been done and what remains to be done' (2010) 45 *Environmental Management* 442–453.
72 Feng et al., 'How ecological restoration alters ecosystem services' 4.
73 R. Yin and G. Ying, '*China's Ecological Restoration Programs: Initiation, Implementation and Challenges in An Integrated Assessment of China's Ecological Restoration Programs*' (Springer 2009) 1.

law and policy can play in achieving social uptake of the kind of behaviours that can contribute to landscape level restoration. Clearly individual economic incentives can trigger community-wide efforts but the law may also need to provide additional economic incentives to maintain restoration efforts or some form of deterrence. Additionally, care needs to be taken when restoration interventions are designed to ensure that acceptable ecological standards form the basis of legal standards. Otherwise, as researchers have suggested in the case of the Chinese reforestation programme, the restoration programme may meet its social and economic targets but fail to achieve its environmental targets.

Brazil: experimenting with new legal restoration strategies

As the fifth largest country in the world in terms of population and size and the home to a large portion of the Amazonian River basin that has become heavily fragmented under pressures by agribusiness, timber and the mining industry, Brazil is a significant player in ecological restoration. Governments at both the federal and subnational level have created a variety of laws and collaborative relationships to bolster restoration efforts.

Restoration is not a new concept for the Brazilian government. In 1861, the government, in order to protect drinking water sources, undertook forest restoration projects around natural springs and streams in the Rio de Janeiro area that had been damaged by coffee plantation practices.[74] In 1934, the government promulgated the Forest Act requiring or encouraging restoration work by private landholders.[75] This Act governs actions by private landowners and its regulatory sweep is broad since over 70 per cent of Brazilian lands are in private ownership.[76]

Recent changes to the Forest Act may have negative consequences for restoration efforts. As a result of a 2012 revision to the definition of a 'watercourse' lobbied for by industrial interests, the current Forest Act's jurisdiction has been reduced by an average of 50 per cent, leading to less mandatory restoration for large tracts of land governed by the Forest Act.[77] The effect of the revised Act on

---

74 S. Pinto et al., 'Governing and delivering a biome-wide restoration initiative: the case of Atlantic Forest Restoration Pact in Brazil' (2014) 5 *Forests* 2212–2229.

75 Brazil Forest Code in 1934 (Decree # 23793/1934); Brazil Forest Code (Law #4771/1965) (further revised in 1989). Restoration was specifically provided for in the 1965 law with the provision for 'Areas of Permanent Preservation' that needed to be restored if heavily degraded and 'Legal Reserves' designating minimal levels of forest cover for each property.

76 L.C. Garcia et al., 'Restoration challenges and opportunities for increasing landscape connectivity under the New Brazilian Forest Act' (2013) 11(2) *Brazilian Journal of Nature Conservation* 181–185.

77 Law of Native Vegetation Protection (Federal Law # 12651/2012); see also Garcia et al., 'Restoration challenges and opportunities' 182. (Under the 2012 revised code, 'Areas of Permanent Preservation' are now measured according to regular seasonal watercourses rather than largest seasonal watercourse. As a result fewer areas of private land are being identified as 'Areas of Permanent Preservation' that mandate restoration efforts).

restoration work may be particularly profound in wetland areas where adjacent upland flora and fauna may no longer be required to be restored because they have not been designated within a regular seasonal watercourse.[78] Brazil has already faced difficulties in achieving compliance with needed restoration efforts because the government has not enforced full compliance.[79] Equally problematic for restoration goals, the new Brazilian Forest Act permits offsets to be made a long distance from the originally impacted lands without regard to whether the offset will create habitat fragmentation rather than connectivity.[80]

In 2011, the national government of Brazil through its National Environmental Council (CONAMA) promulgated a national resolution for the 'recuperation' of 'permanent preservation areas'.[81] The government requires responsible parties to apply restoration methods that either encourage the natural regeneration of native species or where necessary the planting of native species.[82] At a minimum, parties involved in 'recuperation' must individually maintain newly planted or established species for at least two years, adopt measures for fire prevention and control, adopt measures to control and eradicate competing species, introduce erosion control, prevent access by domestic animals and adopt 'measures for the conservation and attraction of native animals that disperse seeds'.[83] While it is unclear from the resolution's language how the law will address non-compliance, the content of the resolution is significant by establishing a uniform best practices approach for all 'preservation areas'.

In response to the ongoing environmental challenges of needing to restore private lands in order to protect landscape integrity, legal entities in Brazil have introduced some significant legal and institutional interventions that rely on innovative framing of the ecological restoration challenges relevant to Brazil. One intervention involves sub-nationally-based restoration legislation that provides a highly detailed approach to restoration work beyond the more general federal requirements.[84] In São Paolo, restoration work is governed by the São Paolo State Resolution on Forest Restoration which legislates provisions on minimum standards for a restoration project that is required to address 'environmental damage resulting from economic activities, violations of environmental laws, or for projects supported by public resources'.[85] Parties regulated by the resolution are expected to undertake restoration planning which includes identifying impacts of human-based disturbances on the

---

78 Ibid., 183.
79 Ibid., 184.
80 Ibid., 183.
81 CONAMA Resolution 429, 28 February 2011, Published in Official Gazette 43 on 2 March 2011, p. 76, available in English at <www.mma.gov.br/port/conama/proces sos/61AA3835/CONAMA-ingles.pdf>.
82 Ibid., Chapter III, Article 3.
83 Ibid., Chapter III, Article 5.
84 J. Aronson et al., 'What roles should government regulation play in ecological restoration? Ongoing debate in São Paolo State, Brazil' (2011) 19(6) *Restoration Ecology* 690–695.
85 Ibid., 692 and Appendix S1 (containing São Paolo State Resolution on Forest Restoration).

areas, locating invasive species, creating restoration priorities and then designing specific restoration strategies for each location to be restored.[86]

It is the next portion of the legislation that distinguishes the São Paolo State Resolution from any other national or sub-national laws regarding restoration efforts. While the Aichi Biodiversity Targets provide specific percentage goals for states to achieve, the São Paolo State Resolution provides even more concrete goals for its regulated parties. For example, as part of the planning process, parties who are restoring in seasonally dry forests, Atlantic rainforests or Savanna woodlands should have as their objective achieving biodiversity defined by the resolution as maintaining at least '80 native woody plant species within a given period of time'.[87] In addition to planting a minimum number of plant species including for most species at least 12 individuals of that species, standards were set on the quality of the restoration:[88] 20 per cent of the species should be animal-dispersed, 5 per cent should be endangered species, and no more than 60 per cent of the species should belong to the same ecological group.[89] While the law as drafted is highly specific, there is disagreement among restoration experts both inside and outside Brazil about the ecological and social efficacy of providing such specific requirements.[90] Some experts consider the detailed requirements of the law as necessary to provide some attainable baseline for projects to achieve;[91] others question whether the designated performance standards are adequate to achieve biological viability or whether the percentages are somewhat arbitrary.[92]

While it remains to be seen whether laws providing for detailed restoration approaches result in more effective restoration projects, one challenge for these laws is uniform enforcement. Even though the law is specific about what needs to be planted, the law does not provide any specific information on enforcement. These laws require instead some willingness on the part of most regulated parties to self-enforce the restoration provisions as part of project management. Otherwise, evaluating each and every potential restoration project across the 95,000 square miles of São Paolo becomes another sizable expense in addition to the reforestation expenses that average around $5000 per hectare for Atlantic Rainforest.[93] Whether or not the percentages set in the law are sufficient to achieve

---

86 Ibid. (citing Articles 3, 5, 9, 10 and 11 of the São Paolo State Resolution on Forest Restoration).
87 Ibid. (citing Article 6).
88 Ibid.
89 Ibid.
90 Ibid., 693.
91 Ibid. (explaining that the standards for number of woody plants was set based on what the forest nurseries of the state deemed to be attainable).
92 Ibid. (describing some parties who argue that evidence shows that forests have achieved a self-perpetuating state after 10–20 years of planting and describing other parties who suggests that 20 years of secondary succession is not sufficient to guarantee long-term sustainability of plantings).
93 Ibid. (São Paolo recognises the sizable expense in doing needed ecological restoration work and has introduced some legislation in its forest remnants programme providing for payments for ecosystem services and other financial incentives).

sustainable ecosystem structure and function, São Paolo's efforts to incorporate meaningful restoration practice into its law must still be applauded for their boldness in establishing discrete restoration standards.

A second notable innovative restoration intervention in Brazil involves institution building across the landscape. In 2009, 160 NGOs, private companies, research institutes and government agencies entered into the Atlantic Forest Restoration Pact to promote restoration of 15 million degraded hectares by 2050 representing 30 per cent of the original rainforest while still protecting existing forest reserves across 15 Brazilian states.[94] The Pact was an attempt to create additional incentives for individual actors to participate in forest restoration efforts because they could now be part of the larger decision-making bodies. The Pact has formed its own governance structure with a Coordinating Board composed of an equal number of members from private companies, research institutions, government agencies and NGOs who oversee several working groups including a group on public policy. The efforts of the Pact are a work in progress but the institution of the Pact offers an encouraging governance model that may be appropriate for other states.

South Africa: restoration practice as community development

In South Africa, the government created the 'Working for Water' programme in 1995 that has employed 20,000 individuals to remove invasive woody plants that disproportionately remove soil moisture and deprive native plants of water increasing risks for wildfire and erosion.[95] The programme is a payment for ecosystem services project. While the programme is designed to restore the biological integrity of the landscape in order to improve water delivery, it is more centrally a programme about initiating livelihood opportunities for individuals. In a country with large-scale poverty and a lack of livelihoods, social welfare outcomes are a high priority. The government of South Africa conceives of the programme as a core community development programme providing short-term contract work for women, youths, disabled individuals and ex-offenders that builds vocational skills.[96] Through the Working for Water programme, the government delivers a number of other social services including HIV/AIDS awareness and financial savings schemes.[97] Based on the 'waste' materials that have been collected from the restoration clearing efforts, the government has promoted a number of value added industries using the invasive biomass including manufacturing screens and

94 M. Calmon et al., 'Emerging threats and opportunities for large-scale ecological restoration in the Atlantic Forest of Brazil' (2011) 19(2) *Restoration Ecology* 154–158; S. Pinto et al., 'Governing and delivering a biome-wide restoration initiative (less than 12 per cent of the original Atlantic Rainforest is left).

95 S.J. Milton, 'Economic incentives for restoring natural capital in Southern African Rangelands' (2003) 1 *Frontiers in Ecology and the Environment* 247–254.

96 Department of Water Affairs, Republic of South Africa, *Working for Water*, <www.dwaf.gov.za/wfw/SocialDev/>.

97 Ibid.

blinds, lights and lamps, indoor and outdoor furniture, fencing, wooden educational toys, firewood, charcoal and woodchips.[98]

Capitalising on the success of the Working for Water programme as a dual community development and restoration programme, the government of South Africa has also launched a Working for Wetlands programme that hires individuals to rehabilitate key wetland areas including existing and proposed Ramsar Wetlands of International Importance.[99] The Working for Wetlands programme is coordinated across a number of different agencies including the Department of Environmental Affairs and Tourism, the Department of Water Affairs and the Department of Agriculture, Forestry and Fisheries as part of an 'expanded public works programme'.[100] In addition to its water and wetlands programme, the government has organised other community-based invasive species eradication programmes including Working on Fire and Working on Woodlands and Woodlots.[101]

In addition to addressing community poverty through restoration work, the South African government also supports private rehabilitation efforts through a series of programmes under the umbrella of 'Landcare'.[102] This programme is financed at the federal level but implemented at the provincial level and is focused primarily on agricultural rehabilitation through a number of sub-programmes including Watercare, Veldcare and Soilcare. While these are not ecological restoration programmes like the Working for Water and Working for Wetlands programme, they might operate as precursors for national environmental restoration work by raising awareness of ecological land values.

Ecuador and Bolivia: restoration as a right

For those states that refer to restoration in their laws, most of the references regard restoration as a functional and remedial activity. Restoration as understood in the legal texts does not redefine the relationships between people and ecosystems. In the past few years, Ecuador and Bolivia have taken a different approach to restoration by incorporating a legal perspective that assigns rights to nature.

In 2008, with active support from non-governmental organisations including the Pacha Mama Foundation, the Community Environmental Legal Defense Fund, ECOLEX, EcoCiencia, CONAIE, IUCN, Frente de Defensa de la Amazonía and Fundación Ambiente y Sociedad, Ecuador drafted and ratified a constitution positioning 'nature' – referred to as Pacha Mama – as central to the Ecuadorian

---

98 Department of Water Affairs, Republic of South Africa, Working for Water, Partnerships, Value Added Industries, <www.dwaf.gov.za/wfw/Partnerships/> (providing links to industries including a company called 'Invader Crafts').

99 South African National Biodiversity Institute, Working for Wetlands Program Overview, <www.sanbi.org/sites/default/files/jobs/documents/Summary%20of%20Rchabplan%20Waterberg.pdf>.

100 Ibid.

101 C. Macaskill, *The National Agricultural Directory 2011* Department of Agriculture, Forestry and Fisheries, Republic of South Africa (2011) 567.

102 Ibid., 570.

state and its people. Based in part on indigenous perspectives on nature, the constitutional 'nature' is embodied with the 'right to integral respect for its [nature's] existence and for the maintenance and regeneration of its life cycles, structure, functions, and evolutionary processes'.[103] As part of this right which can be invoked at any time by 'all persons, communities, peoples, and nations',[104] nature has 'the right to be restored'.[105] The right of restoration exists separate from any duty to compensate individuals who may have also experienced losses due to a loss of ecosystem services.[106] The state has the potential to play an active role. If the environmental damage to nature is deemed to be 'severe or permanent' then the state must establish 'the most effective mechanisms to achieve the restoration and shall adopt adequate measures to eliminate or mitigate harmful environmental consequences'.[107] When this language is interpreted literally, the state can presumably require a private entity to 'eliminate or mitigate harmful environmental consequences' but if these measures fail then the state may have a residual duty to undertake restoration activities on its own. This interpretation is reinforced in Chapter Two of the Constitution providing that in the case of environmental damage 'the State shall act immediately ... to guarantee the health and restoration of ecosystems'.[108]

Several safeguards have been built into the Constitution to support the state's efforts to achieve ecological integrity. First, there is no statute of limitation on prosecution for environmental damage under the Constitution.[109] Second, the state should be able to call upon efforts by its nationals who have the responsibility under Article 83 of the Constitution 'to respect the rights of nature' which includes the 'right to be restored'.[110]

Inspired in part by Ecuador's success in rewriting its Constitution, Bolivia passed a similar law recognising a right to nature. In a 2010 Act called the 'Rights of Mother Earth Act', the Bolivian legislative assembly recognised 'rights of Mother Earth' and 'obligations and duties of the Plurinational State and of the society'.[111] The law articulated several principles including the principle of regeneration which recognises that 'The States (and its different levels) and society ... must ensure the necessary conditions for Mother Earth's living systems to be able to absorb damages, adapt to the alterations, and regenerate without significantly affecting their structural and function characteristics.'[112] The law recognises that 'living systems have limits to their ability to regenerate, and that humanity has

---

103 Ecuador Constitution, Article 71.
104 Ibid.
105 Article 72.
106 Ibid.
107 Ibid.
108 Article 397.
109 Article 396.
110 Article 83.
111 Law 071 of the Plurinational State of Bolivia (2010), Article 1.
112 Ibid., Article 2(3).

limits in its ability to reverse its actions'.[113] All Bolivians, in addition to having duties as members of Bolivian society, also share 'collective rights in the life systems of Mother Earth'.[114] As with the Ecuadorian Constitution, Mother Earth has the right to restoration which is understood as 'the right to timely and effective restoration of the living systems affected, directly or indirectly by human activities'.[115]

The 2010 Act has been further refined by a Law 300 entitled 'The Framework Law of Mother Earth and Holistic Development for Living Well'. The concept of 'living well' (*'vivir bien'*) is a concept defined in the law as 'a civilizational and cultural alternative to capitalism based on the indigenous worldview' that 'signifies living in complementarity, harmony and balance with Mother Earth and societies, in equality and solidarity and eliminating inequalities and forms of domination'.[116] The law signals that *'vivir bien'* is to 'Live Well amongst each other, Live Well with our surroundings and Live Well with ourselves'. Restoration is considered one of the techniques for achieving 'holistic development for living well'.

Among the central principles embodied in the law is restoration. The law provides that Mother Earth has the right to be restored. Specifically the law provides that the state and any individual or collective 'who causes accidental or premeditated damage to the components, parts and life systems of Mother Earth, is required to conduct a comprehensive and effective restoration or rehabilitation of the functionality thereof' so as to return ecosystems to a pre-damage condition.[117] The law provides a very specific definition of restoration describing it as 'the planned process of intentionally modifying a life zone or life system in order to re-establish the diversity of its components, processes, cycles, relationships and interactions and dynamics'.[118] Life zones are defined as unique biogeographical-climatic areas and life systems are defined as ecosystems that include humans and their 'cosmovisions'.[119] The target for restoration is to return the system to the condition that it was in before it became damaged. The law calls for the restored ecosystem to be 'self sustainable in ecological, social, cultural and economic terms'.[120] Under Law 300, the state assumes some costs for restoration as part of its framework law. Specific language is provided requiring restoration of forests,[121]

---

113 Ibid.
114 Ibid., Article 6.
115 Ibid., Article 7(6).
116 Law 300 Ley Macro de la Madre Tierra y Desarrollo Integral para Vivir Bien, Article 5.2.
117 Ibid., Article 4.5; see also Article 11 (El responsable directo del daño ocasionado a los componentes o zonas de vida de la Madre Tierra está obligado a restaurar el mismo, de manera que se aproximen a las condiciones preexistentes al daño, sea directamente o por medio del Estado, cuando corresponda).
118 Ibid., Article 5.10.
119 Ibid., Article 5.16 and Article 5.12.
120 Ibid.
121 Article 25 (Promover y desarrollar políticas de manejo integral y sustentable de bosques de acuerdo a las características de las diferentes zonas y sistemas de vida, incluyendo programas de forestación, reforestación y restauración de bosques, acompañados de la implementación de sistemas agroforestales sustentables, en el marco de las prácticas productivas locales y de regeneración de los sistemas de vida).

restoration of areas damaged by mineral and hydrocarbon removal,[122] water sources[123] and land.[124]

While Ecuador and Bolivia have taken the most expansive view of restoration in their law of any state, by seeking a paradigm shift in environmental practices, only time will tell whether this rights-based approach can be operationalised. As of now, there are no restoration targets in the law and it is unclear whether the government will be seeking contribution to restoration by parties that it might prosecute for failing to respect Mother Earth. If the government of Bolivia was to prosecute the dozens of factories that are currently dumping nitrates and sulphates into the Rocha River under its new law, then the law may indeed have teeth.

Conclusion

National responses to achieving ecosystem restoration outcomes vary greatly depending on the political will, financial resources and technical abilities of a State. While there is a great deal of variety among the domestic laws promoting restoration, ranging from technical approaches to rights-based approaches, there is at least one similarity. Each of the legal systems described recognises that humans have been the source of unnecessary ecological degradation and now seek through a variety of legal mechanisms, from public-private partnerships to standing for nature, to revive self-sustaining ecosystem functions that can be conserved for generations to come.

While there are no definitive lessons to be learned from the domestic law covered in this section about how best to domesticate international obligations to restore, there are two general observations that can be made about how states might structure ongoing and future ecological restoration efforts. First, while states have the prerogative about how best to implement their restoration obligations within their territory or in coordination with other states, most governments would benefit from investing in restoration projects that are both science and community knowledge based. Reconciling the approaches of science and communities to what constitutes a good restoration outcome is not always straightforward. One can imagine a scenario where scientists desire the return of 'dangerous' environments such as wetlands with mosquitos and other species that threaten human well-being but the community prefers a different landscape that addresses their needs as residents.

Second, for many states, there will need to be an internal conversation among state decision-makers about whether restoration projects, required by or possibly

122 Article 26.
123 Article 27.7 (Garantizar la conservación, protección, preservación, restauración, uso sustentable y gestión integral de las aguas fósiles, glaciales, humedales, subterráneas, minerales, medicinales y otras, priorizando el uso del agua para la vida).
124 Article 28 (Establecimiento de instrumentos institucionales, técnicos y jurídicos para verificar que el uso de la tierra y territorios se ajusten a las características de las zonas y sistemas de vida, incluyendo la vocación de uso y aprovechamiento, condiciones para la continuidad de los ciclos de vida y necesidades de restauración).

financed by national governments, reflect adequate coordination across a national landscape in order to achieve the types of restoration-related goals set by governments, such as protecting areas of ecological significance, reviving damaged ecosystem services or adapting for climate change. Quality of restoration matters, of course, but the physical scale of restoration is likely to become increasingly important as states endeavour to revive or maintain self-sustaining ecosystems. For large-scale restoration goals to be achieved, restoration must be regarded not just as an implementation matter for a given state's Ministry of Forestry or a Department of Conservation, but also a cross-cutting priority for other government agencies including administrative bodies responsible for transportation, commerce, education, defence, health and housing. While at this point, this level of proposed integrated decision-making is at best aspirational, this approach might ensure that ecological gains from restoration efforts are not lost due to fragmented future decisions.

This chapter has examined an array of national legal responses with examples chosen to be illustrative of the diversity of responses that states have taken to implementing restoration efforts. There are also many other inspirational efforts under way that may accelerate implementation of restoration obligations if the efforts are capable of mainstreaming restoration goals across landscapes and social institutions. For example, two additional efforts that have the potential to be transformational because of their scale include India's National Green Mission[125] and the Rwandan Forest and Landscape Restoration programme.[126]

What is clear from this review of a variety of approaches is that states understand that they must take some coordinated legal or policy action to restore ecological values within their territory. What motivates the action of each state depends in part on the situation within each state. China restores its ecosystems in hopes of reviving livelihoods that are in jeopardy. South Africa restores in order to enhance its social development programme for some of the most marginalised of its citizens. The United States seeks leverage of its restoration work through partnerships with states, municipalities, NGOs and businesses. Bolivia and Ecuador restore because of

---

125 National Mission for a Green India, Government India, Ministry of Environment & Forests, Foreword, <www.moef.gov.in/sites/default/files/GIM_Mission%20Docum ent-1.pdf> (calling for a holistic view on 'greening' that will go beyond trees and plantations to encompass both protection and restoration. Emphasis will be placed on restoration of degraded ecosystems and habitat diversity, e.g. grasslands and pastures (more so in arid/semi-arid regions), mangroves, wetlands and other critical ecosystems).

126 This programme is evolving. For more information about restoration opportunities in Rwanda, see Republic of Rwanda, Ministry of Natural Resources, *Forest Landscape Restoration Opportunity Assessment for Rwanda* (September 2014) p. 27 and Appendix 3 at p. 50, <https://portals.iucn.org/library/sites/library/files/documents/ 2014-077.pdf> (one of the key generic success factors for forest landscape restoration is whether a 'Law requiring restoration exists and is enforced'. The report observes that there is no specific formal law related to restoration for Rwanda and that unfortunately the ability to improve laws in Rwanda in general is low due to a lack of enforcement because of a lack of budget and a lack of human resources).

their citizens' powerful spiritual beliefs about what relationship should exist between the state and the environment.

A review of a limited number of state legislative and policy practices illustrates that individual states are deeply engaged in creating laws and policies to further restoration work. In international law, there is a theoretical curiosity as to whether international law informs domestic law or domestic law underpins changes in international law. Here, it is unclear what the relationship between domestic restoration legislation is with the existing international treaty law. While states have not been explicit in articulating that their national legislation is a response to international obligations, in most cases the domestic laws underlying ecological restoration efforts such as those discussed in this chapter support and implement the international restoration objectives described in Chapters 4–6.

Part III

Thematic issues on ecological restoration

9 Private non-state actors and ecological restoration

Up to this point, the chapters have focused exclusively on the actions of state actors in negotiating international treaties, domesticating international obligations and designing national programmes. Yet, restoration is not limited to governance by the states. In fact, much of the area of emerging governance in ecological restoration work is the product of personal investments of time and money by private individuals or communities who envision a better world one acre or one reef at a time. For example, in North America Aldo Leopold's name is synonymous with restoration because of his personal contribution of taking his hard-scrabble acreage in Wisconsin and transforming it into a biodiverse thriving forest.[1] In New Zealand, the Karori Sanctuary/Zealandia project providing for the long-term ecological restoration of a valley in Wellington, New Zealand is the vision of a committed group of conservationists and bird lovers.[2] Nobel Prize winner Wangari Maathai with the help of thousands of community members initiated the Green Belt Movement as a community empowerment movement to restore forests to desertified landscapes.[3]

While it is not possible to discuss all of the contributions that non-state actors have made to achieving restoration priorities, the objective of this chapter is to discuss a few themes regarding interaction between non-state actors, the law and ecological restoration. The chapter starts with non-state actors as 'objects' of the law with a variety of legal provisions mandating non-state actors to engage in restoration as a form of compensation. The chapter progresses by examining how law may operate as a barrier to ecological restoration for private actors. The chapter then examines how several private actors have become key contracting partners in a variety of agreements to promote ecological restoration work. After exploring the evolving norm of ecological restoration as articulated by civil society and on the basis of corporate social responsibility (CSR) initiatives, the chapter concludes by suggesting that it may be time for private actors across an array of

---

1 C. Meine, *Aldo Leopold: His Life and Work* (University of Wisconsin Press 2010) xxiv.
2 J. Lynch, *History of Zealandia* (2007), <www.visitzealandia.com/what-is-zealandia/our-history/one-mans-vision/>.
3 W. Maathai, *The Green Belt Movement: Sharing the Approach and the Experience* (Lantern Books 2004).

interests to agree on a negotiated ecological restoration code that reflects core obligations and standards that non-state actors agree to follow in their ecological restoration efforts.

Statutory mandates to private actors to restore

In many cases, private actors engage in ecological restoration because they are being held accountable by the state as compensation to offset a project impact or to remedy intentional or accidental damage such as the *Exxon Valdez* oil spill or the British Petroleum Deepwater Horizon blowout.[4] For example, a company involved in coal mining in the United States is expected to pay a reclamation fee to be used for 'the restoration of land and water resources and the environment that have been degraded by the adverse effects of coal mining practice'.[5] Likewise, a company that removes wetlands in the United States must compensate through compensatory mitigation efforts that may involve restoration.[6] Other privately funded restoration efforts are triggered by the judicial system ordering a polluter or a party that has damaged resources to provide restoration. For example, in Uganda, the government has the authority to issue an environmental restoration order requiring 'the person to restore the environment as near as it may be to the state in which it was before the taking of the action which is the subject of the order'.[7]

In the case of catastrophic events, some countries have designated specific remedies that encompass ecological restoration. For example, when the *Exxon Valdez* hit the reef in Alaska in 1989, the US Congress reacted by passing the Oil Pollution Act.[8] This Act enabled the US government to collect damages from responsible parties based on 'the cost of restoring, rehabilitating, replacing or acquiring the equivalent of, the damaged natural resources.'[9] The agency practice of assessing and quantifying natural resource damage was codified into regulations in 1996.[10] Damages are assessed based on a baseline which has been statutorily defined as 'the condition of the natural resources and services that would have

---

4　See Chapter 10 for a discussion of how restoration is a legal compensation strategy for protected areas.

5　Surface Mining Control and Reclamation Act, 30 USC 1233 (a)(1)(B).

6　Clean Water Act, Section 404(h)(1); Memorandum of Agreement between the US Environmental Protection Agency and Department of the Army ('Appropriate and practicable compensatory mitigation is required for unavoidable adverse impacts which remain after all appropriate and practicable minimization has been required'); see Chapter 10 for the distinction that is made by the Ramsar Convention and the EU between compensation and mitigation within protected areas, which highlights the potential confusion of using terms such as 'compensatory mitigation', compensation and mitigation.

7　The National Environmental Statutes of 17 May 1995 (Statutes Supplement to the Uganda Gazette, No. 21, Volume LXXXVIII) Part IV, Section 68(1)(a).

8　33 USC §§ 2701 et seq.

9　33 USC § 2706 (d)(1).

10　15 CFR Part 990.

existed had the incident not occurred'.[11] Defining a baseline can generate controversy. When an oil spill has occurred and after emergency restoration efforts are implemented, natural resource trustees are required to draft a longer-term restoration plan that will include primary restoration and possible compensatory restoration projects (covering public losses of a resource amenity before recovery is completed) that will be subject to environmental impact review.[12] This process can take several years because of the complexity of quantifying damages within a complex ecosystem. Once a programme has been approved, the plan will be implemented and monitored.

After the British Petroleum Deepwater Horizon accident, BP was ordered to pay natural resource damages. In an effort to improve its corporate reputation as a proactive company and potentially reduce its costs, BP proposed entering a legal agreement with the US government providing a 'framework for early restoration'.[13] Early restoration is considered to be a form of compensatory remediation and allows for a responsible party to begin restoration activities before a natural resource assessment has been finalised.

Given the expense associated with compensatory restoration, a new market has emerged to respond to the statutory requirement for restoration called 'restoration banking'. In the United States, entrepreneurs are now restoring areas to sell natural resource credits to responsible parties to be used for ecological restoration projects.[14] If a party is found responsible, it will have readily available compensatory mitigation credits to offer to cover damages. While these type of restoration banking projects demonstrate a highly functional motivation for private actors undertaking ecological restoration, these projects provide the necessary funding for undertaking the potentially high-quality restoration projects and ensuring long-term monitoring.

In these cases, restoration work undertaken by private actors for either anticipated environmental harms (e.g. restoration banking) or actual environmental harms (e.g. primary restoration for an oil spill) functions in the context of a corrective justice theory of restoration. In theory, the specific harm to society will be remedied by the restoration efforts and society will be made whole by the replanting of damaged sea grasses or the reintroduction of seabirds. In theory, statutory mandated restoration can make a public space whole again depending on what the state demands of the private actor, but it really depends on the scale of the damaged resources. As the post-*Exxon Valdez* and post-Deepwater Horizon

11 15 CFR § 990.30.
12 US National Oceanic and Atmospheric Administration, Office of Response and Restoration, <http://response.restoration.noaa.gov/environmental-restoration/na tural-resource-damage-assessment.html> (providing an overview of the Damage Assessment, Remediation, and Restoration Program).
13 British Petroleum and US Government Natural Resource Trustees, Framework for Early Restoration, <www.restorethegulf.gov/sites/default/files/documents/pdf/fram ework-for-early-restoration-04212011.pdf>.
14 Bluefield Holdings, <http://bluefieldholdings.com/> (describing the availability of natural resource damage credits along the Duwamish River for a variety of sites).

restoration efforts unfortunately illustrate, holistic ecological restoration can be elusive when damage to the environment is extensive. In relation to the *Exxon Valdez* restoration efforts, the National Oceanic and Atmospheric Administration's Office of Response and Restoration opines that ecological recovery defined as 'a return to conditions as they were before the spill' may not be viable in a dynamic intertidal zone.[15] Even though many decades have passed, oil remains below the surface of beaches indicating that at least the beaches in the regions of the original spill have not returned to a baseline condition.[16]

The *Exxon Valdez* case raises an interesting question regarding the responsibility of private actors in the context of liability for ecological restoration. In the case of the *Exxon Valdez*, the government settled with the Exxon Company for $900 million with annual payments stretched over a ten-year period and the possibility of a 'reopener' where the governments could make a claim for up to an additional $100 million for restoration costs.[17] Should the law put limits on what society can expect from a private actor when restoration has not yet been achieved? Is there a set of ecological parameters that can be applied to signal when a private actor has done enough to return a system to a baseline condition? Or should there be continued payments until comprehensive ecological restoration has been achieved? These are liability questions that judges may need to grapple with to implement the polluter pays principle and to achieve ecological restoration objectives of attempting to return an ecological system to a pre-incident baseline.

Laws as hurdles for restoration workers

In the years to come, ecological restoration has the potential of becoming a profitable business enterprise with estimates of global investments in restoration of up to $18 billion.[18] What this means is that many more individuals may become involved in ecological restoration work. In spite of all of the good intentions associated with ecological restoration work, restoration like other social activities is bounded by existing law. Ecological restoration firms like most other businesses will need insurance to cover potential tort liabilities associated with restoration efforts. For example, an ecological restoration firm who may have failed to plan for impacts on neighbouring landowners, such as flooding caused by restoring a river channel, might be held liable under traditional legal doctrines of negligence.

15 US National Oceanic and Atmospheric Administration, Office of Response and Restoration, Has Prince William Sound Recovered from the Spill?, <http://response. restoration.noaa.gov/oil-and-chemical-spills/significant-incidents/exxon-valdez-oil-sp ill/prince-william-sound-recovered.html>.

16 US National Oceanic and Atmospheric Administration, Office of Response and Restoration, Is the Oil Gone?, <http://response.restoration.noaa.gov/oil-and-chem ical-spills/significant-incidents/exxon-valdez-oil-spill/oil-gone.html>.

17 United State of America v. Exxon, Agreement and Consent Decree (25 September 1991), <www.evostc.state.ak.us/static/PDFs/agreement_consent_decree093091.PDF>.

18 M.H. Menz et al., 'Hurdles and opportunities for landscape-scale restoration' (2013) 339 *Science* 526–527.

Restoration professionals need to be aware of the standard of care associated with various projects.

Existing laws can also create unintended barriers for restoration efforts on both public and private lands. For example, burning is a common practice for restoring prairies, grasslands and meadows. Depending on the location of the project and whether there are residential units or other potentially sensitive facilities in the vicinity such as schools or hospitals, certain restoration activities may be prohibited within a particular land zoning classification or, even where they are permitted, errant smoke from grassland burning might be identified as a private nuisance or a trespass. In others cases, restoration workers will need access to properties across private lands. If there has not been an access easement negotiated, restoration workers may find themselves needing to consult a lawyer if they are refused access in order to avoid a trespass violation.

In addition to the potential for there to be zoning or common law violations, ecological restoration practitioners may find themselves needing to become knowledgeable about regulatory approval for certain types of restoration efforts such as dredging and filling to restore certain types of habitat. In some cases, if a restoration worker wants to reintroduce a species on a large scale, they may find that they are required by state, provincial or federal law to undertake an environmental impact assessment or to engage in a lengthy permit process.

A restoration case study undertaken reflects some of the challenges associated with implementing restoration. The case study focuses on the removal of three stream barriers preventing steelhead from returning to a watershed in Central California. In order to remove the barriers, the landowners who were voluntarily undertaking the project required permitting approval by the California Department of Fish and Game, the National Marine Fisheries Service, the US Fish and Wildlife Service, the Regional Water Quality Control Board and the County of San Mateo. Coordinating across these groups proved so difficult that it took eight years before one of the three barriers was removed.[19] The second two stream passage barriers were not removed because the landowner who had originally agreed to their removal withdrew from the project due to the lengthy process, escalating costs, poor relationship with the agencies and concern over losing water diversions.[20] Ultimately, the 'regulatory obstacles faced during the course of the project resulted in the loss of the 4-mile Apanolio Creek as strong viable habitat for steelhead'.[21]

Recognising the potential regulatory challenges of performing restoration work, there have been some efforts to coordinate review of projects and reduce certain

---

19 AgInnovations, Permitting Restoration Case Study: Apanolio Creek Fish Passage Project, <www.aginnovations.org/uploads/result/1431289151-93b99a9dd9a0442df/Apa nolio_Creek.pdf>.

20 Ibid., 2 (observing that restoring 150 feet of the creek ultimately cost $500,000 due to construction and permitting costs and observing that the landowner was informed that the voluntary restoration project could trigger a reassessment of the appropriateness of the landowner's current water diversions).

21 Ibid.

regulatory burdens. For example, in California, the California Roundtable on Agriculture and the Environment has proposed a number of suggestions for both state agencies and restoration proponents to avoid the situation where restoration projects become prioritised 'based on the feasibility of permitting rather than on the magnitude of environmental benefits'.[22] As the Roundtable participants observed in 2010, California agencies 'do not necessarily distinguish between permit applications for environmental restoration projects which seek to provide a public benefit through environmental enhancement and those for conventional development projects (e.g., residential, commercial, industrial)'.[23] The Roundtable suggested that an inter-agency permit coordination task force be designed to help landowners with an expedited permit review for voluntary restoration projects.[24] Different jurisdictions might benefit from taking an approach such as the European Union has taken where environmental assessment is not necessary for plans and projects directly related to the management of a protected area.[25]

Private actors as partners to ecological restoration law contracts

In many cases, private actors are active stakeholders in formal partnerships. In keeping with the CBD COP Decision calling upon more involvement from the private sector in ecological restoration,[26] many groups including companies are entering into public-private partnerships to contribute to the funding and planning of restoration activities. For example, in the Eldorado National Forest in California, the Coca-Cola Company is partnering with the National Forest Service and the non-profit National Fish and Wildlife Foundation to restore a 500-acre meadow at the headwaters of a significant California river.[27] The success of these public-private collaborations in terms of achieving ecological restoration objectives may depend upon the ability of the partners to negotiate long-term partnership agreements because many of the projects will require long-term commitments of resources. One such example of a public-private partnership agreement is the Master Stewardship Agreement for the Cornerstone Project among the managers of the Stanislaus National Forest in California, the Eldorado National Forest in California and the Amador Calaveras Consensus Group that includes watershed improvements and habitat restoration projects as part of a larger effort for

---

22 California Roundtable on Agriculture and the Environment, Permitting Restoration: Helping Agricultural Land Stewards Succeed in Meeting California Regulatory Requirements for Environmental Restoration Projects (November 2010) 2, <www.aginnovations.org/uploads/result/1431289151-93b99a9dd9a0442df/CRAE_Permitting_Restoration.pdf>.
23 Ibid., 6.
24 Ibid., 1.
25 EU Habitats Directive, Article 6(3).
26 Decision XII/9. Engagement with subnational and local governments, UNEP/CBD/COP/DEC/XII/9 (2014) para. 4(c).
27 Region 5, Regionwide Listing of Ecological Restoration Projects, <www.fs.usda.gov/detail/r5/landmanagement/?cid=stelprdb5382852>.

sustainable timber product harvesting.[28] This agreement is the result of the US government-sponsored Collaborative Forest Landscape Restoration Program that invests in multiparty efforts to restore forest landscapes and monitor for restoration while also encouraging the use of forest by-products such as thinned trees from restoration treatments as a source of income for local communities. The challenge for these programmes is the difficulty in developing a competitive market for restoration by-products that may require more processing to create a product than other forest products.

In the coming years with the emphasis on investing in projects that support climate change adaptation, there may be the formation of many new public-private partnerships to achieve restoration outcomes. For example, South Korea announced in 2009 that it would seek to partner with private actors to restore four important watersheds in order to protect against flooding.[29] A critical aspect of these public-private partnership will be some degree of harmonisation of ecosystem restoration standards to ensure that the projects meet ecological quality control criteria. Presumably some of the public-private partnerships that might emerge as part of national ecological infrastructure projects such as Korea's projects are subject to a public procurement process. These procurement processes should clearly describe what ecological parameters and objectives will constitute performance by a private partner. Otherwise, there is the potential for accountability problems if private performance falls short of public expectations.

Another form of agreement that might be characterised as public-private is negotiated payments for ecosystem services.[30] The South African Working for Water programme described in Chapter 8 is one example of a payment for ecosystem service programme. The strategy behind these programmes is to ensure that protection of ecosystems services such as clean water, healthy habitat or carbon sequestration is valued and assigned a financial return. Ideally, a programme offering a payment offers the payment for a benefit that would not have happened under a 'business as usual' model. While these programmes that are often underwritten by public-sector agencies, businesses and non-governmental organisations provide a positive financial incentive for investing in long-term restoration, the challenge is ensuring adequate ongoing financing to provide the appropriate payments. Additional challenges might exist in relation to verification that ecosystem services are delivered.

---

28 Collaborative Forest Landscape Restoration Project Amador Calaveras Consensus Group (ACCG) Cornerstone (CFLR015) Eldorado and Stanislaus National Forest (2013), <www.fs.fed.us/restoration/documents/cflrp/2013Reports/AmadorCala verasCornerstone_FY13.pdf>.
29 The Economics of Ecosystems and Biodiversity for National and International Policy-makers, Chapter 9 Investing in Ecological Infrastructure, Box 9.16. <www.cbd.int/doc/case-studies/inc/cs-inc-teeb.Chapter%209-en.pdf>.
30 CIFOR, 'What are "payments for environmental services"?', <www.cifor.org/pes/_ref/a bout/index.htm> (a payment for an ecosystem service is understood as a voluntary transaction where an ecosystems service buyer provides payments to an ecosystem service provider as long as the provider continues to supply the negotiated service).

In addition to public-private partnership agreements, there are a number of important private-private partnership agreements that are designed to support restoration efforts. For example, Perrier Vittel, concerned with protecting the aquifers that they used for water, signed long-term contracts with local farmers in the Rhine-Meuse watershed to invest in certain services including specific watershed protection and restoration activities.[31] In states where the government has been historically weak on environmental protection, civil society has replaced some of the governance functions of the state by actively engaging in rebuilding environmental resources. For example, in regions with widespread poverty but an abundance of environmental capital, some NGOs have organised contractual programmes to promote restoration work among communities based on payment from the NGOs to communities for ecosystem projects that support global objectives such as delivering carbon sequestration or protecting biodiversity and freshwater supplies. These programmes are anchored in basic contract law with significant legal questions possibly arising from either the perspective of the NGO or the community over adequacy of performance. The success of these programmes as vehicles for restoration depends on the honesty of the parties engaged in the contracts and the ability to reduce transaction costs associated with developing the original contracts.

One example of a privately operated payment for ecosystem services programme is Wetlands International's 'biorights' programme that creates a de facto market for communities who engage in conservation and restoration work through a 'rights trading system'.[32] The NGO provides micro-credits to local communities to assist them in developing sustainable income generating activities. In order to repay their loan and interest on the loan, the communities offer in-kind services either in the form of refraining from unsustainable environmental activities or in the form of labour on a variety of conservation and restoration projects. If the conservation and restoration projects are deemed successful by the NGO, then the micro-credits are converted into a combination of cash payments and community-based sustainable development funds.[33]

At the contract stage, Wetlands International negotiates multi-year contracts between rural communities and 'investment' parties who may include sustainable forest product companies or ecotourism operators.[34] The contracts generally include a description of the roles of each party (including buying materials for restoration or implementing sustainable development measures), the roles of intermediaries, a description of the ecosystem service to be delivered, specific contract measures, payment amounts, project monitoring obligations, project

31 United Nations Environment Programme, Payments for Ecosystem Services: Getting Started: A Primer, 7, <www.iied.org/NR/forestry/documents/Vittelpaymentsfor ecosystemservices.pdf>.
32 P. van Eijk and R. Kumar, 'Bio-rights in theory and in practice' (2009) 21, <www. wetlands.org/Portals/0/publications/Report/WI_Bio-rights%20in%20theory%20and %20practice.pdf>.
33 Ibid., 6.
34 Ibid., 23.

duration and terms on legal enforcement of contract provisions.[35] The contracts are formally registered within the legal system and include a number of specific indicators of adequate performance.[36] Past contracts have included indicators such as seedling survival rates or specific reductions in ecological degradation rates.[37] The party 'buying' the conservation or restoration services assumes certain risks through a force majeure clause, the risk for natural disasters and acts of war.[38] In the case of a resource that is extremely valuable to exploit in the short term, contracts for the delivery of ecosystem services may be negotiated not just as multi-year contracts but also ongoing renewable contracts.[39]

One aspect of a number of these public-private and private-private contracts that raises a number of potentially challenging legal questions is the role of 'adaptive management'. Even though performance of parties may have been negotiated under one set of circumstances, parties may need to do more than they negotiated to do in order to achieve original ecological objectives. This could require the reopening of a contract negotiation with all of its associated transaction costs. It may benefit all parties to a long-term payment for ecosystem services agreements to include provisions within a contract that allow for renegotiation of specific terms of the contract under pre-determined circumstances without fully renegotiating the entire contract.

Other legal issues including the security of land tenure are implicit in the conservation and restoration contracts. Because many of the communities with whom parties wish to contract for restoration services lack legally recognised land-tenure rights over the lands that they occupy, contracts may need to be enlarged to include private landowners or government landowners.[40] To ensure enforceability of a contract in case one of the parties fails to perform, contracts must conform with local legislative requirements.[41] Depending on who is investing in the conservation or restoration of ecosystem services, there may be different levels of monitoring to ensure that the community has delivered adequate services. For some community enhancement projects, the local community and project manager can provide monitoring services. For other projects that involve global markets for carbon reduction, outside auditors may be required.[42]

From 1998 to 2008, the NGO Wetlands International worked with several thousand parties through contracts to restore mangrove and peatland ecosystems.[43] The first project in Java provides a useful example of the potential for civil society to foster restoration activities as economic development opportunities. In 1998, the Wetlands International Indonesia Programme in cooperation with a

35 Ibid., Annex IV, 125–126.
36 Ibid., 22.
37 Ibid.
38 Ibid., 23.
39 Ibid., 24.
40 Ibid.
41 Ibid., 26.
42 Ibid., 36.
43 Ibid., 7.

local NGO worked with communities to restore mangrove systems that had been removed a couple decades earlier to facilitate a number of failed aquaculture projects. The programme started with one group of five people but eventually supported the restoration activities of several hundred people. An interesting outcome of the project was that post-implementation, many of the groups formed to support the restoration activities still exist and have formally registered themselves as corporations under Indonesian law working on various sustainable development projects.[44] Perhaps most encouraging from the perspective of an NGO seeking to change behaviour, the community groups as of 2008 had continued to meet regularly to discuss maintenance of existing environmental restoration and proposals for new projects even though the bio-rights contracts had formally ended in 2005.[45]

In addition to the private-private agreements between NGOs and communities designed to support ecological restoration outcomes, there are an increasing number of agreements being concluded among landowners. In one recent example, soft law is creating ongoing partnerships between private parties who desire both to restore landscapes and to continue economic development activities. The Wildlands Network in the United States has supported the formation of the Western Landowners Alliance who provides direct support to a number of important private restoration efforts by the ranching community on both private lands and federally leased lands used by the landowners for grazing.[46] In some cases members have agreed to coordinate ecological stewardship efforts across private property boundaries. For example, the Chama Peak Land Alliance includes a Memorandum of Cooperation around future oil and gas exploration on private lands which includes members of the alliance informing each other about proposed leases and evaluating the impact of these leases on the land alliance membership as a whole.[47] The rules that these alliances establish for their activities carry some legal weight in terms of creating a standard of care among alliance members. Assuming a joint restoration project across private landowners' lands, terms for such a project might be memorialised in a memorandum of understanding or a contract.

One of the most interesting models for a land alliance promoting restoration is the EKA Legacy Partners, LLC that runs The Earthkeeper Alliance. Founded by a group of real estate investors, the group pursues restoration through impact investing in what they term 'undevelopment'. Their model for 'undevelopment' entails aggregating land that has the potential to be high-quality conservation land, investing in some low-impact development, attaching a permanent conservation easement and then selling the land as a legacy investment.[48] The Alliance has created for itself a set of rules that it intends to govern land use for the lands transferred as legacy investment. It is unclear whether these rules have been

44 Ibid., 88.
45 Ibid., 95.
46 <www.westernlandownersalliance.org/>.
47 <http://chamapeak.org/programs/energy/oil-and-gas/landowner-mou/>.
48 EKA Legacy Partners, LCC/The EarthKeeper Alliance, Undevelopment, <www.earth
    keeperalliance.com/faq>.

memorialised in a document but, conceivably, the Alliance can assert a form of private governance over its lands as long as the lands are not managed in a way that violates federal or state law.

Potentially, important for restoration outcomes is the engagement of individual large landowners in managing their lands with a restoration ethic. The 100 largest landowners in the United States own collectively 2 per cent of the United States land mass, or about 33 million acres, with much of this land in the form of large undeveloped forests tracts or open rangeland.[49] Some of these landowners continue to invest in rural and undeveloped land which may be suitable for cost-effective ecosystem restoration. At least one of the top 100 landowners in the US, Ted Turner, is actively involved in restoration on his lands.[50] Smaller landowners such as M.C. Davis have been restoring Southeast longleaf pine forest ecosystems by investing around $1 million a year to replant longleaf pine forest habitat in the spaces in between existing islands of longleaf forests.[51] Today, his project is the largest forest restoration project in the US east of the Mississippi.

Individual investment by families such as the Tompkins has been critical to the conservation and restoration of a number of large tracts of land in Argentina and Chile including the Esteros del Ibera, a large marshland and Pumalin Park.[52] For each of their landholdings, the family has committed itself to a major landscape restoration initiative including restoring millennia-old trees in the temperate rainforest and restoring overgrazed grasslands to productive wildlife habitat.[53] The family has also introduced a 'rewilding' programme to return species that had disappeared from the landscape including giant anteaters, pampas deer, huemul deer, jaguars, peccaries, tapirs and giant otters.[54]

Law plays an important supporting role in enhancing these large-scale conservation and restoration projects through basic real estate contracts and ecological restoration service contracts. Under contract law, these large landowners can require specific performance standards from the parties with whom they contract to do restoration work. These performance standards would legalise ecological planning. Setting performance standards is of course as much art as science since restoration science is still evolving. Efforts to develop performance standards

---

49 Behind the 2013 Land Report 100: America's Largest Landowners Double Down, <www.landreport.com/2013/10/behind-the-2013-land-report-100-americas-larges t-landowners-double-down/>.
50 Ted Turner Enterprises, Vermejo Park Rank, <http://www.tedturner.com/turner-ra nches/turner-ranch-map/vermejo-park-ranch-new-mexico/> (describing forest and riparian restoration projects on the 590,823-acre ranch plus the effort to restore endangered black footed ferrets).
51 T. Hiss, 'Can the world really set aside half of the planet for wildlife?', Smithsonian. com (2014), <www.smithsonianmag.com/science-nature/can-world-really-set-a side-half-planet-wildlife-180952379/?no-ist>.
52 Tompkins Conservation, Landscape Restoration, <http://tompkinsconservation.org/ landscape_restoration.htm>.
53 Ibid.
54 Tompkins Conservation, Wildlife Recovery, <http://tompkinsconservation.org/wild life_recovery.htm>.

between a private landowner and an ecological restoration firm offer the opportunity for an iterative dialogue about the objectives of a given restoration effort and the tools of restoration.

Law can also create barriers for private landowners to make investments in restoration. After Douglas Tompkins acquired the Esteros del Ibera, lawmakers where the wetlands were located amended the municipal constitution to prevent non-Argentinians from buying land in strategic resources.[55] Since that decision, Argentina now has a law limiting the sale of rural lands, defined as all lands outside of an urban area, to foreigners because of concerns about food security. Under the 2011 law, which is not retroactive, foreigners may not hold more than 24,710 acres of land.[56] The Tompkins hold about 370,000 acres in Esteros del Ibera that were formerly cattle ranching land.[57]

Evolving norms for private restoration efforts

Non-state actors are actively negotiating norms for what constitutes best practices for restoration. Some of the norms proposed by civil society groups or incorporated into corporate strategies may become the basis for state and municipal understandings of what comprises a legal obligation to restore a particular site. This sub-section reviews efforts by civil society groups and corporations to define good ecological restoration practice.

Civil society

One policy initiative with potential long-term normative impact in defining appropriate restoration practices is the Society for Ecological Restoration's (SER's) International Primer that was drafted and utilised by SER for restoration professionals.[58] The SER Primer's definition of restoration as the 'process of assisting the recovery of an ecosystem that has been degraded, damaged, or destroyed' has been widely adopted including by local municipalities such as the Auckland Regional Council and the Chicago Metropolitan Agency for Planning as well as by international organisations such as the Food and Agriculture Organization and state and federal government agencies including the British Columbia Ministry of Environment, the US National Parks Service, the US National Ocean and Atmospheric Administration, the US Environmental Protection Agency, the US Forest

55 S. Romig, 'American buys slices of South America', *The Washington Post* (9 June 2007) (indicating that the community was concerned that the Tompkins would control one of South America's largest freshwater reserves and would not cede the land back to the state).
56 Government of Argentina, Consejo Interministerial de Tierras Rurales, Ley No. 26.737, Articulo 10 (22 December 2011).
57 EcoAmericas, 'Panther to Make Return in Esteros del Ibera' (August 2015).
58 As of 2016, the International Primer is currently under revision. Many of the basic principles of the Primer will remain unchanged.

Service, the US Bureau of Land Management, the US Geological Service and the New Zealand Department of Conservation.[59]

The Primer proposes a number of processes and outcomes that may eventually become the source of legal customary rules if adopted by sufficient restoration practitioners across the globe.[60] The Primer introduces a number of key ideas that have the potential to define what type of restoration will be achieved under the various international and national laws mandating restoration. Chief among these ideas is the principle that 'Historic conditions are … the ideal starting point for restoration design.'[61] For example, the Primer refers to reference ecosystems as actual ecosystems or descriptions of ecosystems that can serve as a model for restoration efforts as well as a baseline against which to monitor restoration outcomes. A successful restoration is deemed to be 'any state … as long as it is comparable to any of the potential states into which its reference could have developed'.[62] The Primer also provides nine attributes for measuring the success of a given restoration project:

1   The restored ecosystem contains a characteristic assemblage of the species that occur in the reference ecosystem and that provide appropriate community structure.

2   The restored ecosystem consists of indigenous species to the greatest practicable extent. In restored cultural ecosystems, allowances can be made for exotic domesticated species and for certain non-invasive species.

3   All functional groups necessary for the continued development and/or stability of the restored ecosystem are represented or, if they are not, the missing groups have the potential to colonise by natural means.

4   The physical environment of the restored ecosystem is capable of sustaining reproducing populations of the species necessary for its continued stability or development along the desired trajectory.

5   The restored ecosystem apparently functions normally for its ecological stage of development, and signs of dysfunction are absent.

6   The restored ecosystem is suitably integrated into a larger ecological matrix or landscape, with which it interacts through abiotic and biotic flows and exchanges.

7   Potential threats to the health and integrity of the restored ecosystem from the surrounding landscape have been eliminated or reduced as much as possible.

59 Society for Ecological Restoration, Adoption of SER Primer's Definition of Ecological Restoration, <www.ser.org/resources/resources-detail-view/adoption-of-ser-p rimer's-definition-of-er> (see Section 2 of SER International Primer).
60 Society for Ecological Restoration International Science & Policy Working Group, *The SER International Primer on Ecological Restoration* (2004), <www.ser.org/resources/ resources-detail-view/ser-international-primer-on-ecological-restoration>.
61 Ibid., Section 1.
62 Ibid., Section 5.

8   The restored ecosystem is sufficiently resilient to endure the normal periodic stress events in the local environment that serve to maintain the integrity of the ecosystem.
9   The restored ecosystem is self-sustaining to the same degree as its reference ecosystem, and has the potential to persist indefinitely under existing environmental conditions.[63]

A project need not contain all of these elements but these characteristics become valuable for measuring whether a party has achieved a successful ecological restoration project. By proposing these attributes as potential metrics, the Society for Ecological Restoration is providing a comprehensive restoration framework against which legal restoration efforts can be compared. For example, if a given restoration project fails to have 'a characteristic assemblage' of species operating in an appropriate community structure or lacks key functional groups or is incapable of sustaining reproducing populations, then it might be possible for a state or a community that was to benefit from a restoration project to demand additional performance from a party who has a restoration obligation. The nine attributes of restored ecosystems can serve as bench markers for ecologically adequate restoration.

The International Union for the Conservation of Nature (IUCN) has designed a series of principles for restoration of protected areas.[64] The protected areas that are the focus for the IUCN include strict nature reserves, wilderness areas, national parks, national monuments, habitat/species management areas, protected landscapes/seascapes and cultural areas with a focus on sustainable use of natural resources.[65] The IUCN relies upon the SER's definition of restoration and then calls upon practitioners to be effective, efficient and engaging. To be effective a restoration 're-establishes and maintains the values of a protected area'.[66] IUCN guidelines call for parties to:

•   'Do no harm' by first identifying when restoration is the best option.[67]
•   Re-establish ecosystem structure, function and composition.[68]
•   Maximise the contribution of restoration actions to enhancing resilience (e.g. to climate change).[69]
•   Restore connectivity within and beyond the boundaries of protected areas.[70]

---

63   Ibid., Section 3.
64   K.A. Keenleyside et al., *Ecological Restoration for Protected Areas: Principles, Guidelines and Best Practices* (IUCN 2012).
65   Ibid.
66   Ibid., 16.
67   Ibid., 17.
68   Ibid.
69   Ibid., 18 (referring to the need to potentially restore in order to provide refuges for species forced to move due to climate shifts).
70   Ibid., 19.

- Encourage and re-establish traditional cultural values and practices that contribute to the ecological, social and cultural sustainability of the protected area and its surroundings.[71]
- Use research and monitoring, including from traditional ecological knowledge, to maximise restoration success.[72]

Efficient ecological restoration 'maximizes beneficial outcomes while minimizing costs in time, resources and effort'.[73] To achieve these goals, the IUCN proposes guidelines encouraging parties to:

- Guideline 2.1 Consider restoration goals and objectives from system-wide to local scales.[74]
- Guideline 2.2 Ensure long-term capacity and support for maintenance and monitoring of restoration.[75]
- Guideline 2.3 Enhance natural capital and ecosystem services from protected areas while contributing to nature conservation goals.[76]
- Guideline 2.4 Contribute to sustainable livelihoods for indigenous peoples and local communities dependent on the protected areas.[77]
- Guideline 2.5 Integrate and coordinate with international development policies and programming.[78]

Finally, ecological restoration operating within an IUCN protected area should be 'engaging' by providing for collaboration and participation from partners and stakeholders.[79] This includes efforts to:

- Guideline 3.1 Collaborate with indigenous and local communities, neighbouring landowners, corporations, scientists and other partners and stakeholders in planning, implementation and evaluation.[80]
- Guideline 3.2 Learn collaboratively and build capacity in support of continued engagement in ecological restoration initiatives.[81]

---

71 Ibid. (recognising that in some cases effective ecological restoration may require the recovery of traditional, ecologically sustainable cultural practices and in other cases ecological restoration may require the end of a traditional cultural practice).
72 Ibid.
73 Ibid., 16.
74 Ibid., 19 (noting that management approaches can differ greatly with some projects managed to resist change while other projects are managed to encourage change).
75 Ibid.
76 Ibid., 20 (noting that restoration work in protected areas must focus primarily on conservation values and not on restoring for ecosystem services).
77 Ibid.
78 Ibid (restoration work can contribute to long-term poverty-reduction programmes and enhance efforts to deliver services for health, waste management, water supply, disaster mitigation and food security).
79 Ibid., 16.
80 Ibid., 20.
81 Ibid., 21.

- Guideline 3.3 Communicate effectively to support the overall ecological restoration process.[82]
- Guideline 3.4 Provide rich experiential opportunities, through ecological restoration and as a result of restoration, that encourage a sense of connection with and stewardship of protected areas.[83]

Taken together the SER Primer and IUCN guidelines offer a number of shared norms that apply to both how a restoration project should proceed and what outcomes define a successful project. For example, there appears to be a clear procedural norm that information about restoration projects must be made available to partners and stakeholders.

Restoration as corporate social responsibility

In addition to the individual and community efforts to restore habitat and ecosystems, a growing number of large companies who do not directly provide ecological restoration services to the market as part of their core business are now engaged in restoration activities. The growing corporate involvement is a response to a combination of state environmental regulations and self-regulation. Many of the larger corporate-managed restoration projects appear to be motivated by an ethos of corporate social responsibility (CSR). For purposes of this discussion, CSR is understood as internal policies adopted by senior decision-makers that balance fiduciary responsibilities to shareholders with responsibilities to the community and the environment.

For example, the CEMEX company, a multinational cement and concrete corporation, has become involved in a multi-year ambitious transborder restoration project as part of its commitment to achieving CSR. Concerned about desertification and fragmentation of land along the US and Mexico border, the company has restored 4000 hectares of habitat for biodiversity values that had been overgrazed and poorly managed.[84] The company explained its motivation in restoring lands as both a demonstration of 'its ability to be a real conservation player' and as a move to 'ensure the continuity of its business'.[85] What is interesting about the CEMEX project is that it appears to be part of a much larger corporate strategy. The company has purchased 120,000 hectares of land south of Big Bend National Park and has conservation agreements with adjacent landowners covering an additional 60,000 hectares. The company is further committed to supporting 'the establishment of protocols within existing policy and/or drawing up new agreements to facilitate the free flow of information, expertise, and material resources across the US-Mexican border that is specifically in support of conservation, ecological restoration, and research related to sustainable land use in the border

82 Ibid.
83 Ibid.
84 World Business Council for Sustainable Development, CEMEX, The Santa Maria Ecological Restoration Initiative, Case Study (2008).
85 Ibid.

region'.[86] In the long-term CEMEX hopes that its private transboundary conservation and biodiversity programme will strengthen restoration efforts at the landscape level.

The trigger for engaging CSR activities can be internal such as the leadership of a highly motivated chief executive office, but more often is likely to originate in response to external pressures critical of corporate activities. Some of this pressure is civil society based. Many of the larger companies that voluntarily pursue conservation or restoration activities have been encouraged to make these investments under civil society pressure. For example, under pressure from the World Wildlife Fund, Greenpeace and Rainforest Alliance, the Asia Pulp and Paper company based in Indonesia agreed to conduct some form of restoration on the forest lands it owns.[87] Questions remain about what will constitute adequate restoration. The paper company has indicated that it will focus on restoring degraded natural forests in its holding back to a healthier condition than the current forest. The World Wildlife Fund would like to see the company re-convert some of its existing tree plantations back to natural forest, a much more costly endeavour. These types of tensions concerning the quality of a restoration effort are common. What typically results in these cases is an ongoing negotiation between the company and civil society about what constitutes a reasonable level of effort. When negotiations break down, civil society groups may attempt to lobby for new legislation for more expansive restoration obligations by resource users. The threat of additional regulation may lead to greater voluntary concessions by a corporation than would otherwise be expected under the existing legal regime.

To the extent that conservation has become mainstreamed in CSR, restoration is also becoming mainstreamed among corporate actors as part of corporate rule-setting within certain sectors. An example of this is the recent Memorandum of Understanding adopted by The Sustainable Trade Initiative (IDH) based in the Netherlands, The Forests Dialogue based at Yale University and the World Business Council for Sustainable Development (WBCSD) to collaborate on developing and mainstreaming sustainable landscape management approaches that help businesses foster ecosystem restoration.[88] As a membership organisation, the WBCSD has participation from some of the largest multinationals in the world with control over large acreages of land. The agreement suggests that an emerging norm is the recognition by corporate firms that successful ecological landscape restoration will require involvement from all social actors including resource-dependent businesses. Even though this may seem obvious, this has not been the practice of firms and the voluntary engagement of corporate actors in the 2015 memorandum of agreement suggests the possibility of a new direction in corporate rule-setting.

86 Ibid.
87 R. Butler, 'A new leaf in the rainforest: longtime villain vows reform' (2014) *Yale Environment* 360. <http://e360.yale.edu/feature/a_new_leaf_in_the_rainforest_long time_villain_vows_reform/2745/>.
88 World Business Council for Sustainable Development, New Global Partnership Will Intervene in Landscapes at Risk (21 April 2015), <www.wbcsd.org/Pages/eNews/ eNewsDetails.aspx?ID=16480&NoSearchContextKey=true>.

While it is possible that the memorandum is merely a strategic business decision on the part of some of the world's largest resource-dependent firms to help 'greenwash' their corporate activities, an equally plausible theory is that the memorandum is a formal recognition that resource-dependent companies must engage in some degree of ecosystem restoration not only to support their business interests but also to regain the trust of the public that companies can steward resources.

While CSR documents only really govern the company that has the policy, CSR documents can reflect emerging norms. While it is beyond the scope of this chapter to empirically review CSR statements from across a range of actors, companies have through a range of CSR commitments agreed to invest in ecological restoration. What the nature of this investment is, based on the CSR commitment, is less than clear. In some cases, private companies support ecological restoration through financial philanthropic contributions. For example, in 2011, Nestle Waters North America donated funds to The Nature Conservancy to restore damaged salmon habitat.[89] In other cases, companies have set corporate goals for themselves that involve investment in restoration. For example, the Coca-Cola company since 2007 has sponsored the 'National Program for Reforestation and Water Harvesting' in Mexico designed to ensure reforestation of 70 million trees in Mexican watersheds to replenish 100 per cent of the water used in the company's products.[90]

Finance law as an enabler

In addition to the partnerships that are led by public actors, there are also a variety of efforts by private actors such as foundations to support restoration activities as a public good. For example, the Restore the Earth Foundation relies upon private management of a combination of private and public funding to purchase and restore land.[91] Proper leveraging of these combined funds may contribute to funds for additional restoration.

For-profit corporations are increasingly engaged in financing the capital needed for restoration work. For example, Beartooth Capital Partners operates a fund for land conservation and habitat restoration particularly on ranches.[92] The company identifies large tracts of land with both conservation and commercial value that non-governmental organisations would not be able to purchase directly. Each

89 Nestle Waters North America, Press Release CSRwire, Nestle Waters North American Marks World Water Day by Highlighting its Commitment to Improving Watersheds Across the US (21 March 2012).

90 Coke and the World's 'Most Important' Ecological Restoration Program (13 August 2015), <www.coca-colacompany.com/stories/coke-and-the-worlds-most-important-e cological-restoration-program/>.

91 Restore the Earth Foundation, Public-Private Partnerships, <www.restoretheearth.org/ what-we-do/public-private-partnerships> (observing that $1 of private funding is generating $3 of public funding to achieve restoration projects).

92 Beartooth Capital Partners, Manta Consulting Inc., Financing Fisheries Conservation, Beartooth Capital Partners (January 2011) 31–34, <www.conservation.org/publica tions/Documents/Manta-Consulting-Financing-Fisheries-Change.pdf>.

equity investment begins with creating a multi-stakeholder partnership focused on achieving restoration and management goals. Beartooth profits by selling conservation easements and restored parcels to buyers concerned about environmental land values such as local governments or non-profits. For risk-mitigation purposes, the fund has a ten-year life span and investors will receive up to 8 per cent returns.

Additional upfront financing for restoration may become available through innovative financial tools such as 'environmental impact bonds'. Like a 'social impact bond' used to reduce recidivism in jails, an 'environmental impact bond' would be a contract between private investors and the public sector where the public sector might agree to a particular payment in exchange for a specific environmental outcome to be measured over a period of time such as the restoration of a given ecosystem service such as pollination.[93] The advantage for governments utilising these tools is that the risk of the initial investment is carried entirely by the private investor. If a proposed restoration project financed through the environmental bond mechanism fails, then the private investor would recover no returns. If the project succeeds and delivers the desired results, the government will provide a requisite bond payment plus interest to the institutional investor. The bonds reward investors with known technical approaches who are financially sophisticated to understand the level of short-term restoration investment needed to return a long-term profit.

Role for a standardised harmonised code to guide restoration actors?

While restoration efforts will vary in terms of the objectives of a project proponent and the specific site where the restoration activities are undertaken, one initiative that deserves further exploration is the negotiation of a technical code that can be used for similar projects as a baseline document to define an appropriate standard of care that both private landowners and ecological restoration practitioners can rely on. While the existing SER Primer and IUCN guidelines provide a significant global framework for furthering restoration objectives in terms of articulating restoration as a socio-ecological effort, there may also be additional room for even greater technical detail that can explicitly guide the efforts of project proponents.

Laws and regulations may have additional roles to play in supporting coordination among volunteer restoration groups by, for example, setting at least minimal benchmarks for ecosystem restoration. While there are 'relatively few generally accepted measurements of integrity or conditions at the ecosystem or landscape level', there may be opportunities in the form of a legal instrument to establish certain practices for groups that otherwise do not have access to the advice and guidance of ecological professionals.[94] The legal instrument can be a soft law

---

93 J. Hartley, Social Impact Bonds are Going Mainstream (15 September 2014), <www.forbes.com/sites/jonhartley/2014/09/15/social-impact-bonds-are-going-ma instream/>.
94 J. Cairns Jr., 'Rationale for restoration' in M. Perrow and A. Davy (eds), *Handbook of Ecological Restoration* (Vol. 1, Cambridge University Press 2002) 19.

instrument such as a code which could assist individuals and communities in designing restoration projects that have a better chance of achieving goals such as returning certain historical ecosystem functions and services to an area and minimising risks such as introducing new invasive species that serve desirable ecosystem functions but outcompete other species.

A Restoration Code might operate like the existing Fire Prevention Code that incorporates specific minimum requirements for fire control and fire extinguishing systems and equipment.[95] Like the National Fire Prevention Code, a Restoration Code would include sub-standards and be regularly updated by a committee based in part on public input and comments.[96] If such a Code was to exist, public contracting might incorporate the compliance with the Code as a basic condition for satisfying the contract.

The need for a standardised Code becomes increasingly important as the ecological restoration economy grows. As of 2015, the US domestic ecological restoration sector is estimated to directly employ 126,000 workers, indirectly employ 95,000 other workers, generate $9.5 billion in direct economic sales and support an addition $15 billion in indirect economic output.[97] These estimates include larger employment numbers than the oil and gas industry when measured by number of individual employed by million dollars invested.[98]

Some private groups such as the Restore the Earth Foundation whose mission is to 'restore one million acres of degraded land in the Mississippi River Basin – North America's Amazon – to its natural state' are applying global standards to their operations.[99] Specifically, their restoration work is based on global metrics articulated in the UN Millennium Ecosystem Assessment, The Economies of Ecosystems and Biodiversity project, the Global Reporting Initiative and International Standards Organization 26000 standard.[100] This desire for standardisation reflects a need for continuity across project designs to provide some calibration for

---

95 National Fire Prevention Association, Fire Prevention Code, <www.nfpa.org/Assets/ files/AboutTheCodes/1/1_A2014_FDReport.pdf>.

96 Ibid. The National Fire Prevention Association provides numerous technical guidelines such as NFPA 664 'Standard for the Prevention of Fires and Explosions in Wood Processing and Woodworking Facilities' and NFPA 701 'Standard Methods of Fire Tests for Flame Propagation of Textiles and Films'. This type of technical guideline model might be imported into restoration efforts with 'standards' for improving water quality for certain types of wetlands or for decontaminating certain types of soils.

97 T. BenDor et al., 'Estimating the size and impact of the ecological restoration economy' (2015) 10 (6) *PLoS One* 1–15 <http://journals.plos.org/plosone/article?id=10. 1371/journal.pone.0128339>.

98 Ibid.

99 Restore the Earth Foundation, Mission and Vision, <www.restoretheearth.org/a bout/mission-and-vision>.

100 Restore the Earth Foundation, Global Standards, <www.restoretheearth.org/our-impa ct/global-standards> ('Restore the Earth Foundation's approach to ecosystem restoration and assessment is designed using these international standards. This ensures that corporate and governmental funders, NGO partners, and impact investors will experience reliable, comparable, and measurable returns on their restoration investments').

quality control including ongoing monitoring across projects. As explained in the conclusion of the book, there is the start of a movement under foot to articulate and disseminate shared standards for restoration work with efforts such as Australasia Society for Ecological Restoration's National Standards for the Practice of Ecological Restoration in Australia. To the extent that it is possible to create standards that can be applied by an array of social actors across a variety of ecosystems, law, whether in the form of a code, guidelines or standards, may guide future landscape level restoration efforts.

10 Protected areas and ecological restoration

Since protected areas are at the core of many international and national nature conservation laws, and many restoration activities are realised within protected areas, this chapter focuses in particular on protected areas[1] and their relationship to restoration. First, the concept of 'protected areas' will be defined, their significance explained and the need for restoration of protected areas will be demonstrated. Second, the chapter will concentrate on evaluating the restoration obligations, criteria or guidelines of a number of key international conventions involving protected areas including the Biodiversity Convention,[2] the Ramsar Convention,[3] the World Heritage Convention,[4] the UNESCO MAB Programme[5] and the European Bern Convention.[6] A third part will focus on the restoration obligations for protected areas under the EU Birds Directive[7] and the Habitats Directive[8] as this is a well-worked out regime, including some very interesting case law by the European Court of Justice. The last two sections will deal with restoration of connectivity between protected areas, and restoration as part of compensation regimes for loss of protected areas.

---

1 On terrestrial protected areas in general see: A. Cliquet and H. Schoukens, 'Terrestrial protected areas' in J. Razzaque and E. Morgera (eds), *Biodiversity and Nature Protection Law, Encyclopedia of Environmental Law* (Edward Elgar Publishing, in press).
2 Convention on Biological Diversity, 5 June 1992, <www.cbd.int>.
3 Convention on Wetlands of International Importance especially as Waterfowl Habitat, Iran, 2 February 1971, <www.ramsar.org>.
4 Convention Concerning the Protection of the World Cultural and Natural Heritage, Paris, 16 November 1972, <http://whc.unesco.org>.
5 UNESCO, Programme on Man and Biosphere – MAB, <www.unesco.org/new/en/na tural-sciences/environment/ecological-sciences/man-and-biosphere-programme/>.
6 Convention on the Conservation of European Wildlife and Natural Habitats, Bern, 19 September 1979, <www.coe.int/en/web/bern-convention/home>.
7 Directive 2009/147/EC of the European Parliament and of the Council of 30 November 2009 on the conservation of wild birds, *OJ L* 20, 26 January 2010, replacing the original Birds Directive, Directive 79/409/EEC of 2 April 1979 on the Conservation of Wild Birds, *OJ L* 103, 25 April 1979 (Birds Directive).
8 Directive 92/43/EEC of 21 May 1992 on the Conservation of Natural Habitats and of Wild Fauna and Flora, *OJ L* 206, 22 July 1992 (Habitats Directive).

Introduction: the role of protected areas in ecological restoration

The concept of a protected area

According to the Biodiversity Convention, or Convention on Biological Diversity (CBD), a 'Protected area' means 'a geographically defined area which is designated or regulated and managed to achieve specific conservation objectives'.[9] This is a very broad definition of protected areas,[10] and there can be enormous differences between protected areas depending on the geographical size of the area, the designation criteria and process, the goals of the protected area and, above all, the management strategy for the area that may or may not permit human activities within an area. The IUCN has designed a general classification scheme for protected areas, including six categories of protected areas: I. Strict Nature Reserve/Wilderness Area, II. National Park, III. Natural Monument, IV. Habitat/Species Management Area, V. Protected Landscape/Seascape and VI. Managed Resource Protected Area.[11] Although this classification scheme has not been adopted in international or national law,[12] in all of these six categories, ecological restoration might be appropriate when the values, for which the areas are established, have been impaired. The degree and type of management intervention and restoration will depend on the management goals of the particular area.[13]

Importance of protected areas

Protected areas provide valuable and numerous benefits that extend spatially far beyond their boundaries, including cultural, ecological, spiritual and scientific benefits.[14] These areas are critical to preserving global biodiversity and stemming the extinction crisis.[15] According to the Biodiversity Synthesis report from the Millennium Ecosystem Assessment,[16] protected areas are extremely important

9 Article 2, Biodiversity Convention.
10 See on this issue A. Gillespie, *Protected Areas and International Environmental Law* (Martinus Nijhoff Publishers 2007) 27–46.
11 See <www.iucn.org/about/work/programmes/gpap_home/gpap_quality/gpap_pa categories/>.
12 Gillespie, *Protected Areas and International Environmental Law* 30–31.
13 K. Keenleyside, N. Dudley, S. Cairns, C.M. Hall and S. Stolton, *Ecological Restoration for Protected Areas: Principles, Guidelines and Best Practices* (IUCN 2012) 9.
14 K.J. Mulongoy and S.B. Gidda, *The Value of Nature: Ecological, Economic, Cultural and Social Benefits of Protected Areas* (Secretariat of the Convention on Biological Diversity 2008) 6–7.
15 N. Lopoukhine, 'Protected areas – For Life's Sake' in Secretariat of the Convention on Biological Diversity, *Protected Areas in Today's World: Their Values and Benefits for the Welfare of the Planet* (CBD Technical Series no. 36, 2008) 2; on the sixth extinction crisis, see G. Ceballos, P.R. Ehrlich, A.D. Barnosky, A. García, R.M. Pringle and T.M. Palmer, 'Accelerated modern human-induced species losses: entering the sixth mass extinction' (2015) 1(5) *Science Advances* 1–5, DOI: 10.1126/sciadv.1400253.
16 Millennium Ecosystem Assessment, *Ecosystems and Human Well-being: Biodiversity Synthesis* (World Resources Institute 2005) 10; for the full report, see K. Chopra et al.

especially in environments where biodiversity loss is sensitive to changes in key drivers. Protected area systems are most successful if they are designed and managed in the context of an ecosystem approach, with due regard to the importance of corridors and interconnectivity of protected areas and to external threats such as pollution, climate change and invasive species.

In recent years specific attention has been given to the economic value of protected areas. The most noticeable initiative in this regard is The Economics of Ecosystems and Biodiversity (TEEB) study.[17] It demonstrates that there are also socio-economic benefits to manage protected areas properly at both the global and local level. When the full range of ecosystem services is taken into account the benefits of protected areas often exceed the costs for protected areas.[18]

A report from the European Environmental Agency confirmed these global findings and stressed the importance of protected areas in Europe. The report refers to research on the costs and benefits of the ecological network of the European Union, the Natura 2000 network. According to the research the annual costs of implementing the Natura 2000 network were approximately € 5.8 billion for the (then) EU-27. However, based on a number of case studies, the benefits of the network can be between three and seven times the costs based on the delivery of a wide range of ecosystem services.[19] According to a report prepared for the European Commission, the value of the benefits of the (terrestrial) Natura 2000 network range between €200 and €300 billion per year at present (or 2–3 per cent of EU GDP).[20] Some of the beneficial ecosystem services that were identified include carbon storage, protection against natural hazards, tourism, water purification and food security.

Number and status of protected areas

Since 1962, the UN has made up a list of protected areas worldwide, which is updated approximately every ten years. The UN list is based on the information

(eds), *Ecosystems and Human Well-Being: Policy Responses: Findings of the Responses Working Group of the Millennium Ecosystem Assessment* (Island Press 2005) 125–131; about the study and the reports, see <www.maweb.org/>.

17 On the study and the reports, see <www.teebweb.org/>.

18 TEEB, *The Economic of Ecosystems and Biodiversity for National and International Policy Makers,Summary: Responding to the Value of Nature* (2009) 20–21; TEEB, *The Economics of Ecosystems and Biodiversity in National and International Policy Making*, Edited by P. ten Brink (Earthscan 2011) Chapter 8: 'Recognizing the value of protected areas'.

19 European Environment Agency, *Protected Areas in Europe – an Overview* (EEA 2012); see for the original report S. Gantioler, M. Rayment, S. Bassi, M. Kettunen, A. McConville, R. Landgrebe, H. Gerdes and P. ten Brink, *Costs and Socio-Economic Benefits Associated with the Natura 2000 Network* (Final report to the European Commission, DG Environment on Contract ENV.B.2/SER/2008/0038, Institute for European Environmental Policy/GHK/Ecologic 2010).

20 P. ten Brink, *The Economic Benefits of the Natura 2000 Network* (Final Synthesis Report to the European Commission, DG Environment 2011).

available in the World Database on Protected Areas.[21] The latest version of the list dates from 2014 and contains 209,429 protected areas covering a total area of 32,868,673 km^2.[22] This translates into 14 per cent of the world's terrestrial areas currently protected. If Antarctica is excluded from the global statistics coverage, the percentage of the total terrestrial area protected is 15.4 per cent (20.6 million km^2). In the marine environment, however, only 3.41 per cent of the total marine area is protected.[23] The coverage of marine protected areas within areas of national jurisdiction is 8.4 per cent of all marine areas within national jurisdiction; but only 0.25 per cent of marine areas beyond national jurisdiction are within protected areas.[24] These global figures also include the internationally designated protected areas, such as the World Heritage sites, Ramsar sites and the Natura 2000 sites, although there is a considerable overlap with nationally designated sites.[25]

The global number and extent of protected areas has increased substantially over the past century. However, the number of protected areas tells us nothing about the level of protection, their conservation status, or their need for restoration. Protected areas should be ecologically representative, well managed and well connected. In spite of their importance for biodiversity conservation, the status of protected areas is not overall positive and shortcomings exist on coverage, representativeness, connectivity and management.

The current system of protected areas is not sufficient to conserve biodiversity, due to insufficient coverage and management.[26] A CBD Conference of the Parties (COP) decision has noted that existing systems of protected areas are neither representative of the world's ecosystems, nor do they adequately address conservation of critical habitat types, biomes and threatened species.[27] According to the Global Biodiversity Outlook 4, even though the protected area network is becoming more representative of the world's diverse ecological regions, many regions have only a small portion of their area protected.[28] Furthermore, today's

21 <www.protectedplanet.net/>.
22 M. Deguignet, D. Juffe-Bignoli, J. Harrison, B. Macsharry, N. Burgess and N. Kingston, *2014 United Nations List of Protected Areas* (UNEP-WCMC 2014).
23 Ibid., 12.
24 D. Juffe-Bignoli, N.D. Burgess, H. Bingham, E.M.S. Belle, M.G. de Lima, M. Deguignet, B. Bertzky, A.N. Milam, J. Martinez-Lopez, E. Lewis, A. Eassom, S. Wicander, J. Geldmann, B. van Soesbergen, A.P. Arnell, B. O'Connor, S. Park, Y.N. Shi, F.S. Danks, B. MacSharry and N. Kingston, *Protected Planet Report 2014* (UNEP-WCMC 2014) 7.
25 B. Bertzy et al., *Protected Planet Report 2012: Tracking Progress towards Global Targets for Protected Areas* (IUCN and UNEP-WCMC 2012) 5.
26 Chopra et al. (eds), *Ecosystems and Human Well-Being*, 125.
27 Decision VII/28. Protected areas, UNEP/CBD/COP/DEC/VII/28 (2004) para. 16 and annex I, para. 2.
28 Around one quarter of terrestrial regions and more than half of marine regions have less than 5 per cent of their area protected.

protected areas will not be adequate to conserve many species whose distributions will shift in the future due to climate change.[29]

According to the Millennium Ecosystem Assessment (MEA) Biodiversity Synthesis report, protected areas need to be better located, designed and managed to deal with problems like lack of representativeness, impacts of human settlement within protected areas, illegal harvesting of plants and animals, unsustainable tourism, impacts of invasive species and vulnerability to global change.[30] In many instances, protected areas are just small islands of protection amidst zones of heavily degraded nature.[31] The lack of proper connectivity between the existing protected sites leads to further deterioration of the status of the remaining habitats and species. Moreover, the majority of the protected areas are located in so-called residual areas of the world, where the least risk exists of interference with extractive activities such as agriculture, mining or forestry.[32]

The benefits to biodiversity from protected areas depend critically on how well these areas are managed. Indeed, many studies show that protected areas are, given the greater richness and/or abundance of species they mostly harbour, the cornerstone to preventing species extinction and habitat degradation.[33] However, in the absence of adequate management, protected areas merely function as 'paper parks', that is protected sites without any physical presence – which are by nature unable to stem the ongoing biodiversity decline.[34] It is therefore important to assess the management effectiveness. A management effectiveness evaluation is the assessment of how well the protected area is being managed – primarily the extent to which it is protecting values and achieving goals and objectives.[35] An analysis of protected area management effectiveness at a global level showed that only 22 per cent of sites were soundly managed, 37 per cent had basic management in place, 28 per cent had major deficiencies and 13 per cent were deficient in terms of management.[36]

29 Secretariat of the Convention on Biological Diversity, *Global Biodiversity Outlook 4* (2014) 83.

30 Millennium Ecosystem Assessment, *Ecosystems and Human Well-being: Biodiversity Synthesis* (World Resources Institute 2005) 10–11.

31 E. Rees, 'Do protected areas for wildlife really work?' (2012) *The Ecologist*, <www.theecologist.org/News/news_analysis/1304082/do_protected_areas_for_wildlife_really_work.html>.

32 B. Pressey and E. Ritchie, 'We have more parks than ever, so why is wildlife still vanishing?' (2014) *The Conversation*, <http://theconversation.com/we-have-more-parks-than-ever-so-why-is-wildlife-still-vanishing-34047>.

33 S.E. Lester, B.S. Halpern, K. Grorud-Colvert, J. Lubchenko, B.I. Ruttenberg, S.D. Gaines, S. Airamé and R.R. Warner, 'Biological effects within no-take marine reserves: a global synthesis' (2009) 384 *Marine Ecology Progress Series* 33–46.

34 E. Di Minin and T. Toivonen, 'Global protected area expansion: creating more than paper parks' (2015) 65(7) *BioScience* 637–638.

35 M. Hockings, S. Stolton, F. Leverington, N. Dudley and J. Courrau, *Evaluating Effectiveness: A Framework for Assessing Management Effectiveness of Protected Areas* (2nd edn, IUCN 2006) xiii.

36 F. Leverington, K. Lemos Costa, J. Courrau, H. Pavese, C. Nolte, M. Marr, L. Coad, N. Burgess, B. Bomhard and M. Hockings, *Management Effectiveness Evaluation in Protected Areas – A Global Study* (2nd edn, The University of Queensland 2010) 692.

In 2004, Goal 4.2 of the CBD Programme of Work on Protected Areas specifically called for countries to assess management effectiveness of at least 30 per cent of each party's protected areas by 2010.[37] In 2010, the Conference of the Parties agreed to:

> Continue to expand and institutionalize management effectiveness assessments to work towards assessing 60 per cent of the total area of protected areas by 2015 using various national and regional tools and report the results into the global database on management effectiveness maintained by the World Conservation Monitoring Centre of the United Nations Environment Programme (UNEP WCMC).[38]

By 2013, only 45 countries had assessed the management effectiveness for 60 per cent or more of the total area of their marine and terrestrial protected areas. Where the quality of management has been assessed, the majority of protected areas had either only basic management or major deficiencies; only 24 per cent of the areas had sound management.[39]

Studies also point to biodiversity loss within protected areas, in spite of their legally protected status. Reasons for degradation include climate change, invasive species, wider landscape changes, as well as illegal encroachment of people into the protected area, poaching and weak management.[40] Government budget constraints, conflicts with human development and a growing human population further undermine the effectiveness of protected areas.[41]

In the EU, Member States have monitoring and reporting obligations, both under the Birds and the Habitats Directives.[42] This allows for a systematic assessment of the conservation status of EU-protected species and habitats. While there have been improvements in the knowledge and evidence base, even in the EU the number of unknown assessments for species and habitats is still substantial.[43] The 2015 'State of Nature' assessment report from the European Environment Agency (EEA)[44] shows that only 16 per cent of the assessments of

---

37 Decision VII/28. Protected areas (Articles 8(a) to (e)), UNEP/CBD/COP/DEC/VII/28 (2004).
38 Decision X/31. Protected areas, UNEP/CBD/COP/DEC/X/31 (2010) para. 19, a.
39 Juffe-Bignoli et al., *Protected Planet Report 2014* 30.
40 Keenleyside et al., *Ecological Restoration for Protected Areas* 11.
41 C. Mora and P.F. Sale, 'Ongoing global biodiversity loss and the need to move beyond protected areas: a review of the technical and practical shortcomings of protected areas on land and sea' (2011) 434 *Marine Ecology Progress Series* 251–266.
42 Articles 10 and 12, Birds Directive and Articles 11 and 17, Habitats Directive.
43 European Environment Agency, *State of Nature in the EU: Results from Reporting under the Nature Directives 2007–2012* (EEA Technical report No 2/2015, European Environment Agency 2015) 42 and 48.
44 European Environment Agency, *State of Nature in the EU*.

Annex I habitats of the Habitats Directive[45] are in a favourable conservation status. Of the EU protected species (other than birds) only 23 per cent assessments are favourable.[46] The report also specifically evaluates the contribution of Natura 2000 in achieving a favourable conservation status for EU protected species and habitats. The EEA report assesses the coverage of the Natura 2000 network, as well as the management of the network. The report recognises that measuring the ecological effectiveness of a network of protected areas is difficult due to the lack of baseline data and difficulty in assigning any measured change to a particular designation or other conservation measure.[47]

Even so, research indicates that the Birds Directive has had a positive effect on protected bird species. Species listed in Annex I of the Birds Directive, for which Member States must implement special conservation measures, including the designation and management of Special Protection Areas, show more positive population trends than species not mentioned in Annex I.[48] For the habitats and species protected under the Habitats Directive it is not so easy to give an overall assessment on the management effectiveness. The State of Nature report states that the impact of Natura 2000 in maintaining or restoring Annex I habitats at favourable conservation status is not clear, with contradictory studies existing, showing the improvement of the conservation status in some areas, but not in other cases. The report acknowledges that restoring habitats often takes many years, especially for habitats such as forests. In some cases, protected areas, although effective in preventing change of land use, cannot address pressures such as climate change on habitats. Even when a functioning habitat has been restored, it may not have the same species composition as undisturbed habitats.[49] In order to be fully effective as a coherent European ecological network, sites must be managed appropriately, and there is evidence to suggest that this is not always the case for a variety of reasons including lack of data, inadequate resources, conflicts between conservation and economic interests and lack of appropriate management.[50]

45 Annex I habitats are natural habitat types of community interest whose conservation requires the designation of special areas of conservation.
46 See Chapter 7.
47 European Environment Agency, *State of Nature in the EU*, 137.
48 P.F. Donald, F.J. Sanderson, I.J. Burfield, S.M. Bierman, R.D. Gregory and Z. Waliczky, 'International conservation policy delivers benefits for birds in Europe' (2007) 317(5839) *Science* 810–813; see also F. Sanderson, R. Pople, C. Ieronymidou, I. Burfield, R. Gregory, S. Willis, C. Howard, P. Stephens, A. Beresford and P. Donald, 'Assessing the performance of EU nature legislation in protecting target bird species in an era of climate change' (2015) *Conservation Letters* DOI: 10.1111/conl.12196.
49 European Environment Agency, *State of Nature in the EU* 138.
50 Ibid., 141.

The need for restoration in protected areas

As a result of the unfavourable conservation status in many protected areas, restoration efforts within protected areas are necessary. We have already seen an evolution from 'hands-off' or laissez-faire management in protected areas to more active management and restoration. The combination of previous degradation with continuing external pressures and climate change will make restoration in protected areas even more necessary.[51] Because of threats such as climate change, restoration is becoming increasingly important both inside and outside protected areas. Restoration also needs to address the challenge of maintaining connectivity in the wider landscape.[52]

Restoration of protected areas is fundamental to addressing several goals that are important for biodiversity conservation and human well-being. Restoration in protected areas can help to maintain and recover threatened species, maintain valuable ecosystems, and regain lost or degraded ecosystems.[53] Furthermore, protected areas fulfil a crucial role in both mitigation and adaptation to climate change.[54]

The IUCN has drafted guidelines and principles for restoration in protected areas. There are three fundamental principles to which ecological restoration in protected areas should adhere: restoration should be effective, efficient and engaging. Effective restoration is restoration that re-establishes and maintains protected area values. Efficient restoration maximises beneficial outcomes while minimising costs in time, resources and effort. Engaging restoration collaborates with partners and stakeholders, promotes participation and enhances visitor experience.[55] These guidelines are valuable for practitioners and governments who are engaged in ecological restoration, even though they are non-binding. When it comes to obliging governments or landowners to restore protected areas, we will have to look for legal obligations. Hereafter, we will first discuss the legal obligations in several conventions for restoration in protected areas. Second, we will look at the EU legislation and related case law.

Restoration in protected areas and international law

Biodiversity Convention

Chapter 6 previously discussed obligations and policy targets with regards to restoration under the Biodiversity Convention. This chapter will specifically look into the obligations of the Biodiversity Convention with regards to restoration within protected areas.

---

51 See Keenleyside et al., *Ecological Restoration for Protected Areas* 4–5.
52 IUCN WCPA, *Next Steps: Convention on Biological Diversity Programme of Work on Protected Areas* (2010) 12.
53 See Keenleyside et al., *Ecological Restoration for Protected Areas* 11–12.
54 See Chapter 11.
55 See Keenleyside et al., *Ecological Restoration for Protected Areas* 15–22.

General obligations for restoration and protected areas

Under the Biodiversity Convention State Parties have the obligation to establish a system of protected areas or areas where special measures need to be taken to conserve biological diversity.[56] Although this obligation does not explicitly contain the obligation to restore protected areas, this can be deduced from the definition of conservation. 'In-situ conservation' means 'the conservation of ecosystems and natural habitats and the maintenance and recovery of viable populations of species in their natural surroundings'.[57] The Convention text also includes an explicit obligation to restore degraded ecosystems,[58] which covers both protected and non-protected areas.

Although an obligation for restoration is present in the Convention itself, it is only recently that specific attention has been paid to restoration by State Parties. In a COP Decision of 2004 on protected areas the parties underlined the importance of conservation of biological diversity not only within but also outside protected areas by promoting the sustainable use of natural resources to achieve a significant reduction of the rate of biodiversity loss by 2010. Parties called for 'increased efforts to integrate biodiversity conservation and restoration aspects into sectoral policies and programmes'.[59] The Programme of Work on Protected Areas which was part of the same decision[60] also addresses restoration. In its first paragraph it states that 'protected areas, together with conservation, sustainable use and restoration initiatives in the wider land- and seascape are essential components in national and global biodiversity conservation strategies'.[61] Under Goal 1.2 of the Programme of Work[62] a suggested activity for State Parties is to 'rehabilitate and restore habitats and degraded ecosystems, as appropriate, as a contribution to building ecological networks, ecological corridors and/or buffer zones'.[63] Under Goal 1.5 of the Programme of Work on Protected Areas[64] parties are encouraged to 'establish and implement measures for the rehabilitation and restoration of the ecological integrity of protected areas'.[65]

In a 2010 COP Decision on protected areas, restoration of ecosystems and habitats is mentioned as an issue that needs greater attention. State Parties are

56 Article 8(a), Biodiversity Convention.
57 Article 2, Biodiversity Convention.
58 Article 8(f): States shall as appropriate 'Rehabilitate and restore degraded ecosystems and promote the recovery of threatened species, inter alia, through the development and implementation of plans or other management strategies'.
59 Decision VII/28. Protected areas (Articles 8(a) to (e)), UNEP/CBD/COP/DEC/VII/28 (2004) para. 12.
60 Decision VII/28. Protected areas (Articles 8(a) to (e)), UNEP/CBD/COP/DEC/VII/28 (2004).
61 Ibid., Annex (Programme of work), para. 1.
62 Goal 1.2: To integrate protected areas into broader land- and seascapes and sectors so as to maintain ecological structure and function.
63 Programme of Work on Protected Areas, 1.2.5.
64 Goal 1.5: To prevent and mitigate the negative impacts of key threats to protected areas.
65 Programme of Work on Protected Areas, 1.5.3.

urged to increase the effectiveness of protected area systems in biodiversity con-servation and enhance their resilience to climate change and other stressors, 'through increased efforts in restoration of ecosystems and habitats and including, as appropriate, connectivity tools such as ecological corridors and/or conservation measures in and between protected areas and adjacent landscapes and seascapes'.[66] State Parties are also urged 'to include restoration activities in the action plans of the programme of work on protected areas and national biodiversity strategies'.[67]

Specifically with regards to climate change, the State Parties were invited to achieve Goal 1.2 of the Programme of Work on protected areas by 2015, through concerted efforts to integrate protected areas into wider landscapes and seascapes and sectors, 'including through the use of connectivity measures such as the development of ecological networks and ecological corridors, and the restoration of degraded habitats and landscapes in order to address climate change impacts and increase resilience to climate change'.[68]

Targets for restoration under the Convention

Most concrete work on pursuing restoration in protected areas has been done since 2010, namely with the Aichi Targets,[69] as well as subsequent work that was done by the Conference of the Parties, the Subsidiary Body on Scientific, Technical and Technological Advice (SBSTTA) and the Executive Secretary.

TARGET 11 ON PROTECTED AREAS

For restoration of protected areas, states agreed to Target 11:

> By 2020, at least 17 per cent of terrestrial and inland water areas, and 10 per cent of coastal and marine areas, especially areas of particular importance for biodiversity and ecosystem services, are conserved through effectively and equitably managed, ecologically representative and well connected systems of protected areas and other effective area-based conservation measures, and integrated into the wider landscapes and seascapes.

Although the target does not explicitly mention restoration of protected areas, the words 'conservation' and 'effectively managed' encompass restoration activities, particularly if a protected area is in an unfavourable conservation status. 'Effec-tively managed' is the degree to which protected area management protects bio-logical and cultural resources and achieves the goals and objectives for which the protected area was established.[70] Effective management includes planning

---

66 Decision X/31. Protected areas, UNEP/CBD/COP/DEC/X/31 (2010) para. 26, a.
67 Ibid., para. 26, b.
68 Ibid., para. 14.
69 See Chapter 6.
70 Explanatory Guide on Target 11 of the Strategic Plan for Biodiversity, 16, <www.cbd. int/protected/tools/default.shtml>.

measures to ensure ecological integrity and the protection of species, habitats and ecosystem processes.[71]

The use of the terminology 'well-connected systems of protected areas' in Target 11 is also relevant to restoration as a reference to ecological networks. According to the CBD Secretariat's guide on implementing the Aichi Targets, creating or restoring functional linkages between protected areas and their surrounding regions is essential to strengthen ecological coherence and resilience for both biodiversity conservation and sustainable development.[72] In order to implement the Programme of Work on Protected Areas, states must have an action plan to improve its protected area network including both restoring degraded protected areas and establishing ecological corridors.[73]

While it is clear from Target 11 that government-facilitated restoration may be required, the target lacks concrete guidelines about what constitutes achievement of the target. For example, the target does not indicate to what level protected areas should be restored and where restoration should be done.

TARGETS 14 AND 15 ON RESTORATION

Although not specific for protected areas, Aichi Targets 14 and 15 can be relevant for restoration of protected areas. Target 14 aims to restore and safeguard by 2020, ecosystems that provide essential services, including services related to water, and which contribute to health, livelihoods and well-being. Target 15 is the 'restoration target':

> By 2020, ecosystem resilience and the contribution of biodiversity to carbon stocks has been enhanced, through conservation and restoration, including restoration of at least 15 per cent of degraded ecosystems, thereby contributing to climate change mitigation and adaptation and to combating desertification.

Even though Target 15 is certainly not limited to protected areas, the danger exists that governments will tend to undertake restoration measures within already protected areas, in order to reach the 15 per cent target. However, very often, there are pre-existing obligations for restoration within protected areas either based on international and/or national laws. Arguably, government-facilitated restoration efforts to achieve Target 15 should not replicate efforts under other laws and targets such as Aichi Target 11 but instead focus national attention on areas outside of protected areas in order to strengthen existing protected areas by providing needed buffer zones and connectivity between protected areas and

71 CBD, Quick guide to target 11 of the Aichi Biodiversity Targets (2012), <www.cbd. int/doc/strategic-plan/targets/T11-quick-guide-en.pdf>.
72 Explanatory Guide on Target 11 of the Strategic Plan for Biodiversity, 16.
73 Ibid., 23.

restoring ecological functioning of the wider landscape. Major improvements and efforts are needed to restore and manage the ecosystems that comprise the 80–90 per cent of the planet that is outside protected areas,[74] in order to achieve Targets 14 and 15.

Still, there remain uncertainties with regards to implementing Targets 14 and 15. As with Target 11, there are no concrete guidelines regarding which ecosystems need be restored or to what quality standards ecosystems should be restored. States understood this uncertainty because at the Conference of the Parties in 2010, when the Aichi Targets were decided, states indicated that at the next Conference of the Parties in 2012, states would need to identify the ways and means to support ecosystem restoration.[75] In practice, this meant that even though states had set for themselves targets, states did not have a working understanding of what it might take to implement those targets. According to one researcher, Target 15 was written in such a way that the goal cannot be reached because progress towards the goal cannot be measured.[76] Since the setting of the targets in 2010, further work on restoration has been undertaken in the framework of the SBSTTA[77] and the COPs, including information documents and notes by the Executive Secretary[78] and COP decisions.[79] In the COP XI Decision on ecosystem restoration of 2012 no distinction is made between prioritising restoration inside protected areas versus outside protected areas. The decision urges parties to make concerted efforts to achieve the Aichi Targets 14 and 15, and contribute to the achievement of all other Aichi Biodiversity Targets through ecosystem restoration. This includes identifying degraded ecosystems that have the

---

74 SBSTTA, Ways and means to support ecosystem restoration. Note by the Executive Secretary, UNEP/CBD/SBSTTA/15/4 (2011) 15.
75 Decision X/9. The multi-year programme of work for the Conference of the Parties for the period 2011–2020 and periodicity of meetings, UNEP/CBD/COP/DEC/X/9 (2010) (a) ix.
76 D. Jørgensen, 'Ecological restoration in the Convention on Biological Diversity targets' (2013) 22 *Biodiversity and Conservation* 2977–2982; see Chapter 6.
77 SBSTTA, Ways and means to support ecosystem restoration. Note by the Executive Secretary, UNEP/CBD/SBSTTA/15/4 (2011); SBSTTA, Recommendation XV/2. Ways and means to support ecosystem restoration, UNEP/CBD/SBSTTA/REC/XV/2 (2011); SBSTTA, Progress report on ecosystem restoration and related Aichi targets. Note by the Executive Secretary, UNEP/CBD/SBSTTA/17/7 (2013); SBSTTA, Report on issues in progress: ecosystem conservation and restoration. Note by the Executive Secretary, UNEP/CBD/SBSTTA/18/14 (2014).
78 Ecosystem restoration. Note by the Executive Secretary, UNEP/CBD/COP/11/21 (2012); Available guidance and guidelines on ecosystem restoration. Note by the Executive Secretary, UNEP/CBD/COP/11/INF/17 (2012); Available tools and technologies on ecosystem restoration. Note by the Executive Secretary, UNEP/CBD/COP/11/INF/18 (2012); Most used definitions/descriptions of key terms related to ecosystem restoration. Note by the Executive Secretary, UNEP/CBD/COP/11/INF/19 (2012); Ecosystem Conservation and restoration. Note by the Executive Secretary, UNEP/CBD/COP/12/22 (2014).
79 Decision XI/16. Ecosystem restoration, UNEP/CBD/DEC/XI/16 (2012).

potential for ecosystem restoration bearing in mind that such areas may be occupied by indigenous and local communities.[80]

Progress on Aichi Targets

Because of a lack of clear definitions, it might be difficult to measure the progress on the restoration targets. In the Global Biodiversity Outlook (GBO) 4 report,[81] the authors report on the progress towards achieving the Aichi Targets. For Target 11 on protected areas, curiously nothing specific is mentioned about restoration in protected areas. With regards to the quantitative target on reaching 17 per cent (for terrestrial) and 10 per cent (for marine) ecosystems designated as protected areas, the report mentions that 'the terrestrial area of the planet protected for biodiversity is increasing steadily, and designation of marine protected areas is accelerating. At the current rate of growth, the percentage targets would be met for terrestrial areas by 2020 and this is reinforced by existing commitments to designate additional terrestrial protected areas. Overall, the extrapolations suggest that the marine target is not on course to be met. However, progress is higher in coastal areas, while open ocean and deep sea areas, including the high seas, are much less covered.'[82]

The report also notes that the designation of new protected areas may not always be adequate to conserve many species whose distributions will shift in the future due to climate change.[83]

As for the quality of the management of protected areas where restoration efforts are being implemented, progress is much harder to establish. According to the GBO 4 report only a minority of protected areas currently enjoy sound management, although management appears to be improving based on the limited information available. National actions reported to the CBD indicate that most countries have targets relating to improvement of protected area coverage. Only a few states have set for themselves quantitative targets and even fewer states focus on ecological representativeness, connectedness or management effectiveness.[84]

On achieving restoration under Target 15, the GBO 4 states that restoration is under way for some depleted or degraded ecosystems, especially wetlands and forests, and is sometimes undertaken on an ambitious scale, as in China. Many countries, organisations and companies have pledged to restore large areas

---

80 Ibid., 1 (c).
81 Secretariat of the Convention on Biological Diversity, *Global Biodiversity Outlook 4* (2014).
82 Ibid., 83; SBSTTA, Draft executive summary with the main messages of the fourth edition of the Global Biodiversity Outlook. Note by the Executive Secretary, UNEP/CBD/SBSTTA/18/2 (2014) para. 33.
83 Secretariat of the Convention on Biological Diversity, *Global Biodiversity Outlook 4* (2014) 83.
84 Ibid.

Abandonment of farmland in some regions including Europe, North America and East Asia is enabling 'passive restoration' on a significant scale. In spite of these positive developments, there is still a net loss of forests, suggesting no overall progress on this component of the target. The GBO 4 concluded that while the combined initiatives currently under way or planned may be on track to restore 15 per cent of degraded ecosystems by 2020, it will be hard to assess.[85] The evaluation of GBO 4 was confirmed and strengthened in an analysis by researchers that determined that the progress on Target 15 cannot be measured as there are no indicators available for extrapolation.[86]

In a document prepared by the Secretariat for the 12th COP,[87] as well as the GBO 4, some key actions are proposed to enhance the implementation of the Aichi Targets. States are encouraged with regards to Target 11 to consider expanding protected area networks and other effective area-based conservation measures in order to be more representative of the planet's ecological regions, especially for marine and coastal areas (including deep-sea and ocean habitats), inland waters and areas of particular importance for biodiversity. States should also consider improving and regularly assessing management effectiveness of protected areas and other area-based conservation measures.[88]

The key actions to achieve Target 15 on restoration include that governments need to identify opportunities and priorities for restoration. These priority areas should include highly degraded ecosystems, areas of particular importance for ecosystem services and ecological connectivity, and areas undergoing abandonment of agricultural or other human-dominated use. Strategies to restore at least 15 per cent or more of degraded areas should be expanded and further developed, including through environmental permitting procedures and market instruments such as wetland mitigation banking, payments for ecosystem services and other mechanisms. The contribution of biodiversity to carbon sequestration through state or private sponsored passive and active afforestation programmes, such as the REDD+ mechanism must be increased.[89] Where feasible, restoration should be made an economically viable activity, by coupling income generation to restoration

---

85 Ibid., 101; SBSTTA, Draft executive summary with the main messages of the fourth edition of the Global Biodiversity Outlook. Note by the Executive Secretary, UNEP/CBD/SBSTTA/18/2 (2014) para. 39; Indicators for assessing progress in the implementation of the strategic plan for biodiversity 2011–2020 and draft terms of reference for a meeting with the ad hoc technical expert group. Note by the Executive Secretary, UNEP/CBD/COP/12/9/Add.2 (2014).

86 D. Tittensor et al., 'A mid-term analysis of progress toward international biodiversity targets' (2014) 346(6206) *Science* 241–244.

87 Key actions to enhance implementation of the strategic plan for biodiversity 2011–2020. Note by the Executive Secretary, UNEP/CBD/COP/12/9/Add.1 (2014); Secretariat of the Convention on Biological Diversity, *Global Biodiversity Outlook 4* (2014) 85.

88 Ibid.

89 On REDD+, see Chapter 11.

activities.[90] The proposed key actions do not distinguish between protected areas and other areas, but they can contribute to restoration of protected areas, inter alia through restoring connectivity between core protected areas and creating new protected areas on abandoned lands.

In preparation for the 13th meeting of the Conference of the Parties in December 2016, the Subsidiary Body on Scientific, Technical and Technological Advice prepared a short-term action plan on ecosystem restoration.[91] One of the guiding principles in the document is that ecosystem restoration is a complement to conservation and can greatly enhance the value of protected areas.[92] Under the plan, State Parties are expected to identify and prioritise areas where restoration would contribute most significantly to achieving those national level targets that contribute to the Aichi Biodiversity Targets. Priority restoration areas might include key biodiversity areas, areas that provide key ecosystem services, areas that would enhance the integrity of protected areas, and areas that would enhance the integration of protected areas into wider landscapes and seascapes.[93] In the Appendix on guidance for integrating biodiversity considerations into ecosystem restoration, the following guidance is relevant for protected areas and restoration:

> Site-based actions should be taken in the context of integrated land- and seascape management practices. For example: priority can be given to restoring ecosystem services within a mosaic of land uses; or promoting ecological connectivity and biodiversity conservation through ecosystem restoration in proximity to species refugia (e.g. protected areas, key biodiversity areas, important bird and biodiversity areas, and Alliance for Zero Extinction sites) creating buffer zones, or connectivity corridors between them.[94]

This guidance confirms that restoration under Target 15 is seen as a way of strengthening the protected areas, rather than duplicating the restoration obligations that already exist in protected areas. Restoration obligations within protected areas already exist in several international conventions that establish international designated sites. These will be discussed in the subsequent sections.

---

90 Secretariat of the Convention on Biological Diversity, *Global Biodiversity Outlook 4* (2014) 102; see also: SBSTTA, Draft executive summary with the main messages of the fourth edition of the Global Biodiversity Outlook. Note by the Executive Secretary, UNEP/CBD/SBSTTA/18/2 (2014) paras 41–42; Key actions to enhance implementation of the strategic plan for biodiversity 2011–2020. Note by the Executive Secretary, UNEP/CBD/COP/12/9/Add.1 (2014).
91 SBSTTA, Protected areas and ecosystem restoration. Note by the Executive Secretary, UNEP/CBD/SBSTTA/20/12 (2016) Annex.
92 Ibid., Annex, para. 8.
93 Ibid., Annex, para. 13(2).
94 Ibid., Appendix.

Ramsar Convention

The Ramsar Convention, which was created in 1971, does not include an explicit reference to restoration in the Convention text. However, the need for restoration has been addressed under the framework of the Convention.[95] Since Ramsar sites are protected sites, this chapter will review the specific restoration obligations for designated Ramsar sites. Within the strategic plans of the Convention, wetland restoration has been recognised as one of its strategic objectives, especially restoration of Ramsar-designated sites.[96] In the latest strategic plan for 2016 to 2024, states are expected to restore or maintain the ecological character of Ramsar sites through effective planning and integrated management.[97]

Designation of Ramsar sites

State Parties to the Ramsar Convention are required to designate suitable wetlands within their territory for inclusion in a List of Wetlands of International Importance (Ramsar List). The State Parties have to designate at least one area. The criteria for inclusion are included in the Convention[98] and are further worked out in guidelines established by the COP.[99] The criteria require sites to contain representative, rare or unique wetland types or are based on species and ecological communities. Already in the designation of Ramsar sites, the possibility of restoration can play a role. According to the Ramsar Handbook on designation of sites, wetlands being considered for designation need not be pristine areas which have not been subjected to impacts from human activities. In fact, Ramsar designation can be used to confer a special type of recognition on these areas by virtue of elevating them to the status of sites recognised as internationally important. In this way, 'Ramsar designation could represent the starting point for a process of recovery and rehabilitation of a particular site', provided the site meets the criteria for listing under the Convention when nominated.[100]

---

95 See also Chapter 5.
96 Strategic Plan 1997–2002, Resolution VI.14. The Ramsar 25th Anniversary Statement, the Strategic Plan 1997–2002, and the Bureau Work Programme 1997–1999 (1996), Operational Objective 2.6. To identify wetlands in need of restoration and rehabilitation, and to implement the necessary measures; Strategic Plan 2003–2008, Resolution VIII.25. The Ramsar Strategic Plan 2003–2008 (2002), Operational objective 4. Restoration and rehabilitation; Strategic Plan 2009–2015, Resolution X.1. The Ramsar Strategic Plan 2009–2015 (2008) and adjusted for the 2013–2015 triennium by Resolution XI.3 (2012), Strategy 1.8. Wetland restoration.
97 Resolution XII.2. The 4th Strategic plan 2016–2024 (2015), Target 5.
98 Article 2(2), Ramsar Convention.
99 <www.ramsar.org/sites/default/files/documents/library/ramsarsites_criteria_eng.pdf>.
100 Ramsar Convention Secretariat, *Designating Ramsar Sites: Strategic Framework and Guidelines for the Future Development of the List of Wetlands of International Importance. Ramsar Handbooks for the Wise Use of Wetlands* (4th edn, vol. 17, Ramsar Convention Secretariat 2010) 22; see also Strategic Framework and guidelines for the future development of the List of Wetlands of International Importance of the Convention on Wetlands (Ramsar, Iran, 1971), third edition, as adopted by Resolution VII.11 (1999) and amended by Resolutions VII.13 (1999), VIII.11 and VIII.33 (2002), IX.1, Annexes A and B (2005) and X.20 (2008) para. 50.

Management of Ramsar sites

Restoration is also an important consideration in light of the obligations under Article 3 (1)–(2) of the Convention providing that 'The Contracting Parties shall formulate and implement their planning so as to promote the conservation of the wetlands included in the List, and as far as possible the wise use of wetlands in their territory' and 'each Contracting Party shall arrange to be informed at the earliest possible time if the ecological character of any wetland in its territory and included in the List has changed, is changing or is likely to change as the result of technological developments, pollution or other human interference'.

The basic management scheme for Ramsar designated sites is to avoid loss. No matter how feasible future restoration may be in a given case, when there is a prospect for a potential loss of natural wetlands, the first priority is to avoid any such loss.[101] Restoration or creation of wetlands cannot replace the loss or degradation of natural wetlands.[102] When change in ecological character at a site is deemed likely, the response obligation at that point is to maintain the character under Article 3(1) of the Convention. If change is occurring or has occurred, the 'maintain' obligation should be interpreted as continuing in effect, 'which would mean an obligation to restore the interests in question, in situ'.[103]

The assessment of the ecological character of a site is thus important to establish the need for restoration actions. The Ramsar handbook on the management of wetlands includes criteria for evaluating ecological character features. Criterion 7 is used to assess the potential for improvement or restoration. Severely degraded features may have varying degrees of potential for improvement; some will have none at all, while others will have potential for total recovery if given appropriate management. The need to identify this potential is crucial. There can be no justification for wasting resources in attempting to manage a degraded feature when the underlying reasons for the damage cannot be reversed.[104]

When a change in ecological character occurs, parties have committed themselves to take restoration measures.[105] Resolution VIII.8 calls upon Contracting Parties to maintain or restore the ecological character of their Ramsar sites, including utilising all appropriate mechanisms to address and resolve as soon as practicable the matters for which a site may have been the subject of a report pursuant to Article 3(2).[106] Sites that undergo ecological change can be put on a

---

101 Recommendation 4.1, Resolution VII.17, para. 10, and Resolution IX.6, Annex, para. 12.
102 Resolution VII.17, para. 10 and Resolution VIII.16, para. 10.
103 Ramsar COP 10 DOC. 27 Background and rationale to the Framework for processes of detecting, reporting and responding to change in wetland ecological character (2008) paras 131–132.
104 Ramsar Convention Secretariat, *Managing wetlands: Frameworks for Managing Wetlands of International Importance and Other Wetland Sites. Ramsar Handbooks for the Wise Use of Wetlands* (4th edn, vol. 18, Ramsar Convention Secretariat 2010) 38.
105 See flowchart in Resolution X.16. A Framework for processes of detecting, reporting and responding to change in wetland ecological character (2008).
106 Resolution VIII.8. Assessing and reporting the status and trends of wetlands, and the implementation of Article 3.2 of the Convention (2002) para. 20.

list, the so-called Montreux Record.[107] If a site is placed on the Montreux Record, the goal is to have the threat mitigated and the values for why it was originally inscribed restored.[108] Since the Montreux Record was created in 1990, 80 sites were entered onto the record. Currently, there are 47 sites on the Montreux Record,[109] while 32 sites that had been listed on the Montreux Record have since been removed from it.[110] However, it appears that restoration efforts have not been sufficient to remove the remaining Ramsar designated sites from the Montreux Record.[111]

A further step is necessary when the efforts to maintain, restore or rehabilitate are not successful. This is not an easy judgement to make, either in ecosystem management terms or in legal terms, and thus informed assessments and a precautionary approach will both be necessary. When the judgement is made that a site can no longer be restored or rehabilitated, it constitutes a decision that the situation has moved beyond the scope of the Article 3(1) requirement, and the question of compensation becomes relevant.[112]

Removal of sites from the Ramsar List

According to the Ramsar Convention a restriction in boundary or removal from the Ramsar List is only possible because of urgent national interests,[113] for which compensation is required.[114] However, the Conferences of the Parties identified other situations in which sites can be removed from the List. According to practice under the Ramsar Convention, a protected wetland can be removed from the Ramsar List when parties and the Ramsar Secretariat agree that there is no possibility of extension, enhancement or restoration of its functions or values.[115]

The potential for restoration becomes the test for whether a state can delist a wetland. Under the 2005 Guidance of Resolution IX.6,[116] states are expected to

107 The Montreux Record was established by Recommendation 4.8. Change in ecological character (1990).
108 Gillespie, *Protected Areas and International Environmental Law* 243.
109 <www.ramsar.org/sites-countries/change-in-ecological-character>.
110 Ramsar Convention Secretariat, *The Ramsar Convention Manual: A Guide to the Convention on Wetlands* (Ramsar, Iran, 1971) (6th edn, 2013) 57, < www.ramsar. org/sites/default/files/documents/library/manual6-2013-e.pdf>.
111 Gillespie, *Protected Areas and International Environmental Law* 243.
112 Ramsar COP 10 DOC. 27 Background and rationale to the Framework for processes of detecting, reporting and responding to change in wetland ecological character (2008) paras 134–135; on compensation, see last section of this chapter.
113 Article 2(5), Ramsar Convention.
114 Article 4(2), Ramsar Convention; see section on compensation in this chapter.
115 Resolution V.3. Procedure for initial designation of sites for the List of Wetlands of International Importance (1993) Annex. Review procedure for listed sites which may not qualify under any of the Criteria established by Recommendation 4.2.
116 Resolution IX.6. Guidance for addressing Ramsar sites or parts of sites which no longer meet the Criteria for designation (2005) Annex, IV; this resolution builds on Resolution VIII.22. Issues concerning Ramsar sites that cease to fulfil or never fulfilled the Criteria for designation as Wetlands of International Importance (2002); see also Ramsar Convention Secretariat, *Addressing Change in Wetland Ecological Character: Addressing*

assess whether the change in ecological character that has led to the site, or part of the site, ceasing to qualify as a Ramsar site, is truly irreversible. If the ecological change appears to have a chance of reversibility, a state must define the conditions under which the change may be reversed, the management actions including restoration that are needed to secure this outcome and the likely timescales needed to permit the recovery of the character of the site. States may find themselves needing to address changes to a Ramsar site in response to damage caused by a natural disaster, the natural inter-annual variability of the size of waterbird or other populations and/or management interventions including restoration or rehabilitation on the site. If there is potential for reversibility of ecological changes, the key ecological features of the site must be monitored for the time period identified as necessary for recovery and the status of the site must be reassessed for its qualification as a protected area under the criteria. States must report recovery of the site, submit an updated Ramsar Information Sheet which clearly identifies the changes that have occurred to the Secretariat and, if appropriate, request the removal of the site from the Montreux Record. If the loss of part or all of the listed site is irreversible, and the attempts at recovery or restoration have failed in terms of its qualification for the Ramsar List, a report must be prepared on the restriction of the site's boundary or its removal from the List, as appropriate.[117] This procedure highlights an important role for ecological restoration. Generally, a state must demonstrate that a Wetland of International Importance cannot be restored, before it can be removed from the List.

World Heritage Convention

The World Heritage Convention does not mention restoration in its text. Several articles in the Convention mention 'rehabilitation'. Article 5 includes the obligation for State Parties to take the appropriate legal, scientific, technical, administrative and financial measures necessary for the identification, protection, conservation, presentation and rehabilitation of this heritage.[118] The World Heritage Committee can receive requests from State Parties for international assistance with respect to property forming part of the cultural or natural heritage, situated in their territories, and included or potentially suitable for inclusion in the World Heritage List or World Heritage List in Danger. The purpose of such requests may be to secure the protection, conservation, presentation or rehabilitation of such property.[119] International assistance by the World Heritage Committee and by the World Heritage Fund can be for reasons of rehabilitation of World Heritage.[120]

Change in the Ecological Character of Ramsar Sites and Other Wetlands. Ramsar Handbooks for the Wise Use of Wetlands (4th edn, vol. 19, Ramsar Convention Secretariat 2010) E. IV, Procedures to apply should deletion or restriction be contemplated.

117 See Resolution IX. 6. Guidance for the consideration of the deletion or restriction of the boundaries of a listed Ramsar site (2005) Annex, IV, paras 24–29.
118 Article 5(2), World Heritage Convention.
119 Article 13(1), World Heritage Convention.
120 See Articles 22 and 23, World Heritage Convention.

Also the operational guidelines pay attention to restoration of World Heritage. In response to monitoring of World Heritage sites that are under threat, when the World Heritage Committee considers that the property has seriously deteriorated, but not to the extent that its restoration is impossible, it may decide that the property be maintained on the List, provided that the State Party takes the necessary measures to restore the property within a reasonable period of time. The Committee may also decide that technical cooperation be provided under the World Heritage Fund for work connected with the restoration of the property, proposing to the State Party to request such assistance, if it has not already been done; in some circumstances State Parties may wish to invite an Advisory mission by the relevant Advisory Body(ies) or other organisations to seek advice on necessary measures to reverse deterioration and address threats.[121]

Sites can also be placed on the List of World Heritage in Danger. When considering the inscription of a property on the List of World Heritage in Danger, the Committee shall develop, and adopt, as far as possible, in consultation with the State Party concerned, a desired state of conservation for the removal of the property from the List of World Heritage in Danger and a programme for corrective measures.[122] Although the operational guidelines do not specifically mention restoration explicitly, the 'corrective' measures can include restoration of sites. If the restoration measures prove to be successful, the site can be removed from the List of World Heritage in Danger, based on a regular review of the state of conservation of properties on the List of World Heritage in Danger.[123]

A deletion of a site from the World Heritage List can be done for the following reasons: where the property has deteriorated to the extent that it has lost those characteristics that determined its inclusion in the World Heritage List; and where the intrinsic qualities of a World Heritage site were already threatened at the time of its nomination by human action and where the necessary corrective measures as outlined by the State Party at the time have not been taken within the time proposed.[124] Also under this Convention, restoration measures play an important role. In general, a site can only be removed from the World Heritage List if restoration measures have failed.

Biosphere reserves

The Statutory Framework of the World Network of Biosphere Reserves[125] does not mention restoration. However, there is a review procedure for biosphere reserves. This review is carried out every ten years and is based on a report that has

121 UNESCO, Operational Guidelines for the Implementation of the World Heritage Convention (WHC.15/01 2015) para. 176, b.
122 Ibid., para. 183.
123 Ibid., paras 190–191.
124 Ibid., para. 192.
125 The Statutory Framework of the World Network of Biosphere Reserves, 28C/Resolution 2.4, UNESCO General Conference 1995, <http://unesdoc.unesco.org/images/0010/001038/103849Eb.pdf>.

to be submitted by the national authority. The Advisory Committee for biosphere reserves provides recommendations to the International Coordinating Council (ICC). If the ICC concludes that a site no longer satisfies the biosphere reserve criteria, it may recommend that the state take measures to ensure conformity with the provisions.[126] It may recommend to restore the site.[127] If the ICC finds that the biosphere reserve still doesn't satisfy the criteria, it can remove a site from the network.[128]

Restoration in protected areas and regional law: the EU as a case study[129]

While Chapter 7 discusses the general policy and legal framework on restoration in the European Union, this chapter will specifically focus on the role that ecological restoration plays in the designation and management of the Natura 2000 network of protected areas in the EU.

Designation of Natura 2000 sites

According to the Habitats Directive 'A coherent European ecological network of special areas of conservation shall be set up under the title Natura 2000.'[130] This network shall be composed of Special Areas of Conservation (SACs), hosting the natural habitat types listed in Annex I[131] and habitats of the species listed in Annex II[132] of the Habitats Directive. The designation process under the Habitats Directive consists of three phases: 1) the Member States propose a list of sites to the Commission; 2) the Commission establishes a list of Sites of Community Importance (SCIs) for each biogeographical region; and 3) the Member States designate the sites as Special Areas of Conservation.[133] The Natura 2000 network shall also include the Special Protection Areas (SPAs) classified by the member

126 Statutory framework, Article 9.5.
127 Gillespie, *Protected Areas and International Environmental Law* 249; for some examples see UNESCO, *Biosphere Reserves: Special Places for People and Nature* (2002) 70–71.
128 Statutory framework, Article 9.6.
129 See also A. Cliquet, K. Decleer and H. Schoukens, 'Restoring nature in the EU: the only way is up?' in C.-H. Born, A. Cliquet, H. Schoukens, D. Misonne and G. Van Hoorick (eds), *The Habitats Directive in its EU Environmental Law Context: European Nature's Best Hope?* (Routledge 2015) 265–284.
130 On site designation, see H. Schoukens and H. Woldendorp, 'Site selection and designation under the Habitats and Birds Directives: a Sisyphean task?' in C.-H. Born, A. Cliquet, H. Schoukens, D. Misonne and G. Van Hoorick (eds), *The Habitats Directive in its EU Environmental Law Context: European Nature's Best Hope?* (Routledge 2015) 31–55.
131 Annex I includes natural habitat types of community interest whose conservation requires the designation of special areas of conservation.
132 Annex II includes animal and plant species of community interest whose conservation requires the designation of special areas of conservation.
133 Article 4, Habitats Directive.

states pursuant to Directive 79/409/EEC.[134] The Natura 2000 network, encompassing SACs and SPAs, now comprises 18.36 per cent of the EU land territory;[135] but only 4 per cent of the total marine area under the jurisdiction of EU Member States has a Natura 2000 designation.[136]

According to Article 4 of the Habitats Directive, designation of SACs should be done according to the criteria of Annex III of the Habitats Directive. Sites undergoing restoration or proposed for restoration can be considered for Natura 2000 designation.[137] The Standard Data Form, submitted by the Member States when designating the site, includes not only information on the habitat types present on the site but also information on the degree of conservation of the existing structure and functions of the natural habitat type concerned and any possibilities for restoration. The sub-criterion on restoration possibilities is used to evaluate to what extent the restoration of a habitat type concerned on the site in question could be possible. The first question to assess is the scientific feasibility of restoration. This requires a comprehensive knowledge of the structure and functions of the habitat type and the concrete management plans and prescriptions needed to restore it. There may be need to stabilise or increase the percentage of area covered by that habitat type, to re-establish the specific structure and functions that are necessary for its long-term maintenance and to maintain or restore a favourable conservation status for its typical species. After this scientific assessment, an assessment must be made whether restoration is cost-effective from a nature conservation point of view. This assessment must take into consideration the degree of threat and the rarity of the habitat type. Using 'best expert judgment', it must be decided whether restoration is 'easy', 'possible with an average effort', or 'difficult or impossible'.[138]

The importance of restoration in designating a protected area under EU law is confirmed in a European Court of Justice case where Ireland failed to properly designate a Natura 2000 site on the basis of its capacity to be restored.[139] In the case at hand, Ireland has challenged the need to classify the Cross Lough area as a Special Protection Area under the Birds Directive, even though it was, up to quite recently, an important breeding ground for the sandwich tern (*Sterna sandvicensis*). The Commission observed that the disappearance of the sandwich tern colony can be traced to predatory activity on the part of the American mink (*Mustela vison*) but that no measures were ever put in place to protect the colony. According to the Commission, with appropriate restoration measures, the sandwich tern might resettle this important long-standing breeding ground. Ireland

134 Article 3(1), Habitats Directive.
135 European Commission, *Natura 2000 Nature and Biodiversity Newsletter* (July 2014) 8.
136 European Environment Agency, *Protected Areas in Europe – an Overview* (2012) 74.
137 See criteria in Annex III, A, c and Annex III, B, b, Habitats Directive.
138 Commission Implementing Decision of 11 July 2011 2011/484/EU concerning a site information format for Natura 2000 sites (notified under document C (2011) 4892), *OJ L* 198, 30 July 2011, explanatory notes, 3.1.
139 Case C-418/04 *Commission v Ireland* [2007] ECR I-10947.

should not be allowed to benefit from the fact that it failed to ensure classification and protection of the Cross Lough area in a timely manner.[140] In her opinion the Advocate General stated:

> However, there would be no point in classifying the area as an SPA if it could not be restored to an area most suitable for the protection of birds. In that case classification of it as an SPA would also be unnecessary. However, as the Commission submits, without contradiction, there is a genuine chance that the sandwich tern may resettle the area of Cross Lough (Killadoon). This species frequently changes its colony site and continues to use sites near the area. Therefore, if appropriate measures were taken against the mink, renewed use of the area would be possible.[141]

The Court agreed with the opinion of the Advocate General and ruled that Ireland breached its obligations under the Birds Directive for not designating the site as a protected area. Although the Court did not explicitly refer to the lack of restoration measures, the Court used a similar reasoning to the Advocate General. According to the Court, Ireland ought, at the very least, to have adopted appropriate measures in order to avoid pollution or deterioration of the habitats in the Cross Lough area or any disturbances affecting the sandwich tern. Having failed to take such measures for that area, Ireland had not provided proof that the area would no longer be suitable for designation even if protection measures had been taken. Moreover, according to the results of scientific studies and observations submitted by the Commission during the proceedings, protection measures were possible. Relying on these observations, which confirm the potential for recolonisation of areas by the sandwich tern, the Commission stated that there is a genuine chance that the sandwich tern may resettle in the area. In those circumstances, the Court found that the action was well founded in respect of the Cross Lough area.[142]

Management of Natura 2000 sites

The Natura 2000 network 'shall enable the natural habitat types and the species' habitats concerned to be maintained or, where appropriate, *restored* at a favourable conservation status in their natural range'.[143] The words 'where appropriate' are not meant to leave a discretionary choice for Member States to decide whether or not to take restoration measures, but will rather depend on the conservation status of the habitats concerned. Habitats in a favourable conservation status must be maintained as such. Habitats that are not in a favourable conservation status must be restored.

---

140 Ibid., paras 79–80.
141 Opinion AG Kokott Case C-418/04 *Commission v Ireland* [2007] ECR I-10947, para. 61.
142 Ibid., paras 85–89.
143 Article 3(1), Habitats Directive (emphasis added).

Once a Site of Community Importance has been adopted by the Commission, the state has to establish its priorities 'in the light of the importance of the sites for the maintenance or *restoration*, at a favourable conservation status, of a natural habitat type in Annex I or a species in Annex II and for the coherence of Natura 2000, and in the light of the threats of degradation or destruction to which those sites are exposed'.[144] The Habitats Directive obliges Member States to establish the necessary conservation measures, including – if need be – appropriate management plans and appropriate statutory, administrative or contractual measures that correspond to the ecological requirements of the habitat types and the species present on the sites.[145] As the definition of conservation encompasses restoration, the conservation obligations of Article 6(1) also implicitly include restoration obligations.

It is also important to know what level of restoration is required within a Natura 2000 site. This will depend on the amount of restoration needed to achieve a favourable conservation status for the habitats and species for which the site has been designated. The overall aim of the Habitats Directive to achieve a favourable conservation status, included in Article 2 of the Habitats Directive, is considered as a result obligation.[146] Conservation and restoration measures of Article 6(1) should correspond to the 'ecological requirements' of the habitat types in Annex I and the species in Annex II present on the sites.[147] The ecological requirements rest on a scientific basis and can only be determined on a case-by-case basis.[148] The detailed assessment of what qualifies as a favourable conservation status is thus a decision for each Member State. This strategy has risks because a Member State can use a low conservation threshold to define favourable status so that restoration measures will not be deemed necessary. This could prove counterproductive if the favourable conservation status is not actually achieved for a habitat or species.

Another question is for which species and habitats restoration measures are required on a Natura 2000 site. According to the Commission guidelines, Article 6(1) establishes a general conservation regime which applies to all SACs of the Natura 2000 network without exception and to all the natural habitat types of Annex I and the species of Annex II present on the sites, except those habitats and species identified as non-significant in the Natura 2000 Standard Data Form.[149] For the habitats and species identified on the site as non-significant, no restoration measures are required, according to the Commission. However, it could still be important to take restoration measures for non-significant species on that site, for example in order to restore a relict population of a species that is characteristic for that habitat, but highly threatened in that region and has a low dispersal capacity.

144 Article 4(4), Habitats Directive (emphasis added).
145 Article 6(1), Habitats Directive.
146 Cliquet et al., 'Restoring nature in the EU' 276.
147 Article 6(1), Habitats Directive.
148 European Commission, *Managing Natura 2000 sites: The provisions of Article 6 of the 'Habitats' Directive 92/43/EEC* (Office for Official Publications of the European Communities 2000) 18.
149 Ibid., 16.

The habitats and species, for which (restoration) measures are required, can evolve. The Standard Data Form must be updated in case of new ecological information, in case that information was initially missing, or in case of new species occurring in a site, due to natural changes or climate change. It will, however, be difficult to remove habitats or species from the Standard Data Form and for those species and habitats mentioned in the Standard Data Form conservation (and restoration) measures are obliged. In analogy with Article 9 of the Habitats Directive, which allows for the declassification of an SAC because of natural developments, one could argue that you could delist habitats and species from the Standard Data Form, but only because of natural developments. This is not the case for the destruction of habitats or disappearance of species from the site due to a lack of restoration measures by the Member State. A similar reasoning can be found with the Advocate General in the *Cascina Tre Pini* case (see below).

Restoration and the obligation to avoid deterioration in Natura 2000 sites

Article 6(2) of the Habitats Directive includes the obligation for Member States to 'take appropriate steps to avoid ... the deterioration of natural habitats and the habitats of species as well as disturbance of the species for which the sites have been designated'. The obligation to avoid deterioration also includes the obligation to take restoration measures. In a 2002 case against Ireland, on the deterioration of the habitat of the red grouse in the Owenduff-Nephin Beg Complex Special Protection Area, the Court decided that it was necessary for the authorities 'not only to take measures to stabilize the problem of overgrazing, but also to ensure that damaged habitats are allowed to recover'.[150] In the *Cascina Tre Pini* case,[151] the importance of restoration measures under Article 6(2) was stressed, particularly to avoid declassification of a protected site. The Advocate General, on the question of declassification of a site, concluded in her opinion that 'Article 6(2) ... requires the Member States to protect SCIs against deterioration. A Member State's failure to fulfil those obligations to afford protection does not warrant the withdrawal of protected status ... *Member States should rather take the necessary measures to restore the site.*'[152] In its ruling, the European Court of Justice stated: 'The failure of a Member State to fulfil the obligation of protecting a particular site does not necessarily justify the declassification of that site. On the contrary, it is for that State to take the measures necessary to safeguard that site.'[153] Although the Court did not explicitly mention restoration measures, it implicitly does by referring to the 'measures necessary to safeguard that site'.

150 Case C-117/00 *Commission v Ireland* [2002] ECR I-5335, para. 31.
151 Case C-301/12 *Cascina Tre Pini s.s. v Ministero dell'Ambiente e della Tutela del Territorio e del Mare and Others (Cascina Tre Pini)* (2013).
152 Opinion A.G. Kokott, Case C-301/12 *Cascina Tre Pini s.s. v Ministero dell'Ambiente e della Tutela del Territorio e del Mare and Others (Cascina Tre Pini)* (2013) para. 50 (emphasis added).
153 Case C-301/12 *Cascina Tre Pini s.s. v Ministero dell'Ambiente e della Tutela del Territorio e del Mare and Others (Cascina Tre Pini)* (2013) para. 32.

The Court has not answered the question of whether states have an obligation to undertake restoration measures in the case of partial deterioration of a Natura 2000 site. Assuming that a certain habitat type in one part of a Natura 2000 site is in an unfavourable conservation status due to fragmentation and lack of maintenance and restoration measures but the same habitat type is in a favourable conservation status in another part of the site, then the overall evaluation of the habitat might still be considered favourable. Is it sufficient to keep the 'good' parts in a favourable conservation status and ignore the unfavourable status in other parts of the site? Partial deterioration of a site appears to violate Article 6(2) of the Habitats Directive mandating no deterioration of habitats, not even when the status of conservation is favourable for the whole site. This is especially the case when the site is composed of a patchwork of spatially isolated sub-sites in a matrix landscape with site-specific differences in species communities.

The more tricky question is whether a state must take additional measures to actually improve the status of a habitat. Neither the Directive nor the Commission guidelines provide a precise answer. One can reasonably assume, however, that a lack of proper restoration and maintenance measures for the habitat type at hand in a sub-site will most likely lead to extinction of relict species or a decrease in the population sizes of target species at the site level leading to decreased connectivity and meta-population functioning. Many habitats in Europe are so fragmented that many species are only able to survive as a meta-population. So most likely restoration measures will be necessary.

The opinion of the Advocate General in the *Cascina Tre Pini* case seems to also support the view that restoration measures are required in case of partial deterioration of a site. In this case a preliminary question was asked on the need for a regular review of the SCIs. The Advocate General stated that 'such reviews are to be undertaken when there are signs that an SCI *or certain parts of it* no longer meet nature conservation requirements'.[154] The Advocate General referred to Article 11 of the Habitats Directive, which includes the obligation to undertake ongoing monitoring of the conservation status of the natural habitats and species of community interest. The fact that Article 11 is included in the section of the Directive which concerns site conservation is evidence of the particular interest in SCIs. The monitoring of SCIs must in particular be suited to ensuring compliance with the conservation priorities established according to Article 4(4) including the maintenance at or restoration to a favourable conservation status of species and habitats in the SCI concerned and the conservation obligations under Article 6(1)–(2). Therefore, states before considering declassification of a site should take 'supplementary measures' to protect the site 'and *to restore the elements which have been adversely affected*'.[155] The Advocate General's analysis seems to suggest an

154 Opinion A.G. Kokott, Case C-301/12 *Cascina Tre Pini s.s. v Ministero dell'Ambiente e della Tutela del Territorio e del Mare and Others (Cascina Tre Pini)* (2013) para. 57 (emphasis added).
155 Ibid., paras 60–63 (emphasis added).

obligation for states to undertake restoration of all parts of a Natura 2000 site even if the overall conservation status of the Natura 2000 site is favourable.

In a 2013 judgment in a preliminary ruling the Court also touched on the destruction of a part of a Natura 2000 site.[156] In this case, a planned road would lead to the permanent loss of approximately 1.47 ha of limestone pavement within the Lough Corrib SCI in Ireland. Preliminary questions were asked to understand the proper interpretation of 'an adverse effect on the integrity of the site' in the Habitats Directive.[157] The planning board in Ireland was of the opinion that the damage to the site was not necessarily incompatible with the finding that there was no adverse effect 'on the integrity of the site'. The Court, however, ruled that in order for the integrity of a site as a natural habitat not to be adversely affected for the purposes of the Habitats Directive, the site needs to be kept in a favourable conservation status. According to the Court, this entails 'the lasting preservation of the constitutive characteristics of the site concerned that are connected to the presence of a natural habitat type whose preservation was the objective justifying the designation of that site'.[158] If after an appropriate assessment the authority concludes that the plan or project will lead to the lasting and irreparable loss of the whole or part of a priority natural habitat type, the plan or project is understood to adversely affect the integrity of that site.[159]

While this case is about the assessment of a project on the integrity of a site, and not explicitly about restoration, there are some interesting lessons for restoration. From the Court's ruling, it is possible to extrapolate that if a project prevents a site or part of a site from being restored (i.e. the 'irreparable loss'), then the project or plan will affect the integrity of a site, and cannot be granted a permit.[160] From the Opinion of the Advocate General it seems that a temporary adverse effect, which can be fully restored after a project has been implemented, might not amount to an adverse effect on the integrity of the site.

The *Sweetman* case also suggests that a favourable conservation status must be achieved at a site level or sub area level for all protected habitats and species.

---

156 Preliminary ruling in Case C-258/11 *Peter Sweetman and Others v An Bord Pleanála (Sweetman)* (2013); see H. Schoukens, 'The Ruling of the Court of Justice in Sweetman: how to avoid a death by a thousand cuts?' (2014) *ELNI* 2–12.

157 Article 6(3) Habitats Directive states: 'Any plan or project not directly connected with or necessary to the management of the site but likely to have a significant effect thereon, either individually or in combination with other plans or projects, shall be subject to appropriate assessment of its implications for the site in view of the site's conservation objectives. In the light of the conclusions of the assessment of the implications for the site and subject to the provisions of paragraph 4, the competent national authorities shall agree to the plan or project only after having ascertained that it will not adversely affect the integrity of the site concerned and, if appropriate, after having obtained the opinion of the general public.'

158 Preliminary ruling in Case C-258/11 *Peter Sweetman and Others v An Bord Pleanála (Sweetman)* (2013) para. 39.

159 Ibid., para. 46.

160 A plan or project with a negative effect on the site can still be allowed under the exception procedure of Article 6(4) of the Habitats Directive.

According to the European Commission, the favourable conservation status of a natural habitat or species has to be considered across its natural range, that is at biogeographical and, hence, Natura 2000 network level.[161] Since, however, the ecological coherence of the network will depend on the contribution of each individual site to it and, hence, on the conservation status of the habitat types and species it hosts, the assessment of the favourable conservation status at site level will always be necessary.[162] In practice, this might mean that restoration measures are required not just at each Natura 2000 site, but also for each distinct sub area of a Natura 2000 site. Although in the *Sweetman* case the habitat type that would be lost was a priority habitat type (limestone pavement), it can be assumed that the Court would come to a similar conclusion for non-priority habitat types.[163]

Restoration of connectivity between protected areas

Connectivity and protected areas

Restoration beyond protected areas is important to restore links between protected areas. This can include the restoration of large corridors, as well as restoring ecological stepping stones.[164] Restoring connectivity can have various functions including to allow movement of species for genetic interchange and to help species adapt to climate change.[165] Restoring ecological stepping stones between protected areas can help to ensure that migratory species have resting and feeding locations in order to ensure their safe passage. A landscape/seascape mosaic can link various habitats into a viable and more functional ecosystem.[166] The scale of connectivity can thus vary from small stepping stones and linear elements to large (international) corridor areas.

---

161 European Commission, *Managing Natura 2000 sites: The provisions of Article 6 of the 'Habitats' Directive 92/43/EEC* (Office for Official Publications of the European Communities 2000) 18.
162 Ibid.
163 Schoukens, 'The ruling of the Court of Justice in Sweetman', 9.
164 On the importance of connectivity see inter alia: A.F. Bennett, *Linkages in the Landscape: The Role of Corridors and Connectivity in Wildlife Conservation* (IUCN 1998, 2003); G. Bennett and K.J. Mulongoy, *Review of Experience with Ecological Networks, Corridors and Buffer Zones* (CBD Technical Series No. 23, 2003).
165 See Chapter 11; see also N. Heller and E. Zavaleta, 'Biodiversity management in the face of climate change: a review of 22 years of recommendations' (2009) 142 *Biological Conservation* 14–32; M.S. Cross, J.A. Hilty, G.M. Tabor, J.J. Lawler, L.J. Graumlich and J. Berger, 'From connect-the-dots to dynamic networks: maintaining and enhancing connectivity as a strategy to address climate change impacts on wildlife' in J.F. Brodie, E.S. Post and D.F. Doak (eds), *Wildlife Conservation in a Changing Climate* (The University of Chicago Press 2013) 307–329.
166 Keenleyside et al., *Ecological Restoration for Protected Areas* 14.

The importance of connectivity has already been recognised in international and EU law.[167] Although several international legal instruments mention networks of protected areas, and some also refer to corridors, few have actually tried to implement them.[168] The focus in nature conservation law is mostly on the designation and protection of core protected areas, but often without specific measures for connecting these core areas.

Restoration and connectivity in international law[169]

Nature conservation conventions can contribute to the protection of corridors and flyways. In the framework of several international and regional nature conservation conventions attention has been given to the importance of connectivity. For example, regional agreements concluded under the Bonn Convention on Migratory Species, such as the African-Eurasian Migratory Waterbird Agreement, can protect flyways or other migration routes.[170] In a Resolution on wetland and climate change under the Ramsar Convention, it was recognised that 'the conservation and wise use of wetlands enable organisms to adapt to climate change by providing connectivity, corridors and flyways along which they can move'. Also in a 2012 Resolution,[171] the Ramsar parties recognised 'that the conservation and wise use of wetlands helps biodiversity to adapt to climate change by providing connectivity, corridors and flyways, and other migratory pathways, along which biota can move'.

In Resolution 10.3, parties to the Bonn Convention recognised that 'sites that perform a critical role in a wider system, such as core areas, corridors, restoration areas and buffer zones, may be linked by strategies that, through a concept of ecological networks, address habitat fragmentation and other threats to migratory species'.[172] Parties agreed in cooperation with other relevant organisations to set network-scale objectives for species conservation 'by restoration of fragmented and degraded habitats and removal of barriers to migration' and to enhance the 'connectivity of terrestrial and aquatic protected areas, including marine areas'.[173] In Resolution 10.19, parties agreed to makes efforts to improve the resilience of

---

167 See more extensively on legal aspects of connectivity: B. Lausche, D. Farrier, J. Verschuuren, A.G.M. La Viña, A. Trouwborst, C.-H. Born and L. Aug, *The Legal Aspects of Connectivity Conservation: A Concept Paper* (IUCN 2013).

168 Gillespie, *Protected Areas and International Environmental Law* 151.

169 See more extensively: A. Savaresi, 'The protection of biodiversity and ecological connectivity in the international arena' in M. Alberton (ed) *Towards the Protection of Biodiversity and Ecological Connectivity in Multi-Layered Systems* (Nomos 2013) 11–28; the issue of connectivity specifically with regards to climate change was also dealt with in A. Cliquet, 'International and European law on protected areas and climate change: need for adaptation or implementation?' (2014) 54 *Environmental Management* 720–731.

170 See Chapter 5.

171 Resolution XI.14. Climate change and wetlands: implications for the Ramsar Convention on Wetlands (2012) para. 18.

172 Resolution 10.3. The role of ecological networks in the conservation of migratory species, UNEP/CMS/Resolution 10.3 (2011) preamble.

173 Ibid., paras 6–7.

migratory species and their habitats to climate change in order to 'strengthen the physical and ecological connectivity between sites, aiding species dispersal and colonization when distributions shift' and 'consider the designation of seasonal protected areas' to support migratory species at critical points of their lifecycle.[174]

In 2014, parties to the Bonn Convention proposed further recommendations for advancing the design and implementation of ecological networks to address the needs of migratory species.[175] In particular, the parties focused on how 'connectivity' in an ecological network can contribute to the elimination of obstacles to migration, including physical obstacles, habitat disturbance, habitat fragmentation and discontinuities in habitat quality.[176] Although restoration is not mentioned explicitly as a strategy, removing various obstacles for ecological network connectivity can include restoration of connectivity elements. As part of a Programme of Work on climate change and migratory species, parties agreed to undertake several actions including to:

- Ensure there is physical and ecological connectivity between sites, aiding species dispersal and colonization when distributions shift;
- Integrate protected areas into wider landscapes and seascapes, ensure appropriate management practices in the wider matrix and undertake the restoration of degraded habitats and landscapes/seascapes;
- Identify migratory species that have special connectivity needs – those that are resource, area, and/or dispersal limited.[177]

The Biodiversity Convention also focuses on connectivity. In the Convention, states are obliged to establish a 'system of protected areas',[178] which implies the introduction of connectivity measures to link core protected areas. In 2004, under the Programme of Work on protected areas, states agreed to a goal 'To integrate protected areas into broader land- and seascapes and sectors so as to maintain ecological structure and function',[179] with the aim by 2015 that 'all protected areas and protected area systems are integrated into the wider land- and seascape, and relevant sectors, by applying the ecosystem approach and taking into account ecological connectivity and the concept, where appropriate, of ecological networks'. Among the suggested activities for State Parties is action 1.2.5, which encourages states to 'Rehabilitate and restore habitats and degraded ecosystems, as

---

174 Resolution 10.19. Migratory species conservation in the light of climate change, UNEP/CMS/Resolution 10.19 (2011) para. 8.
175 Resolution 11.25. Advancing ecological networks to address the needs of migratory species, UNEP/CMS/Resolution 11.25 (2014).
176 Ibid., Annex, 10.
177 Programme of work on climate change and migratory species, UNEP/CMS/Resolution 11.26 (2014).
178 Article 8(a), Biodiversity Convention.
179 Convention on Biological Diversity, Programme of Work on Protected Areas: Goal 1.2 (2004).

appropriate, as a contribution to building ecological networks, ecological corridors and/or buffer zones.'[180]

A CBD COP Decision of 2010 on protected areas includes several provisions that are relevant for connectivity.[181] The decision invites parties at the national level to 'Enhance the coverage and quality, representativeness and, if appropriate, connectivity of protected areas as a contribution to the development of representative systems of protected areas and coherent ecological networks that include all relevant biomes, ecoregions, or ecosystems.'[182] At the regional level parties are invited to create an enabling environment for transboundary cooperation in regards to ecological connectivity.[183] At the global level, the IUCN World Commission on Protected Areas, and other relevant organisations are invited to develop technical guidance on ecological restoration, monitoring and evaluation of connectivity and conservation corridors.[184] In order to address climate change impacts and increase resilience to climate change, parties were encouraged to increase efforts to integrate protected areas into wider landscapes and seascapes by using connectivity measures such as ecological networks and ecological corridors and restoring degraded habitats and landscapes.[185]

Connectivity is also part of the Aichi Targets.[186] Although initiatives exist to develop corridors, there is still not sufficient connection between protected areas. Few national biodiversity strategies and action plans (NBSAPs) explicitly address the connection or integration of protected areas into wider landscapes and seascapes.[187]

In the regional Bern Convention, the importance of connectivity has also been recognised in recommendations.[188] In a 1991 Recommendation, parties

---

180 See further technical guidance on the implementation of Goal 1.2.: J. Ervin, K.J. Mulongoy, K. Lawrence, E. Game, D. Sheppard, P. Bridgewater, G. Bennett, S.B. Gidda and P. Bos, *Making Protected Areas Relevant: A Guide to Integrating Protected Areas into Wider Landscapes, Seascapes and Sectoral Plans and Strategies* (CBD Technical Series No. 44, 2010).

181 Decision X/31. Protected areas, UNEP/CBD/COP/DEC/X/31 (2010).

182 Ibid., para. 1, a.

183 Ibid., para. 5.

184 Ibid., para. 8.

185 Ibid., 14, a; see also similar provisions in para. 26; Decision X/33. Biodiversity and climate change, UNEP/CBD/COP/DEC/X/33 (2010) para. 8, d, iii

186 See previous discussion of Aichi Target 11.

187 P.W. Leadley, C.B. Krug, R. Alkemade, H.M. Pereira, U.R. Sumaila M. Walpole, A. Marques, T. Newbold, L.S.L. Teh, J. van Kolck, C. Bellard, S.R. Januchowski-Hartley and P.J. Mumby, *Progress towards the Aichi Biodiversity Targets: An Assessment of Biodiversity Trends, Policy Scenarios and Key Actions* (Secretariat of the Convention on Biological Diversity, Technical Series 78, 2014) 266.

188 Inter alia Recommendation no. 25 on the conservation of natural areas outside protected areas proper (1991); Recommendation no. 135 on addressing the impacts of climate change on biodiversity (2008) paras 22–23; Recommendation no. 143 on further guidance for Parties on biodiversity and climate change (2009) para. III, 4–5; Recommendation no. 180 on improving the conservation of nature outside protected areas proper (2015) para. 1.

encouraged each other to invest in conservation and restoration of ecological corridors.[189] Parties were expected to restore or to compensate for the loss of ecological corridors by providing for crossing routes for example for otters, badgers and deer or closing roads during the spring migrational period for amphibians.[190] Parties were also recommended to restore watercourses and restore vegetation along their banks.[191]

Another example of a regional convention that pays attention to connectivity is the Carpathian Convention and more specifically its Biodiversity Protocol.[192] The main objective of the Biodiversity Protocol is the enhancement of conservation, restoration and sustainable use of biological and landscape diversity in the Carpathians.[193] A specific objective is to establish ecological networks in the Carpathians. Article 9 states that each party shall take measures to improve and ensure continuity and connectivity of natural and semi-natural habitats in the Carpathians in order to allow for the dispersal and migration of wild species populations such as large carnivores, and genetic exchange between such populations. Article 15 on enhancing conservation and sustainable management in the areas outside of protected areas states that each party shall facilitate coordination and cooperation between all relevant stakeholders, so as to enhance conservation and sustainable management in the areas outside of protected areas in the Carpathians, 'with the objective of improving and ensuring connectivity between existing protected areas and other areas and habitats significant for biological and landscape diversity of the Carpathians'.

The Protocol on Conservation of Nature and the Countryside to the Alpine Convention,[194] another regional convention, includes the obligation for parties to 'adopt all the measures necessary for preserving and, to the extent necessary, restoring special structural, natural and near-natural elements of the landscapes, biotopes, ecosystems and traditional rural landscapes'.[195] A specific ecological connectivity

---

189 Recommendation no. 25 on the conservation of natural areas outside protected areas proper (1991).

190 Ibid., Appendix, III, 1.

191 Ibid., Appendix, III, 2.

192 Framework Convention on the Protection and Sustainable Development of the Carpathians, Kiev, 22 May 2003; Protocol on conservation and sustainable use of biological and landscape diversity, Bucharest, 19 June 2008; <www.carpathianconvention.org/>; on the Carpathian Convention and connectivity, see H. Egerer, K. Kuras and G. Luciani, 'The protection of biodiversity and ecological connectivity in the Carpathian Convention' in M. Alberton (ed), *Towards the Protection of Biodiversity and Ecological Connectivity in Multi-Layered Systems* (Nomos 2013) 81–99.

193 Article 1(1), Biodiversity Protocol.

194 Alpine Convention, Salzburg, 7 November 1991; Protocol on the Implementation of the Alpine Convention of 1991 Relating to the Conservation of Nature and the Countryside, Chambéry, 20 December 1994; <www.alpconv.org/>; on the Alpine Convention and connectivity, see M. Onida, 'The protection of biodiversity and ecological connectivity in the Alpine Convention' in M. Alberton (ed) *Towards the Protection of Biodiversity and Ecological Connectivity in Multi-Layered Systems* (Nomos 2013) 57–79.

195 Article 10(1), Conservation of nature and the countryside Protocol.

project undertaken by the Protocol parties is the Alpine Carpathian Corridor project between Austria and Slovakia.[196] The project aims to create a coherent Green Infrastructure which supports restoration of the ecosystems present and enables wild animal populations to move more freely and interact.[197]

Restoration and connectivity in the EU

The EU Habitats Directive requires that states undertake some measures to promote connectivity. Under the Habitats Directive Article 3, states 'shall endeavour to improve the ecological coherence of Natura 2000 by maintaining, and where appropriate developing, features of the landscape that are of major importance for wild fauna and flora as referred to in article 10'. Article 10 provides that 'Member States shall endeavour, where they consider it necessary ... in particular with a view to improving the ecological coherence of the Natura 2000 network, to encourage the management of features of the landscape which are of major importance for wild fauna and flora'. Articles 3 and 10, however, seem weak obligations when compared with other obligations in the Directive using words like 'shall endeavour', 'where they consider it necessary', 'to encourage'. Presumably as a consequence of this rather weak formulation, the implementation of the Habitats Directive has so far mainly been aimed at the conservation of the status quo of habitats and species within the core areas. A more explicit legal framework, for instance by adapting the legal wording of Articles 3 and 10 of the Habitats Directive, has been advocated in literature.[198]

   Combining the provisions of Articles 3 and 10 with other obligations in the Directive requiring states to achieve favourable conservation status, it appears, however, that connectivity measures are already mandatory under the existing legal regime. Because a large number of habitats and species in the EU are in an unfavourable status of conservation due to fragmentation and other human impacts, connectivity measures will be essential to reach a favourable conservation status. What this means is that Member States must adopt measures beyond the Natura 2000 network aimed at ensuring the functional connectivity between the Natura 2000 sites.[199] The monitoring, obliged by Articles 11 and 17 of the

---

196 ‹www.alpenkarpatenkorridor.at/›.
197 European Commission, *Building a Green Infrastructure for Europe* (2013) 19.
198 See J. Verschuuren, 'Connectivity: is Natura 2000 only an ecological network on paper?' in C.-H. Born, A. Cliquet, H. Schoukens, D. Misonne and G. Van Hoorick (eds), *The Habitats Directive in its EU Environmental Law Context: European Nature's Best Hope?* (Routledge 2015); see also L Krämer, 'The protection of biodiversity and ecological connectivity in the European Union' in M Alberton (ed) *Towards the Protection of Biodiversity and Ecological Connectivity in Multi-Layered Systems* (Nomos 2013) 50.
199 See G. Van Hoorick, 'Biodiversity outside protected areas: an outlaw waiting to be saved?' in C.-H. Born, A. Cliquet, H. Schoukens, D. Misonne and G. Van Hoorick (eds), *The Habitats Directive in its EU Environmental Law Context: European Nature's Best Hope?* (Routledge 2015) 459.

Habitats Directive, can point to the shortcomings in the required coherence of the network.

According to the Commission, the concept of favourable conservation status is not limited to the Natura 2000 network or to the species protected by this network.[200] It therefore follows from this that Member States should promote the implementation of connectivity measures where these are required to maintain or restore a favourable conservation status, irrespective of their contribution to the coherence of the Natura 2000 network.[201]

The EU Birds Directive also appears to promote connectivity. The obligation in Article 3(1) of the Birds Directive to take the measures to preserve, maintain or re-establish a sufficient diversity and area of habitats for all wild bird species would require connectivity measures even if connectivity only served the purpose of climate induced latitudinal and altitudinal dispersal.[202] The preservation, maintenance and re-establishment of biotopes and habitats shall include the upkeep and management in accordance with the 'ecological needs' of habitats inside and outside the protected zones.[203] The re-establishment and creation of linear corridors and stepping-stones between habitats can be considered as an ecological need and thus promotes ecological connectivity.[204] Article 4(2) of the Birds Directive provides for the protection of staging posts of migratory birds along their migration routes, by designating them as special protection areas.

In a recent case before the European Court of Justice involving Spain,[205] the Court recognised the importance of connectivity. The case dealt with an open-cast coal mining project in the Natura 2000 site 'Alto sil', causing a loss of habitat for the brown bear in a corridor area. Because of the noise and vibration caused by the mining operations, the bears moved 3.5 to 5 kilometres from the areas of impact. The disturbance caused by the mining operation made it more difficult or even prevented the animals to access the Leitariegos corridor, which is a critical north-south transit route for the western population of that species.[206] Consequently, the noise and vibrations caused by the open-cast mines, and the closure of the Leitariegos corridor by reason of those mines, constituted disturbances of

200 European Commission, *Guidance document on the strict protection of animal species of Community interest under the Habitats Directive 92/43/EEC* (2007).
201 M. Kettunen, A. Terry, G. Tucker and A. Jones, *Guidance on the maintenance of landscape features of major importance for wild flora and fauna – Guidance on the implementation of Article 3 of the Birds Directive (79/409/EEC) and Article 10 of the Habitats Directive (92/43/EEC)* (Institute for European Environmental Policy (IEEP) 2007); <http://ec.europa.eu/environment/nature/ecosystems/docs/adaptation_fra gmentation_guidelines.pdf>.
202 A. Trouwborst, 'Conserving European biodiversity in a changing climate: the Bern Convention, the European Union Birds and Habitats Directives and the adaptation of nature to climate change' (2011) 20 *RECIEL* 62–77.
203 Article 3(2)(b), Birds Directive.
204 K. Wheeler, 'Bird protection & climate changes: a challenge for Natura 2000?' (2006) 13 *Tilburg Foreign Law Review* 283–299.
205 Case C-404/09 *Commission v. Spain* [2011] ECR I-11853.
206 Ibid., para. 188.

the 'Alto Sil' site.[207] Spain was found in violation of its obligations under Article 6(2) of the Habitats Directive. Also, the mining operations were found capable of producing a barrier effect likely to contribute to the fragmentation of the habitat of the capercaillie, a type of grouse, and the isolation of certain sub-populations of that species.[208] This seems to indicate that this provision also protects the sub-populations located outside the site to which the site's populations are connected.[209] Although the case relates to activities within a Natura 2000 site, it is likely that the Court would have arrived at a similar decision if the case had been about populations in non-adjacent Natura 2000 sites, connected through a corridor not lying in a Natura 2000 site.[210] The case shows that in order to safeguard the integrity of a Natura 2000 site, features of the landscape that are essential for migration, dispersal and genetic exchange of wild species for which the site has been designated, must be preserved and, if necessary, restored.[211]

The EU well understands the need for connectivity, as reflected in various initiatives, including Target 2 of the 2011 EU Biodiversity Strategy encouraging states to develop Green Infrastructure by 2020 capable of maintaining and enhancing ecosystems and their services. Green Infrastructure (GI) is defined as:

> a strategically planned network of natural and semi-natural areas with other environmental features designed and managed to deliver a wide range of ecosystem services. It incorporates green spaces (or blue if aquatic ecosystems are concerned) and other physical features in terrestrial (including coastal) and marine areas. On land, GI is present in rural and urban settings.[212]

The work that has already been done by EU members to establish the Natura 2000 network provides a backbone for Green Infrastructure efforts. According to the Green Infrastructure Strategy it is 'a reservoir of biodiversity that can be drawn upon to repopulate and revitalize degraded environments and catalyze the development of GI. This will also help reduce the fragmentation of the ecosystem,

---

207 Ibid., para. 191.
208 Ibid., para. 148.
209 Verschuuren, 'Connectivity' 285–302.
210 A. Trouwborst, 'The Habitats Directive and climate change: is the law climate-proof?' in C.-H. Born, A. Cliquet, H. Schoukens, D. Misonne and G. Van Hoorick (eds), *The Habitats Directive in its EU Environmental Law Context: European Nature's Best Hope?* (Routledge 2015) 303–324.
211 G. Van Hoorick, 'Biodiversity outside protected areas. An outlaw waiting to be saved?' in C.-H. Born, A. Cliquet, H. Schoukens, D. Misonne and G. Van Hoorick (eds), *The Habitats Directive in its EU Environmental Law Context: European Nature's Best Hope?* (Routledge 2015) 461.
212 European Commission, Communication from the Commission to the European Parliament, the Council, the European Economic and Social Committee and the Committee of the Regions, *Green Infrastructure (GI) – Enhancing Europe's Natural Capital*, COM(2013) 249 final.

improving the connectivity between sites in the Natura 2000 network and thus achieving the objectives of Article 10 of the Habitats Directive.'[213]

The Strategy aims to promote Green Infrastructure in other policy areas and the Commission commits itself to explore the opportunities for setting up financing mechanisms to support GI. The Commission's Strategy was supported by a resolution from the European Parliament of 2013.[214] The Commission has funded several studies on Green Infrastructure. One of these studies has to assess the opportunities for developing a TEN-G initiative (Trans-European Network on Green Infrastructure). As of May 2016 it is unclear what the legal implications will be of such an initiative.

Multi-stakeholder initiatives to promote ecological connectivity between protected areas exist in Europe. An example is the European Green Belt initiative which forms a 12,500 km long transcontinental axis of a European ecological network along the former 'Iron Curtain'.[215] In 2014 the European Green Belt Association was formed, with membership from both government and non-governmental organisations.[216] The collaboration is seen as an important contribution towards the realisation of international ecological networks including the Natura 2000 network, the Emerald Network (under the Bern Convention) and the Pan-European Ecological Network, as well as the achievement of obligations under multilateral environmental agreements such as the Ramsar Convention and the Biodiversity Convention.

Restoration as mitigation and compensation for loss of protected areas

Compensation for loss of protected areas

Restoration projects 'on the ground' are frequently the result of compensation obligations provided for in legal instruments. When a protected area will be reduced in size or lose its protected status because of a development or infrastructure project, some laws require that the lost nature area is compensated elsewhere. Compensation can include the designation of a new protected area, or the restoration of a non-protected area, or even additional restoration within another protected area. There are different terms used at the international, regional, and national level, including 'habitat banking' or 'biodiversity offsetting'. In some countries the term 'mitigation' is used, rather than 'compensation'. This chapter will only briefly look at international and EU law on compensation, in relation to the loss of protected areas. It will not deal with national legislation in this regard.

213 Ibid., 7.
214 European Parliament resolution of 12 December 2013 on Green Infrastructure – Enhancing Europe's Natural Capital (2013/2663(RSP)).
215 Fact sheet European Green Belt, <www.europeangreenbelt.org/fileadmin/content/downloads/Fact-sheet_2014_EGB.pdf>.
216 <www.europeangreenbelt.org/association.html>.

It will also not evaluate compensation obligations for environmental damage, for instance in the case of an incident causing environmental pollution.

It is important to stress that compensation should only be considered as a last-resort option. Compensation for the loss of protected areas should not lead to a legal 'licence to trash'. Although there are many successes in restoration for purposes of compensation, there are no guarantees that nature will be restored. Compensation is not a long-term conservation strategy. For some habitat types such as ancient forests and peatlands, compensating for any losses to these habitats is not feasible in a reasonable time frame and conservation on site must be given priority. You can plant trees as compensation, but you cannot create a forest ecosystem within a short time. Woodlands can take hundreds of years to regenerate. The effectiveness of habitat (re)creation and restoration as compensation has been questioned in scientific literature.[217]

Compensation for loss of protected areas in international law

The Ramsar Convention on the protection of wetlands includes a compensation obligation for the loss or restriction of a Ramsar site. According to Article 2(5), a Contracting Party may, because of its urgent national interest, delete or restrict the boundaries of wetlands listed as wetlands of international significance. Article 4(2) states that where a Contracting Party has acted under Article 2(5) it should as far as possible compensate for any loss of wetland resources, and should create additional nature reserves for waterfowl and for the protection, either in the same area or elsewhere, of an adequate portion of the habitat type from the original listing. No Ramsar site has ever been totally removed from the list and parties have only extremely rarely restricted the boundaries of a site on this basis.[218] The parties agreed on further guidance on this issue through several resolutions.[219] The guidance in the resolutions clearly respects the mitigation hierarchy and sees compensation only as a last resort. A three-step approach should be taken to

---

217 See, for example, R.K.A. Morris, I. Alonso, R.G. Jefferson and K.J. Kirby, 'The creation of compensatory habitat – can it secure sustainable development?' (2006) 14 *Journal for Nature Conservation* 106–116; M. Curran, S. Hellweg and J. Beck, 'Is there any empirical support for biodiversity offset policy?' (2014) 24(4) *Ecological Applications* 617–632; D. Moreno-Mateos, V. Maris, A. Béchet and M. Curran, 'The true loss caused by biodiversity offsets' (2015) 192 *Biological Conservation* 552–559.

218 In Belgium 30 hectares of the Belgian Ramsar site Galgenschoor was lost for the construction of a container terminal. This was compensated by extending another Belgian Ramsar site, the Blankaart lake with surrounding grasslands of the Yzervalley (see Ramsar National Report 1993–1995), <https://data.inbo.be/purews/files/276075/170092.pdf>.

219 Resolution VII.24. Compensation for lost wetland habitats and other functions (1999); Resolution VIII.20. General guidance for interpreting 'urgent national interests' under Article 2.5 of the Convention and considering compensation under Article 4.2 (2002); Resolution XI.9. An Integrated Framework and guidelines for avoiding, mitigating and compensating for wetland losses (2012).

respond to current or likely changes in the ecological character of wetlands, whether or not such wetlands are included in the Ramsar List, namely:

a avoiding impacts (e.g. systematic assessment of projected negative changes to ecological character of potentially impacted wetlands through strategic planning to systematically identify potential areas for conservation);
b mitigating on-site for unavoidable impacts (e.g. through minimising project impacts and restoring area after the project); and
c compensating for, or offsetting, any remaining impacts (e.g. off-site restoration).[220]

Resolution XI.9 reaffirms the Contracting Parties commitment to avoid negative impacts on the ecological character of Ramsar sites and other wetlands as the primary step for stemming the loss of wetlands. Where such avoidance is not feasible, states may apply appropriate mitigation and/or compensation/offset actions, including through wetland restoration.[221] The integrated framework includes definitions of mitigation and compensation. Mitigation is defined as:

Mitigating wetland impacts refers to reactive practical actions that minimize or reduce in situ wetland impacts. Examples of mitigation include "changes to the scale, design, location, siting, process, sequencing, phasing, management and/or monitoring of the proposed activity, as well as restoration or rehabilitation of sites" (Resolution X.17 Annex, paragraph 23). Mitigation actions can take place anywhere, as long as their effect is to reduce the effect on the site where change in ecological character is likely, or the values of the site are affected by those changes. In many cases it may not be appropriate to regard restoration as mitigation, since doing so represents an acknowledgement that impact has already occurred: in such cases the term "compensation" may be a truer reflection of this kind of response.[222]

Compensation for wetland impacts refers to actions that are intended to offset the residual impacts on the wetland ecological character that remain after any mitigation has been achieved. An example of compensation would be an on-site or off-site wetland restoration or creation project, provided it adds value beyond what would have happened otherwise (i.e. relying on an already-planned benefit that would not constitute compensation). The term 'restoration' in the Ramsar Convention framework is used in its broadest sense to include both projects that promote a return to or towards original conditions and projects that improve the ecological character of the wetland without necessarily promoting a return to original/reference conditions. Although some Ramsar texts imply a distinction between these two

220 Resolution XI.9. An Integrated Framework and guidelines for avoiding, mitigating and compensating for wetland losses (2012) preamble.
221 Ibid., para. 14.
222 Ibid., Annex, para. 19.

potential scenarios by referring to 'rehabilitation' as well as 'restoration', such a distinction in practice is not precise and the two terms are often used interchangeably. The term 'restoration' applies to locations where wetland habitat has previously existed or where an existing wetland habitat is degraded.[223]

The framework provides decision criteria that should be considered during the development and implementation of compensation measures. Those criteria show that preference should be given to the restoration of a similar wetland type (e.g. habitat type for habitat type). The compensation provided should address the area extent, significant ecosystem components, and the functional performance of the wetland. Ideally compensation should be in close proximity to the impacted wetland and within the same hydrological catchment or coastal zone. Guidelines are also given on how compensation should be achieved. For instance, compensation should be established in advance of, or at least in consideration of, the timing of the proposed impacts. Costs and risks associated with compensation should be considered. Because of the complexity and irreplaceability of certain ecosystems, where the risk of failure of adequate compensation is high, a decision needs to be made as to whether compensation can be an appropriate strategy or whether a party should simply avoid damage by refraining from the damaging activity.[224]

Compensation for loss of protected areas in EU law

There is an implicit obligation for restoration in the compensation obligation under Article 6(4) of the Habitats Directive: if, in spite of a negative assessment of the implications for the site and in the absence of alternative solutions, a plan or project must nevertheless be carried out for imperative reasons of overriding public interest, including those of a social or economic nature, the Member State shall take all compensatory measures necessary to ensure that the overall coherence of Natura 2000 is protected.[225] According to the Commission guidelines on Article 6, compensation measures include re-creating a habitat on a new or enlarged site, to be incorporated into Natura 2000; or improving a habitat on part of the site or on another Natura 2000 site, proportional to the loss due to the project.[226] The guidelines on Article 6(4) specifically mention restoration in existing sites as a compensatory measure.[227] Compensatory measures should thus be additional to the actions that are normal practice under the Habitats and Birds Directives.[228]

---

223 Ibid.
224 Ibid., Annex, para. 40.
225 On the compensation obligations under the Habitats Directive, see G. Van Hoorick, 'Compensatory measures in European Nature Conservation Law' (2014) 10(2) *Utrecht Law Review* 161–171.
226 European Commission, *Managing Natura 2000 sites: The provisions of Article 6 of the 'Habitats' Directive 92/43/EEC* (Office for Official Publications of the European Communities 2000) 45.
227 European Commission, *Guidance document on Article 6(4) of the 'Habitats Directive' 92/43/EEC* (2007/2012) 14.
228 Ibid., 10.

Due to the unfavourable conservation status of many Natura 2000 sites in Europe, restoration will already be obliged in many situations and compensation must clearly been seen as a last-resort option.

Since compensation measures should ensure that the overall coherence of the Natura 2000 network is protected, the compensatory measures proposed should a) address the habitats and species negatively affected, in comparable proportions to which they have been affected; and b) provide functions comparable to those which had justified the selection criteria of the original site, particularly regarding the adequate geographical distribution.[229] The compensation obligation in the EU thus imposes a type-for-type restoration, as is also advised in the Ramsar framework. For certain habitat types, which cannot be restored, or which take a very long time to be restored, the Commission, rather weakly, states that 'the zero option should be seriously considered'. In a Greek case, on the diversion of a river, the European Court of Justice recognised that extensive compensation measures were required.[230] Unfortunately, the Court also agreed that compensation could consist of the conversion of a natural fluvial ecosystem into a largely man-made fluvial and lacustrine ecosystem.[231] The viewpoint should rather be that a project or plan cannot take place if the condition of compensation cannot be fulfilled.

Compensation under Article 6(4) should be an 'active' compensation meaning that compensation requirements should be fulfilled before damage has occurred. Under certain circumstances where this requirement cannot be fulfilled, over-compensation would be required for the interim losses.[232] Member States have to inform the Commission of the compensatory measures adopted in application of Article 6(4) of the Habitats Directive. The Commission finds several shortcomings in the information provided by the Member States. In most of the cases, a time schedule for the implementation of the compensatory measures is provided, but states do not always indicate when the expected compensatory results will be achieved as regards the ecological function that compensation areas should fulfil. The techniques and methods proposed for the implementation of the proposed compensatory measures are mostly not described and the existing conditions in the areas where the compensatory measures are to take place are often not explained, which makes it difficult to assess feasibility and possible effectiveness of the compensation measures.[233] It also remains to be seen whether the

229 Ibid., 13.
230 Case C-43/10, *Nomarchiaki Aftodioikisi Aitoloakarnanias and Others* (2012).
231 Ibid., para. 139; this case has led to criticism; see, for example, Van Hoorick, 'Compensatory measures in European Nature Conservation Law' 169; H. Schoukens, 'Omlegging Griekse rivier: de mythe van "groene" infrastructuurprojecten' (2013) 1 *Tijdschrift voor Omgevingsrecht en Omgevingsbeleid* 67–69.
232 European Commission, *Guidance document on Article 6(4) of the 'Habitats Directive'* 92/43/EEC (2007/2012) 14.
233 European Commission, *Implementation of Article 6(4), first subparagraph, of Council Directive 92/43/EEC (Habitat Directive). Period 2007–2011. Summary report* (2012) 8.

compensation areas can indeed provide the same level of biodiversity and thus contribute to the coherence of the Natura 2000 network.[234]

In the Commission guidelines on Article 6(4) a distinction is made between 'compensation' and 'mitigation'. Mitigation measures are those measures that aim to minimise, or even cancel, the negative impacts on a site that are likely to arise as a result of the implementation of a plan or project. These measures are an integral part of the specifications of a plan or project. Compensatory measures are independent of the project. They are intended to offset the negative effects of the plan or project so that the overall ecological coherence of the Natura 2000 Network is maintained.[235]

Under Article 6(4) of the Habitats Directive, project developers have to demonstrate the overriding public interest of their project, as well as the lack of alternatives and they have to take active compensation measures. This procedure is perceived by many project developers as rather cumbersome. In order to avoid the procedure from Article 6(4), some Member States developed a practice, in which they included restoration works in the planning procedures for project development so that the restoration would qualify as 'mitigation' measures. This approach was thought to minimise or avoid the adverse effects of the project on a protected site so the government could more easily grant the permission for the project to take place, without having to resort to the exception procedure of Article 6(4). In several Member States, national court cases elaborated on the distinction between mitigation and compensation, with different outcomes.[236]

In 2014 the European Court of Justice brought clarification on the distinction in the *Briels* case.[237] The case concerned the granting of a permission to extend a motorway on the basis of habitat creation measures that would counterbalance the damage that would be inflicted upon the nitrogen-sensitive Molinia meadows by the motorway. Molinia meadows are listed as a protected habitat type in Annex I to the Habitats Directive. In a preliminary procedure, the European Court was asked whether Article 6(3) of the Habitats Directive must be interpreted as meaning that a plan or project that has negative implications for a type of natural habitat, and that provides for the creation of an area of equal or greater size of the same natural habitat type within the same site, has an effect on the integrity of that site, and, if so, whether such measures may be categorised as 'compensatory

---

234 See on compensation: D. McGillivray, 'Compensatory measures under Article 6(4) of the Habitats Directive: no net loss for Natura 2000?' in C.-H. Born, A. Cliquet, H. Schoukens, D. Misonne and G. Van Hoorick (eds), *The Habitats Directive in its EU Environmental Law Context: European Nature's Best Hope?* (Routledge 2015) 101–118.

235 European Commission, *Guidance document on Article 6(4) of the 'Habitats Directive' 92/43/EEC* (2007/2012) 10.

236 See on the Flemish cases, H. Schoukens and A. Cliquet, 'Mitigation and compensation under EU nature conservation law in the Flemish region: beyond the deadlock for development projects?' (2014) 10(2) *Utrecht Law Review* 194–215.

237 Case C-521/12 *Briels and Others v Minister van Infrastructuur en Milieu* (2014); for an extensive discussion on this case, see H. Schoukens and A. Cliquet, 'Biodiversity offsetting and restoration under the EU Habitats Directive: balancing between no net loss and deathbed conservation?' *Ecology & Society* (2016) 21(4):10.

measures' within the meaning of Article 6(4).[238] According to the Court such restoration measures are indeed to be categorised as 'compensatory measures' under Article 6(4). The habitat creation measures in question were not aimed either at avoiding or reducing the significant adverse effects for that habitat type caused by the motorway project. Rather, the measures were only intended to compensate for post-construction impacts with no guarantee that the road project would not adversely affect the integrity of the site. Any positive effects of a future creation of a new habitat that is aimed at compensating for the loss of area and quality of that same habitat type on a protected site, even where the new area will be bigger and of higher quality, are highly difficult to forecast with any degree of certainty and, in any event, will be visible only several years into the future.[239] The Court, however, recognised that the restoration measures could still be taken into account as compensation in the context of the derogation clause contained in Article 6(4).

With this ruling the Court clearly used the mitigation hierarchy as an implicit reference criterion in its qualification of the habitat creation measures at stake in *Briels*, even though the Court did not explicitly refer to the mitigation hierarchy in its ruling. Restoration as compensation is clearly seen as a last resort option.

The reluctance of the Court towards endorsing more flexible mitigation strategies under Article 6(3) is justified. If such flexible mitigation strategy were to be accepted, the risk exists that EU states would invest less in strategies of avoidance. The reliance on restoration measures in early stages of the decision-making process could create the impression that states were given developers a 'licence to trash' EU habitats, rather than strictly applying the derogation clauses that are currently present in EU nature conservation law.[240] The derogation is only to be allowed whenever the public interests related to the infrastructure project clearly outweigh the ecological importance of the preservation of the Natura 2000 site. Also, by referring to the uncertainties related to the creation of new habitats, the Court's decision confirms the scientific findings which point to the uncertain outcomes of compensatory restoration. Also, the Court takes a realistic viewpoint that Member States might 'abuse' the restoration measures in order to avoid the exception procedure under Article 6(4). The Court states in this regard:

> the effectiveness of the protective measures provided for in Article 6 of the Habitats Directive is intended to avoid a situation where competent national authorities allow so-called 'mitigating' measures – which are in reality compensatory measures – in order to circumvent the specific procedures provided for in Article 6(3) and authorize projects which adversely affect the integrity of the site concerned.[241]

238 Case C-521/12 *Briels and Others v Minister van Infrastructuur en Milieu* (2014) para. 18.
239 Ibid., paras 28–32.
240 Schoukens and Cliquet, 'Biodiversity offsetting and restoration under the EU Habitats Directive'.
241 Case C-521/12 *Briels and Others v Minister van Infrastructuur en Milieu* (2014) para. 33.

Finally, it should be underscored that restoration measures used for mitigation or compensation purposes should go beyond existing restoration obligations that exist in Article 6(1) of the Habitats Directive. Although the Court itself did not elaborate on this aspect, the Advocate General in her Opinion did not accept that conservation measures that are required under Article 6(1) could also serve at the same time as mitigation for impacts from a plan or project.[242]

Conclusion

Protected areas are a keystone strategy in most nature conservation policies and laws. Many international legal documents require the conservation and restoration of protected areas. Since many protected areas all over the world are in an unfavourable conservation status, states must give more attention to restoring protected areas in order to implement their international restoration obligations. EU nature conservation law is a clear example of regional law that contains restoration obligations for protected areas under EU law. Recent case law from the European Court of Justice confirms the restoration obligations under EU nature conservation law. International and regional law also recognises the importance of connectivity as part of ecological networks and imposes obligations for restoring connectivity elements. Although some of these provisions are soft law, in the form of non-binding COP decisions, the legal basis for ecological networks can be found in several international and regional instruments. Even much more than restoration in core protected areas, the implementation for restoring connectivity on the ground is missing in many states. Restoration as compensation is to be considered as a last resort option, especially in the light of scientific findings which point to the low success rate in biodiversity offsetting projects. Both the framework under the Ramsar Convention and the EU law as interpreted in the *Briels* case at the European Court of Justice rightly confirm that compensation is a last step in the mitigation hierarchy.

---

242 Opinion A.G. Sharpston, Case C-521/12 *Briels and Others v Minister van Infrastructuur en Milieu* (2014) para. 48.

11 Climate change and ecological restoration

Introduction

There are several relationships between restoration and climate change that are of legal relevance. First of all, ecosystems such as forests, oceans and peatlands can play a vital role in mitigating and adapting to climate change by serving as carbon sinks capturing and sequestering greenhouse gases. Ongoing deforestation releases greenhouse gases in the atmosphere and degraded ecosystems lose the capacity for storing greenhouse gases. Restoration can help to maintain and enhance these carbon sinks. Preventing the destruction of ecosystems and restoring degraded ecosystems is thus an important mitigation measure to reduce emissions of greenhouse gases. The protection and restoration of mainly forests has become a focus for the international climate change regime.

Second, restoration of ecosystems can help humans to adapt to the adverse effects of climate change. As mitigation measures have been insufficient and were taken too late to avoid dangerous levels of greenhouse gases entering into the atmosphere, climate change effects are already occurring and will further increase in the future. In addition to adaptation measures, such as building dikes to prevent flooding from sea-level rise, investments in restoration can play a vital role in protecting people from the adverse effects of climate change. For example, restoring ecosystems such as mangrove forests and coral reefs can create natural buffers from severe storms and flooding. Although some attention has been paid in international law and policy to these 'nature-based solutions', the possible advantages of restoring ecosystems as an adaptation measure are still not fully recognised in law or policy.

Finally, climate change is one of the major threats to biodiversity, aggravating the already deplorable state of nature of many species and habitats.[1] Rising

---

1 For the impact of climate change on biodiversity see, e.g., O. Sala et al., 'Biodiversity – Global biodiversity scenarios for the year 2100' (2000) 287 *Science* 1770–1774; J. Travis, 'Climate change and habitat destruction: a deadly anthropogenic cocktail' (2003) 270 *Proceedings of the Royal Society of London B* 467–473; C. Thomas et al., 'Extinction risk from climate change' (2004) 427 *Nature* 145–148; K. Willis and S. Bhagwat, 'Biodiversity and climate change' (2009) 326 *Science* 806–807; Secretariat

temperatures, extreme weather events, changed rainfall patterns, drought and melting sea ice are all factors that can negatively influence the timing of flowering, migration patterns and distribution of species.[2] Additional efforts have to be undertaken to restore areas and species to adapt to these detrimental effects. However, climate change poses additional scientific and legal challenges to restore habitats and species because many nature conservation laws require conservation or restoration to a favourable conservation status. This usually implies that there is a return to a historical reference situation. If global changes, such as climate change fundamentally alter ecological conditions, it might be difficult or even impossible to return to a historical state.[3] Can nature conservation laws be adapted to address these challenges?

The subsequent sections of this chapter will discuss: 1) restoration as climate change mitigation, focusing on the Climate Change Convention; 2) restoration of nature as a way of helping humans adapt to climate change; and 3) the challenges for restoration of habitats and species under climate change.

Restoration as mitigation

As mentioned above, many ecosystems are important carbon sinks. Obviously, avoiding degradation and destruction of ecosystems through conservation activities is the most effective way for storing and sequestering carbon. However, in degraded ecosystems, restoration can enhance the capacity to sequester carbon. By doing so, ecosystems help to mitigate the emission of greenhouse gases into the atmosphere. 'Mitigation' of emissions in the context of climate change is to be distinguished from 'mitigation' measures with the purpose of avoiding damage to nature while undertaking human activities, such as infrastructure works.[4]

Mitigation under the Climate Change Convention

The role of ecosystems as carbon sinks for CO^2 emissions has gradually been recognised in international law. The United Nations Framework Convention on Climate Change of 1992 (UNFCCC) in its early years focused mostly on mitigating climate change by reducing emissions from human activities.[5] The 1992

of the Convention on Biological Diversity, *Global Biodiversity Outlook 3* (CBD 2010); C. Bellard, C. Bertelsmeier, P. Leadley, W. Thuiller and F. Courchamp, 'Impacts of climate change on the future of biodiversity' (2012) 15 *Ecology Letters* 365–377.

2 Secretariat of the Convention on Biological Diversity, *Global Biodiversity Outlook 3* (CBD 2010) 56.
3 See S. Allison, *Ecological Restoration and Environmental Change: Renewing Damaged Ecosystems* (Routledge 2012) 90–92.
4 On 'mitigation' in the sense of avoiding damage, see Chapter 10 on compensation.
5 United Nations Framework Convention on Climate Change, Rio de Janeiro, 9 May 1992 (UNFCCC), <www.unfccc.int>.

Convention text recognised the role of ecosystems in Article 4 by obliging State Parties 'to promote and cooperate in the conservation and enhancement of sinks and reservoirs of greenhouse gases, including biomass, forests and oceans as well as other terrestrial, coastal and marine ecosystems'. 'Enhancement' of sinks could include the restoration of ecosystems, although it is not mentioned explicitly as such.

With the development of the REDD (Reducing Emissions from Deforestation and forest Degradation in developing countries) mechanism under the Convention, attention increased for the role of ecosystems – or at least the role of forests – as carbon sinks. REDD is a mechanism to provide funding for developing countries for climate mitigation activities and sustainable management of forests.[6] The initial focus of REDD was on reducing emissions from deforestation and forest degradation, but was broadened to also include the role of conservation, sustainable management of forests and enhancement of forest carbon stocks in developing countries (known as REDD+). REDD+ includes several activities, including the enhancement of forest carbon stocks. Although ecological restoration is not explicitly mentioned, the inclusion of sustainable management and carbon-stock enhancement enables REDD+ funding for forest restoration activities that reduce emissions, sequester carbon, and provide important benefits to communities and biodiversity.[7]

REDD was first introduced at the Conference of the Parties (COP) of 2005, and the item continued to be discussed at the subsequent COPs.[8] Despite the absence of a formal legal agreement before 2015 on REDD+, the Conference of the Parties in 2007 asked the countries to undertake REDD+ action on a voluntary basis.[9] In order to guide this work, the UN developed the UN-REDD programme in 2008.[10] The programme is supported inter alia by the Forest Carbon Partnership Facility (FCPF)[11] at the World Bank.

In the 2010 Cancun Agreements, references were included to several safeguards that parties should take into account when implementing REDD+ activities that are relevant to restoration work. For example, REDD+ activities should be consistent with the objective of environmental integrity and take into account the multiple functions of forests and other ecosystems. REDD+ proponents should ensure that actions are consistent with the conservation of natural forests and

---

6 See the REDD+ Web platform at UNFCCC: <http://redd.unfccc.int/>; the REDD Web platform was established by COP Decision 2/CP.13. Reducing emissions from deforestation in developing countries: approaches to stimulate action, FCCC/CP/2007/6/Add.1 (2007) para. 10.

7 S. Alexander, C. Nelson, J. Aronson, D. Lamb, D. Martinez, J. Harris, E. Higgs, R. Lewis III, M. Finlayson, K. Erwin, R. Hobbs, W. Covington, C. Murcia, R. Kumar, A. Cliquet and R. De Groot, 'Opportunities and challenges for ecological restoration within REDD+' (2011) 6 *Restoration Ecology* 683–689.

8 For an overview of the relevant COP decisions, see: <http://unfccc.int/files/land_use_and_climate_change/redd/application/pdf/compilation_redd_decision_booklet_v1.1.pdf>.

9 Decision 2/CP.13. Reducing emissions from deforestation in developing countries: approaches to stimulate action, FCCC/CP/2007/6/Add.1 (2007) para. 1.

10 <www.un-redd.org/>.

11 <www.forestcarbonpartnership.org/>.

biological diversity, and that REDD+ activities do not lead to the conversion of natural forests, but instead incentivise the protection and conservation of natural forests and their ecosystem services while enhancing other social and environmental benefits.[12] These safeguards are of particular importance for restoration efforts as the risk exists with afforestation or reforestation that native plants will be replaced by monocultures or exotic tree species at the expense of natural forests. Forest restoration, on the other hand, is aimed at the regeneration of the natural forest.

Parties were also asked to work out a system to provide information on how these safeguards will be taken into account.[13] The summary of the requested information should be provided periodically and be included in national communications, and can also be provided, on a voluntary basis, at the REDD+ Web Platform.[14] The provision of the most recent summary of information on how the safeguards are being addressed and respected is one of the requirements for states to be eligible for results-based payments.[15] In 2013 the COP decided to establish the Lima REDD+ Information Hub on the REDD+ Web Platform as a means to publish information on the results of REDD+ activities, and corresponding results-based payments.[16] As of May 2016 only Brazil has provided information in the Lima REDD+ Information Hub.[17]

The Paris Agreement of 2015, decided at COP 21,[18] gives a formal legal basis to COP decisions relating to REDD+. The Paris Agreement includes several references that are relevant for restoration. The Preamble states: 'Recognizing the importance of the conservation and enhancement, as appropriate, of sinks and reservoirs of the greenhouse gases referred to in the Convention.' For the first

12 Decision 1/CP.16. The Cancun Agreements: Outcome of the work of the Ad Hoc Working Group on Long-term Cooperative Action under the Convention, FCCC/CP/2010/7/Add.1 (2010) para. 69 and Appendix 1. Guidance and safeguards for policy approaches and positive incentives on issues relating to reducing emissions from deforestation and forest degradation in developing countries; and the role of conservation, sustainable management of forests and enhancement of forest carbon stocks in developing countries, para. 1, d and para. 2, e.

13 Decision 1/CP.16. The Cancun Agreements: Outcome of the work of the Ad Hoc Working Group on Long-term Cooperative Action under the Convention, FCCC/CP/2010/7/Add.1 (2010) para. 71, d.

14 Decision 12/CP.17. Guidance on systems for providing information on how safeguards are addressed and respected and modalities relating to forest reference emission levels and forest reference levels as referred to in decision 1/CP.16, FCCC/CP/2011/9/Add.2 (2011) para. 4; Decision 12/CP.19*. The timing and the frequency of presentations of the summary of information on how all the safeguards referred to in decision 1/CP.16, appendix I, are being addressed and respected, FCCC/CP/2013/10/Add.1 (2013) para. 3.

15 Decision 9/CP.19*. Work programme on results-based finance to progress the full implementation of the activities referred to in decision 1/CP.16, para. 70, FCCC/CP/2013/10/Add.1 (2013) para. 4.

16 Ibid., para. 9.

17 <http://redd.unfccc.int/info-hub.html>.

18 Decision 1/CP.21. Adoption of the Paris Agreement, FCCC/CP/2015/10/Add.1 (2015).

time, a formal document under the Climate Change Convention framework also refers explicitly to biodiversity: 'Noting the importance of ensuring the integrity of all ecosystems, including oceans, and the protection of biodiversity, recognized by some cultures as Mother Earth.'[19]

Article 5 of the Paris Agreement addresses the role of ecosystems as carbon sinks. Parties should take action to conserve and enhance, as appropriate, sinks and reservoirs of greenhouse gases as referred to in Article 4 of the Convention, including forests. Parties are encouraged to take action to implement and support, including through results-based payments, the existing framework as set out in related guidance and decisions already agreed under the Convention. These include policy approaches and positive incentives for activities relating to reducing emissions from deforestation and forest degradation, and the role of conservation, sustainable management of forests and enhancement of forest carbon stocks in developing countries. These also include alternative policy approaches, such as joint mitigation and adaptation approaches for the integral and sustainable management of forests, while reaffirming the importance of incentivising, as appropriate, non-carbon benefits associated with such approaches.[20]

Mitigation under other conventions

Other conventions recognise the role of ecosystems in helping to mitigate climate change, including wetlands and the role of soil, as important sinks for greenhouse gases. Under the Biodiversity Convention, the COP Decision on climate change and biodiversity of 2010 invited parties and relevant organisations to consider how to conserve, sustainably use and restore biodiversity and ecosystem services while contributing to climate change mitigation and adaptation. Under 'ecosystem-based approaches for mitigation', parties were invited to implement ecosystem management activities, including the protection of natural forests, natural grasslands and peatlands; the sustainable management of forests considering the use of native communities of forest species in reforestation activities; sustainable wetland management; restoration of degraded wetlands and natural grasslands; conservation of mangroves, salt marshes and seagrass beds; sustainable agricultural practices and soil management. These activities are seen as a contribution towards achieving the objectives of the Convention on Climate Change, the Desertification Convention, the Ramsar Convention and the Biodiversity Convention.[21] Measures were also proposed to reduce the biodiversity impacts of climate change mitigation and adaptation measures.[22] States were asked to promote biodiversity conservation, especially with regard to soil biodiversity, while

---

19 Paris Agreement, preamble.
20 Paris Agreement, Article 5.
21 Decision X/33. Biodiversity and climate change, UNEP/CBD/COP/DEC/X/33 (2010) para. 8, n.
22 Ibid., para. 8, u–z.

conserving and *restoring* organic carbon in soil and biomass, including in peatlands, other wetlands, grasslands, savannahs and drylands.[23]

The Parties to the Biodiversity Convention have also paid attention to forest restoration activities for climate change mitigation and the need to implement safeguards for biodiversity under REDD+.[24] When designing, implementing and monitoring afforestation, reforestation and forest restoration activities for climate change mitigation, parties are asked to consider conservation of biodiversity and ecosystem services. This can, for example, include prioritising local and acclimatised native tree species when selecting species for planting.[25]

Aichi Target 15, which has been dealt with extensively in other chapters, includes a restoration target, and explicitly refers to mitigation and adaption:

> By 2020, ecosystem resilience and the contribution of biodiversity to carbon stocks has been enhanced, through conservation and restoration, including restoration of at least 15 per cent of degraded ecosystems, thereby contributing to climate change mitigation and adaptation and to combating desertification.

Under the Desertification Convention, synergies with the Climate Change Convention have been recognised. In the Strategic Plan operational objective 2, outcome 2.5 states that 'Mutually reinforcing measures among desertification/land degradation action programmes and biodiversity and climate change mitigation and adaptation are introduced or strengthened so as to enhance the impact of interventions.'[26] Restoring land can play an important role in reaching the necessary mitigation to stay below a temperature rise of 2°C. In a brief from the UNCCD Secretariat, the Secretariat proposes that new commitments to achieve Land Degradation Neutrality[27] could entail restoring and rehabilitating 12 million hectares of land per year that in combination with additional measures in nature conservation, ecosystem restoration and sustainable agricultural/livestock management

23 Ibid., para. 8, s (emphasis added).
24 Decision XI/19. Biodiversity and climate change related issues: advice on the application of relevant safeguards for biodiversity with regard to policy approaches and positive incentives on issues relating to reducing emissions from deforestation and forest degradation in developing countries; and the role of conservation, sustainable management of forests and enhancement of forest carbon stocks in developing countries, UNEP/CBD/COP/DEC/XI/19 (2012). The CBD Secretariat also provided technical advice on REDD+; see Secretariat for the Convention on Biological Diversity, *REDD-plus and Biodiversity* (CBD Technical Series No. 59, 2011).
25 Decision X/33. Biodiversity and climate change, UNEP/CBD/COP/DEC/X/33 (2010) para. 8, p, ii.
26 Decision 3/COP.8. The ten-year strategic plan and framework to enhance the implementation of the Convention (2008–2018), ICCD/COP(8)/16/Add.1 (2007).
27 The Land Degradation Neutrality goal has been accepted under the Sustainable Development Goals (Goal 15.3): 'By 2030, combat desertification, restore degraded land and soil, including land affected by desertification, drought and floods, and strive to achieve a land degradation-neutral world.'

practices could bring states closer to the 2°C mitigation target. This could help close the emissions gap by up to 25 per cent in the year 2030.[28]

The Conference of the Parties to the Ramsar Convention at its 8th meeting adopted Resolution VIII/3 on climate change and wetlands, which, inter alia, called on relevant countries to take action to minimise the degradation, as well as promote restoration, and improve management practices of those peatlands and other wetland types that are significant carbon stores, or have the ability to sequester carbon.[29] Resolution X.24, in turn, urged relevant Contracting Parties to take immediate action to reduce the degradation, promote restoration, improve management practices of peatlands and other wetland types that are significant greenhouse gases sinks, and to encourage expansion of demonstration sites on peatland restoration and wise use management in relation to climate change mitigation and adaptation activities.[30] Resolution XI.14 urged Contracting Parties to sequester and store carbon as important responses for climate change mitigation through the maintenance and enhancement of their ecological functions, and to reduce or halt the release of stored carbon that can result from the degradation and loss of wetlands.[31]

International commitments for forest restoration

Recently, some non-binding initiatives have been taken regarding restoration of forests, to speed up several international commitments that have already been agreed upon. Within the Global Partnership on Forest and Landscape Restoration, which is a partnership between governments, international organisations, communities and individuals, the Bonn Challenge was accepted in 2011.[32] The Bonn Challenge calls upon the members to restore 150 million hectares of deforested and degraded lands by 2020. The Bonn Challenge is not a new global commitment but rather a practical means of realising other existing international commitments, including the CBD Aichi Target 15, the UNFCCC REDD+ goal, and the Rio+20 goal to end land degradation. As of May 2016, 64 per cent of the goal (96.13 of 150 million hectares) has been committed to,[33] but it is not clear how many of these commitments have already been realised in practice.

In 2014, an additional commitment was taken in the New York Declaration on Forests[34] to restore an additional 200 million hectares by 2030. The Declaration

28 UNCCD, *Land Matters for Climate Reducing the Gap and Approaching the Target*, <www.unccd.int/Lists/SiteDocumentLibrary/Publications/2015Nov_Land_matters_For_Climate_ENG.pdf>.
29 Resolution VIII.3. Climate change and wetlands: impacts, adaptation, and mitigation (2002) para. 15.
30 Resolution X.24. Climate change and wetlands (2008) para. 32.
31 Resolution XI.14. Climate change and wetlands: implications for the Ramsar Convention on Wetlands (2012) para. 26.
32 <www.bonnchallenge.org/content/challenge>.
33 <www.bonnchallenge.org/>.
34 New York Declaration on Forests, New York, 23 September 2014, <www.un.org/climatechange/summit/wp-content/uploads/sites/2/2014/07/New-York-Declaration-on-Forest-%E2%80%93-Action-Statement-and-Action-Plan.pdf>.

was signed by various governments, companies, civil society and indigenous organisations.[35] In 2015 an assessment framework was proposed and interim results were given on the progress made.[36] Two indicators are used: 1) forest restoration pledges under the Bonn Challenge, in hectares; 2) afforestation, restoration and reforestation pledges as part of the Intended Nationally Determined Contributions (INDCs) of Parties to the UNFCCC, in hectares. The interim report shows under indicator 2 that the total forest restoration, reforestation and afforestation pledges are estimated at 121.7 million hectares (35 per cent of the 350-million-hectare 2030 restoration target).[37] The report acknowledges gaps in data because information on specific restoration activities and initiatives is currently spread across a number of databases, including those provided by the United Nations Environmental Programme World Conservation Monitoring Centre and the Society for Ecological Restoration. Standardising this information and compiling global data should make it possible to present global aggregates with greater confidence.[38]

Challenges for mitigation through restoration

While restoration is one strategy for mitigating greenhouse gases, there are a number of concerns about the quality of restoration work that is being undertaken for climate mitigation. Although both the Climate Change Convention and Biodiversity Convention promote restoration of a variety of ecosystems, most of the policy attention is aimed at forest restoration. Although the increased attention for forest restoration certainly has its merits, other ecosystems such as non-forested peatlands can also play a major role in carbon sequestration.[39] Given the rate of climate change, current policy attention may be too narrowly limited to forests. When restoration policy is too narrowly focused on forests without acknowledging the importance of other ecosystems, a danger exists of policy-makers pursuing ecologically perverse programmes by, for example, planting trees in grassland ecosystems.[40]

Also potentially perverse is that restoration of forests is seen exclusively as a way to enhance carbon stocks, disregarding biodiversity and other ecosystem services that are provided by forests and other ecosystems. The REDD+ mechanism has been criticised for its focus on enhancement of forest carbon stocks, as there is a

---

35 <www.forestdeclaration.org>.
36 Climate Focus, *Progress on the New York Declaration on Forests – An Assessment Framework and Initial Report* (Prepared by Climate Focus, in collaboration with Environmental Defense Fund, Forest Trends, The Global Alliance for Clean Cookstoves, and The Global Canopy Program 2015).
37 Ibid., 27.
38 Ibid., 28.
39 See M. Strack (ed) *Peatlands and Climate Change* (International Peat Society 2008).
40 J.W. Veldman, G.E. Overbeck, D. Negreiros, G. Mahy, S. Le Stradic, G. Wilson Fernandes, G. Durigan, E. Buisson, F.E. Putz and W.J. Bond, 'Tyranny of trees in grassy biomes' (2015) 347 *Science* 484–485.

possibility that other services and social issues could be adversely affected.[41] For example, the species-rich Cerrado woodlands and savannas of Brazil are already being replaced by plantations of Eucalyptus, species native to Australia, as at least one carbon credit project gets under way.[42]

Mechanisms need to be installed to avoid the negative biodiversity consequences of valuing forests and other natural ecosystems only or primarily for their climate mitigation potential.[43] It is therefore crucial that biodiversity concerns are integrated into the REDD+ programmes.[44] It remains a challenge, however, to set up restoration projects that are effective both in sequestering carbon and promoting biodiversity conservation while also economically beneficial to communities and landowners.[45] For example, ecological restoration of degraded forests or deforested areas should only be done by using native tree species and not exotic tree species.

The attention under the Climate Change Convention and Biodiversity Convention to build in safeguards when implementing REDD+ projects is certainly a step forward, but it remains to be seen if and how these safeguards are actually implemented in national programmes and in the field. Several international standards have been developed, including the UN-REDD Programme Social and Environmental Principles and Criteria,[46] the Strategic Environmental and Social Assessment (SESA) used by the Forest Carbon Partnership Facility (FCPF) of the World Bank[47] and the REDD+ Social and Environmental

41 J.M. Bullock, J. Aronson, A.C. Newton, R.F. Pywell and J.M. Rey-Benayas, 'Restoration of ecosystem services and biodiversity: conflicts and opportunities' (2011) 10 *Trends in Ecology and Evolution* 541–549.

42 C. Stickler, D. Nepstad, M. Coe, D. McGrath, H. Rodrigues, W. Walker, B. Soares-Filho and E. Davidson, 'The potential ecological costs and cobenefits of REDD: a critical review and case study from the Amazon region' (2009) 15 *Global Change Biology* 2803–2824.

43 F.E. Putz and K.H. Redford, 'Dangers of carbon-based conservation' (2009) 19 *Global Environ* 400–401.

44 See, for example: C.A Harvey, B. Dickson and C. Kormos, 'Opportunities for achieving biodiversity conservation through REDD' (2010) 3 *Conservation Letters* 53–61; T. Gardner et al., 'A framework for integrating biodiversity concerns into national REDD+ programmes' (2012) 154 *Biological Conservation* 61–71; for more information on REDD+ and biodiversity see, for example, T. Pistorius, *REDD from the Conservation Perspective: Pitfalls and Opportunities for Mutually Addressing Climate Change and Biodiversity Conservation* (Institute of Forest and Environmental Policy – Albert-Ludwigs-University Freiburg 2009); Secretariat of the Convention on Biological Diversity, *REDD-plus and Biodiversity* (CBD technical series no. 59, 2011).

45 S. Alexander, C. Nelson, J. Aronson, D. Lamb, D. Martinez, J. Harris, E. Higgs, R. Lewis III, M. Finlayson, K. Erwin, R. Hobbs, W. Covington, C. Murcia, R. Kumar, A. Cliquet and R. De Groot, 'Opportunities and challenges for ecological restoration within REDD+' (2011) 6 *Restoration Ecology* 686–689.

46 <file:///C:/Users/acliquet/Downloads/UNREDD_PB6_Social%20&%20Environmental%20Principles%20and%20 Criteria%20Version%201%20(1).pdf>.

47 <www.forestcarbonpartnership.org/sites/forestcarbonpartnership.org/files/Documents/PDF/Nov2011/FCPF%20Readiness%20Fund%20Common%20Approach%20_Final_%2010-Aug-2011_Revised.pdf>.

Standards.[48] These can serve as guidance for national REDD+ projects.[49] Monitoring and verification of the REDD+ projects and a greater input from restoration scientists will remain crucial to prevent negative effects on biodiversity from these projects.

Restoration as adaptation

More recently, the role of ecosystems in helping people to adapt to the effects of climate change is recognised.[50] For instance, well-managed ecosystems such as wetlands can buffer against floods, landslides and storms, which will occur more often as a consequence of climate change. Ecosystems can help to safeguard water supply and provide clean water. In the framework of several international conventions, attention has been paid to the role of ecosystems for adaptation.

Under the Climate Change Convention, gradually more attention has been paid to adaption. Within the Cancun Adaptation Framework of 2010, parties were encouraged to increase resilience of socio-economic and ecological systems, including through sustainable management of natural resources.[51] The 2015 Paris Agreement repeats and formally adopts the Cancun Adaption Framework in Article 7(9):

> Each Party shall, as appropriate, engage in adaptation planning processes and the implementation of actions, including the development or enhancement of relevant plans, policies and/or contributions, which may include: ... (e) Building the resilience of socioeconomic and ecological systems, including through economic diversification and sustainable management of natural resources.

The Biodiversity Convention has also incorporated decisions with specific attention to restoration as a form of adaptation. In addition to Aichi Target 15 calling

48 <www.redd-standards.org/>.
49 Decision XI/19. Biodiversity and climate change related issues: advice on the application of relevant safeguards for biodiversity with regard to policy approaches and positive incentives on issues relating to reducing emissions from deforestation and forest degradation in developing countries; and the role of conservation, sustainable management of forests and enhancement of forest carbon stocks in developing countries, UNEP/CBD/COP/DEC/XI/19 (2012) annex, para. 16.
50 See, for example: European Environment Agency, *Exploring Nature-Based Solutions: the Role of Green Infrastructure in Mitigating the Impact of Weather- and Climate Change-Related Natural Hazards* (EEA Technical report No. 12, 2015); N. Dudley, S. Stolton, A. Belokurov, L. Krueger, N. Lopoukhine, K. MacKinnon, T. Sandwith and N. Sekhran (eds), *Natural Solutions: Protected Areas Helping People Cope with Climate Change* (IUCN WCPA, TNC, UNDP, WCS, The World Bank and WWF 2010); World Bank, Environment Department, *Convenient Solutions to an Inconvenient Truth: Ecosystembased Approaches to Climate Change* (2009).
51 Decision 1/CP.16. The Cancun Agreements: Outcome of the work of the Ad Hoc Working Group on Long-term Cooperative Action under the Convention, FCCC/CP/2010/7/Add.1 (2010) para. 14, d.

for the restoration of at least 15 per cent of degraded ecosystems to support adaptation, Decision X.33 has recognised that ecosystems 'can be managed to limit climate change impacts on biodiversity and to help people adapt to the adverse effects of climate change'. State Parties are invited to implement, where appropriate, 'ecosystem-based approaches for adaptation, that may include sustainable management, conservation and restoration of ecosystems, as part of an overall adaptation strategy that takes into account the multiple social, economic and cultural co-benefits for local communities'.[52]

In CBD Decision XI/21 the parties were encouraged to recognise the significant role that protected areas, restored ecosystems and other conservation measures can play in climate-change-related activities. Parties agreed to support the strengthening of inventorying and monitoring of biodiversity and ecosystem services at appropriate scales in order to evaluate the threats and likely impacts of climate change as well as the positive and negative impacts of climate change mitigation and adaptation on biodiversity and ecosystem services. Parties are expected to also consider reviewing land-use planning with a view to enhancing ecosystem-based adaptation to climate change, such as the role of mangroves in adapting to changes in sea level.[53] In relation to disaster risk reduction, CBD parties acknowledged that even though biodiversity and ecosystems are vulnerable to climate change, the conservation and sustainable use of biodiversity and the restoration of ecosystems can play a significant role in climate change mitigation and adaptation, combating desertification and disaster risk reduction.[54]

Under the Ramsar Convention, states have also recognised the role of wetlands for climate adaptation. In a 2002 Resolution, Ramsar parties agreed to focus on managing wetlands so as to increase their resilience to climate change and extreme climatic events and to reduce the risk of flooding and drought in vulnerable countries. To enhance resilience, states were encouraged to promote wetland and watershed protection and restoration.[55] In Resolution X.24 of 2008, Ramsar parties recognised that 'the wise use and restoration of wetlands contributes to building the resilience of human populations to climate change impacts and can attenuate natural disasters expected with climate change, such as the use of restored floodplain wetlands to reduce risks from flooding'.[56] The resolution further recommends that parties promote restoration for several wetland types, including rivers, lakes, coastal wetlands and peatlands.[57] Resolution XI.14 of the

---

52 Decision X/33. Biodiversity and climate change, UNEP/CBD/COP/DEC/X/33 (2010) para. 8, j.
53 Decision XI/21. Biodiversity and climate change: integrating biodiversity considerations into climate-change related activities, UNEP/CBD/COP/DEC/XI/21 (2012) para. 6, d-f.
54 Decision XII/20. Biodiversity and climate change and disaster risk reduction, UNEP/ CBD/COP/DEC/XII/20 (2014) preamble.
55 Resolution VIII.3. Climate change and wetlands: impacts, adaptation, and mitigation (2002) para. 14.
56 Resolution X.24. Climate change and wetlands (2008) para. 18.
57 Ibid., paras 30–32.

11th meeting in 2012 urges Contracting Parties to promote the ability of wetlands to contribute to nature-based climate change adaptation, particularly the role of wetlands in regulating water, including reducing risks from water-related disasters.[58]

Restoration of habitats and species under climate change

Adapting nature to climate change

Climate change is one of the major threats to biodiversity and will further aggravate the already unfavourable state of nature in many parts of the world. Scientific literature gives numerous recommendations for biodiversity management in the face of climate change.[59] One of the key elements in these recommendations is increasing the resistance and resilience of ecosystems. Resistance means the ability of a system to remain unchanged in the face of external forces, whereas resilience means the ability of a system to recover from perturbation.[60] Ecosystems should be made more resilient, in order to be able to sustain the additional pressure from climate change, as resilient systems will most likely be able to continue to function. The restoration of ecosystems and ecosystem functions plus the recovery of species are seen as important strategies for increasing the resilience of ecosystems to recover from the negative effects of climate change.[61]

Protected areas can play a key role in adapting to climate change. Protected areas provide safe havens for species under climate change, and can also allow their dispersal to suitable habitats under changing conditions.[62] Scientists recommend

---

58 Resolution XI.14. Climate change and wetlands: implications for the Ramsar Convention on Wetlands (2012) para. 26.
59 See, for example: N. Heller and E. Zavaleta, 'Biodiversity management in the face of climate change: a review of 22 years of recommendations' (2009) 142 *Biological Conservation* 14–32; J. Mawdsley, R. O'Malley and D. Ojima, 'A review of climate-change adaptation strategies for wildlife management and biodiversity conservation' (2009) 23 *Conservation Biology* 1080–1089; J. Lawler, 'Climate change adaptation strategies for resource management and conservation planning' (2009) 1162 *The Year in Ecology and Conservation Biology* 79–98; J. Lawler et al., 'Resource management in a changing and uncertain climate' (2010) 8(1) *Frontiers in the Ecology and the Environment* 35–43; see also footnote 28 in A. Trouwborst, 'International nature conservation law and the adaptation of biodiversity to climate change: a mismatch?' (2009) 21 *Journal of Environmental Law* 419–442. Specific recommendations with regards to protected areas are given in, e.g.: J. Baron, L. Gunderson, G. Allen, E. Fleishman, D. McKenzie, L. Meyerson, J. Oropeza and N. Stephenson, 'Options for national parks and reserves for adapting to climate change' (2009) 44 *Environmental Management* 1033–1042; L. Hannah, G. Midgley, S. Andelman, M. Araújo, G. Hughes, E. Martinez-Meyer, R. Pearson and P. Williams, 'Protected area needs in a changing climate' (2007) 5 *Frontiers in the Ecology and the Environment* 131–138.
60 Lawler, 'Climate change adaptation strategies' 81.
61 See Mawdsley et al., 'A review of climate-change adaptation strategies'.
62 K.A. Keenleyside, N. Dudley, S. Cairns, C.M. Hall and S. Stolton, *Ecological Restoration for Protected Areas: Principles, Guidelines and Best Practices* (IUCN 2012) 12.

increasing the number and the size of protected areas, creating buffer zones around them, and adding larger protected areas to ecological networks as climate change adaptation mechanisms.[63] Protected areas that have a high level of ecological integrity will be relatively resilient to change because they can better resist change and/or can better tolerate and adapt to new climatic conditions.[64] As protected areas are one of the best ways of protecting biodiversity, restoration of protected areas can help to increase ecological functioning, and will help to avoid some of the negative effects of climate change. To support climate change adaptation, scientists frequently advocate the establishment or restoration of connectivity between protected areas. Corridor areas or stepping stones might create possibilities for species to disperse to other areas in case their original habitat has become unsuitable due to climate change induced changes such as temperatures changes. The legal aspects of connectivity are covered in Chapter 10 on protected areas. Scientists also frequently promote adaptive management as an important approach to address climate change. While the concept of adaptive management is central to all ecological restoration, it is particularly important in the context of rapid change.[65] The next section will examine how far these scientific recommendations have been incorporated into international biodiversity law.

Adapting nature to climate change in international biodiversity law

In most multilateral biodiversity conventions, an explicit reference to climate change in the convention text is missing. This is no surprise as most of these conventions were drafted in a period before climate change was a priority policy concern. Even the Biodiversity Convention, drafted in the same year as the UN Framework Convention on Climate Change, does not contain any explicit provision on climate change. The lack of explicit language in other treaties such as the Biodiversity Convention does not diminish the role that these instruments can play in making ecosystems more resilient to climate change effects. By providing a legal basis for establishing and managing protected areas and protecting species, these treaty regimes can contribute to the long-term adaptation of ecosystems to climate impacts. In particular, states have used COP decisions to address the challenges of climate change and to encourage restoration efforts.[66] For example, the COP to the Biodiversity Convention has taken several decisions on climate change and

---

63 See Lawler, 'Climate change adaptation strategies' 81.
64 Keenleyside et al., *Ecological Restoration for Protected Areas* 12.
65 Ibid., 14.
66 For an overview, see also A. Cliquet, 'International and European law on protected areas and climate change: need for adaptation or implementation?' (2014) 54 *Environmental Management* 721–722; Trouwborst, 'International nature conservation law'; A. Trouwborst, 'Conserving European biodiversity in a changing climate: the Bern Convention, the European Union Birds and Habitats Directives and the adaptation of nature to climate change' (2011) 20 *RECIEL* 62–77; A. Trouwborst, 'Transboundary wildlife conservation in a changing climate: adaptation of the Bonn Convention on migratory species and its daughter instruments to climate change' (2012) 4 *Diversity* 258–300.

biodiversity.[67] Several of these decisions explicitly point to the role of restoration. For example, in a 2006 COP Decision parties were encouraged 'to integrate bio-diversity considerations into all relevant national policies, programmes and plans in response to climate change; taking into account the maintenance and restoration of the resilience of ecosystems which are essential for sustaining the delivery of their goods and services'.[68]

At the 10th COP meeting in 2010, parties were invited to consider the guidance in the decision on ways to conserve, sustainably use and restore biodiversity and ecosystem services while contributing to climate change mitigation and adaptation. This included, inter alia, to 'Reduce the negative impacts from climate change as far as ecologically feasible, through conservation and sustainable management strategies that maintain and restore biodiversity' and 'Strengthening protected area networks including through the use of connectivity measures such as the devel-opment of ecological networks and ecological corridors and the restoration of degraded habitats and landscapes'.[69] In 2014, a COP Decision invited parties to the Convention, to 'promote ecosystem restoration activities, in particular large-scale restoration activities, noting also the cumulative benefits of small-scale restoration activities that can collectively contribute to biodiversity conservation, climate-change adaptation and mitigation, and reducing desertification, in the context of sustainable development'.[70]

Other conventions, such as the Ramsar Convention, have also focused on ecolo-gical restoration in relation to climate change. In a 2012 Resolution, parties focused on the implications of climate change for the Ramsar Convention. Con-tracting Parties were encouraged to maintain or improve the ecological character of wetlands, including their ecosystem services, by enhancing the resilience of

---

67 Decision VII/15. Biodiversity and Climate Change (2004); Decision VIII/30. Biodi-versity and climate change: guidance to promote synergy among activities for biodiversity conservation, mitigating or adapting to climate change and combating land degradation (2006); Decision IX/16. Biodiversity and climate change (2008); Decision X/33. Bio-diversity and climate change (2010); Decision XI/21. Biodiversity and climate change: integrating biodiversity considerations into climate-change related activities (2012); Decision XI/20. Climate-related geoengineering (2012); Decision XI/19. Biodiversity and climate change related issues: advice on the application of relevant safeguards for biodiversity with regard to policy approaches and positive incentives on issues relating to reducing emissions from deforestation and forest degradation in developing countries; and the role of conservation, sustainable management of forests and enhancement of forest carbon stocks in developing countries (2012); Decision XII/20. Biodiversity and climate change and disaster risk reduction (2014); for these and other decisions relevant for biodiversity and climate change, see the CBD website at <www.cbd.int/climate/decision.shtml>.
68 Decision VIII/30. Biodiversity and climate change: guidance to promote synergy among activities for biodiversity conservation, mitigating or adapting to climate change and combating land degradation, UNEP/CBD/COP/DEC/VIII/30 (2006) para. 1.
69 Decision X/33. Biodiversity and climate change, UNEP/CBD/COP/DEC/X/33 (2010) para. 8, c and d, iii.
70 Decision XII/19. Ecosystem conservation and restoration, UNEP/CBD/COP/DEC/XII/19 (2014) para. 4, d.

wetlands as far as possible to climate-driven ecological changes and, where necessary, restoring degraded wetlands.[71]

Parties to the Ramsar Convention have paid particular attention to the protection and restoration of peatlands. The COP to the Ramsar Convention on Wetlands at its 8th meeting adopted Resolution VIII/3 on climate change and wetlands, which called on relevant countries to take action to minimise the degradation as well as promote the restoration of those peatlands and other wetland types that are significant carbon stores or have the ability to sequester carbon. Because of the important role of peatlands, global action guidelines for peatlands have been elaborated under the Ramsar Convention,[72] in which the restoration of peatlands is seen as a priority for Contracting Parties. Emphasising the importance of peatlands, Resolution XII.11 of the 12th meeting of the parties in 2015 encouraged the Contracting Parties to designate as Wetlands of International Importance at least one peatland area that is also suitable for communication, education and awareness raising about the conservation, restoration and wise use of peatlands and the services they provide, such as their role in relation to climate change, the protection of habitats for rare and threatened species and the provision of water supplies.[73]

Parties to the World Heritage Convention consider that World Heritage may be 'in danger' because of certain causes, including serious fires, landslides, changes in water level, floods and tidal waves.[74] As such climate change is not mentioned in the Convention text, but the above-mentioned events can be a consequence of global climate change. The Operational Guidelines contain criteria for the inscription of properties on the List of World Heritage in Danger. In 2008, the World Heritage Committee amended these criteria and agreed that the emphasis on corrective measures for World Heritage sites that are in danger due to climate change should be on 'adaptation' rather than on 'mitigation'.[75] The criteria in the Operational Guidelines now include the 'threatening impacts of climatic, geological or other environmental factors'.[76] World Heritage sites under threat of climate change impacts can give rise to reactive monitoring, including a site on the World Heritage in Danger List, or the removal of a site from the World Heritage List.[77] If climate change fundamentally alters the characteristics of a site that formed the basis of its inclusion in the World Heritage List, the site could be removed from the List.[78] A failure to take restoration measures might lead to the removal of a

---

71 Resolution XI.14. Climate change and wetlands: implications for the Ramsar Convention on Wetlands (2012) para. 26.
72 Resolution VIII.17. Guidelines for Global Action on Peatlands (2002).
73 Resolution XII.11. Peatlands, climate change and wise use: Implications for the Ramsar Convention (2015).
74 Article 11(4), World Heritage Convention.
75 World Heritage Committee, Decision 32COM 7A.32 (2008).
76 UNESCO, Operational Guidelines for the Implementation of the World Heritage Convention, WHC.15/01 (2015) para. 179.
77 See Chapter 10.
78 Although various World Heritage sites are under threat from climate change (see some examples in A. Colette, *Climate Change and World Heritage Report on Predicting and Managing the Impacts of Climate Change on World Heritage and Strategy to*

site from the World Heritage List. In general, however, the work under the World Heritage Convention focuses on adapting World Heritage sites to the effects from climate change.[79]

In the framework of the Convention on Migratory Species, several resolutions from the Conference of the Parties are relevant for climate change. In a Resolution adopted at the 10th meeting in 2011 the parties and the Scientific Council were urged to improve the resilience of migratory species and their habitats to climate change. States were encouraged to reduce known threats besides climate change in order to maintain or increase population size and genetic diversity and to consider ex situ measures such as assisted colonisation or translocation for those migratory species most severely threatened by climate change.[80] States adopted a programme of work on climate change and migratory species in 2014. It contains measures to facilitate species adaptation in response to climate change, including action to 'integrate protected areas into wider landscapes and seascapes, ensure appropriate management practices in the wider matrix and undertake the restoration of degraded habitats and landscapes/seascapes'.[81] While parties adopted this habitat recovery as a long-term objective to be completed within three triennia or longer, it demonstrates the importance for states of ecological restoration as a mechanism to adapt to climate change.

The above-mentioned examples show that under the major international biodiversity conventions, although international convention texts themselves do not explicitly refer to climate change, the need to adapt to the effects of climate change to protect biodiversity is clearly recognised and restoration measures are one approach to adaptation.

Is biodiversity law 'climate proof'?

Given the dynamic nature of ecosystems, some authors argue that nature conservation laws are too static and unfit to deal with a dynamic system.[82] The dynamic nature of ecological processes will become increasingly uncertain in the

---

Assist States Parties to Implement Appropriate Management Responses (World Heritage Report No. 22, UNESCO World Heritage Centre 2007) 20–24), no sites have been removed from the List because of climate change.

79 Guidance is given to Contracting Parties in several reports, inter alia, Policy Document on the Impacts of Climate Change on World Heritage Properties, Document WHC-07/16.GA/10 adopted by the 16th General Assembly of States Parties to the World Heritage Convention (2007); Colette, *Climate Change and World Heritage Report*; J. Perry and C. Falzon, *Climate Change Adaptation for Natural World Heritage Sites: A Practical Guide* (World Heritage Paper Series No. 37, UNESCO 2014).

80 UNEP/CMS/Resolution 10.19. Migratory species conservation in the light of climate change (2011) para. 6.

81 UNEP/CMS/Resolution 11.26. Programme of work on climate change and migratory species (2014).

82 See, for instance: B. van Leeuwen and P. Opdam, 'Klimaatsverandering vergt aanpassing van het natuurbeleid' (2003) 104 *De Levende Natuur* 122–124; H. Woldendorp, 'Dynamische natuur in een statische rechtsorde' (2009) 3 *Milieu & Recht* 134–143.

face of human-induced climate change. Is international law adequate to restore nature to either counter or cope with the negative effects of climate change, or in other words is the law 'climate proof'?[83] The recommendations in scientific literature point to the necessity of making ecosystems more resilient so that they can recover from the negative effects of climate change. To increase resilience, states must have policies in place capable of increasing and restoring available habitat; allowing for dispersal and migration of species; and reducing the negative impacts of other threats to biodiversity, such as habitat fragmentation and habitat deterioration.[84] From the perspective of law, states must have policies designed to designate additional protected areas, or increase the size of existing areas; to increase connectivity between core protected areas by protecting linear elements or stepping stones in order to allow for species dispersal; and to restore both core protected areas and connectivity areas.

Can international law facilitate implementation of the above-mentioned policies by individual states? Two aspects will be further explored: 1) the need to expand protected area networks; and 2) the need for adaptive management in order to cope with ecological dynamics.

Expanding networks of protected areas include both increasing and expanding core protected areas, as well as providing connectivity measures between protected areas, in order to allow for species dispersal. This means that additional areas will have to be designated or existing areas will need to be enlarged. In general, international law allows or even obliges states to designate additional protected areas. The obligations to designate protected areas do not end upon ratification of the conventions or with the transposition of EU law in national laws. The designation is not considered to be a 'one time operation'.[85] This shows clearly from case law of the European Court of Justice on the designation of sites for the Natura 2000 network.[86]

International law is actively driving the designation of protected areas as a proactive measure to cope with climate change. As was discussed above, several international biodiversity conventions urge states to strengthen the ecological networks as a way to face the challenges of climate change. Also, the criteria to designate sites include criteria that can allow for the designation of additional areas to support the restoration of habitats and species. Although most criteria for the designation of protected areas are aimed at the designation of areas because of the

---

83 See, for instance, the analysis by Arie Trouwborst on the Habitats Directive: A. Trouwborst, 'The Habitats Directive and climate change: is the law climate proof?', in C.-H. Born, A. Cliquet, H. Schoukens, D. Misonne and G. Van Hoorick (eds), *The Habitats Directive in its EU Environmental Law Context: European Nature's Best Hope?* (Routledge 2015) 303–324.

84 See also Trouwborst, 'International nature conservation law' 428.

85 See A. Cliquet, C. Backes, J. Harris and P. Howsam, 2009, 'Adaptation to climate change: legal challenges for protected areas' (2009) 1 *Utrecht Law Review* 158–175, 164.

86 European Court of Justice, Case C-209/04 *Commission v Austria* [2006] ECR I-2755.

actual presence of certain habitats or species in a certain place, the designation of additional areas that allow for the dispersal of species is certainly not excluded. For instance, the criteria for the designation of Ramsar sites explicitly include criteria that relate to climate change. Guidelines for the application of Criterion 2 ('A wetland should be considered internationally important if it supports vulnerable, endangered, or critically endangered species or threatened ecological communities') refer to sites that can no longer develop under contemporary conditions (because of climate change or anthropogenic interference, for example).[87] The criteria for designation of Special Areas of Conservation under the EU Habitats Directive include inter alia 'restoration possibilities'.[88] Although the designation of sites to cope with the effects of climate change is not explicitly included in the criteria, it can be taken into account while designating sites.[89] Although legal possibilities exist for additional site designation and expansion, the political will to actually implement the existing laws in this regard might be missing. Measures are needed to increase political and public support. Researchers advocate that more investments are needed including educational and financial support.[90]

As was already stated in Chapter 10, although attention has been paid to connectivity in international and EU law, the law lacks concrete and binding obligations for the establishment or restoration of areas of connectivity. Still many uncertainties exist with regards to connectivity, especially in light of climate change. Connectivity measures that allow for species dispersal are not necessarily beneficial for all species and habitats. There are concerns about invasive species impacts as migrating invasive species outcompete local, endemic species. Also, there are scientific uncertainties with regards to species migration patterns because of climate change, making it more difficult to proactively take measures that enable migration. However, according to Trouwborst the urgency for taking adaptation measures in biodiversity conservation cannot wait until there is full scientific certainty of migrations patterns. In this sense, the precautionary principle can play a role here. For instance, with regards to connectivity, EU Member States should proactively create ecological infrastructure ensuring mobility for all species groups, rather than restricting these measures for cases in which scientific studies have conclusively established that a certain species in a specific site is in trouble due to climate change. On the other hand, the precautionary principle also can play a role in restraining measures for instance with regards to translocation of species.[91]

The second issue that needs further exploration is the alleged static character of existing nature conservation laws as opposed to ecological dynamics. Many

---

87 Resolution IX.1 Annex B. Revised Strategic Framework and guidelines for the future development of the List of Wetlands of International Importance (2005).
88 See the criteria in Annex III of the Habitats Directive.
89 See also Cliquet, 'International and European law on protected areas and climate change' 725–726; Trouwborst, 'The Habitats Directive and climate change' 315.
90 A. Hochkirch et al., 'Europe needs a new vision for a Natura 2020 Network' (2013) 00 *Conservation Letters* 1–6.
91 Trouwborst, 'The Habitats Directive and climate change' 321–322.

conservation laws oblige states to keep certain habitats and species in a favourable conservation status.[92] If species and habitats are in an unfavourable conservation status, for instance because of the effects of climate change, restoration measures will be legally required. The objective of restoration measures may refer to a historical reference situation.[93] It is possible that if an ecosystem is changing in response to climate change it could become difficult, if not impossible, to restore habitats or species to a historical reference situation? However, certain laws contain an obligation to achieve a certain result. Is law based on static objectives capable of dealing with these dynamic changes?

Scientific literature suggests that in some cases efforts should no longer be made to return an ecosystem to a historical reference situation, but decision-makers should embrace new opportunities to introduce species and habitats that will be more resilient to existing climate conditions.[94] Does that mean that a reference in law to restore to a historical reference situation becomes useless? The answer is highly context specific. The effects of environmental and ecological change are not distributed evenly and may vary tremendously at local and regional scales. As a result, some protected areas may be relatively resistant to change and restoration with a focus on historically determined goals will still be viable.[95]

It could also be that ecosystems become 'hybrid ecosystems' that have some characteristics of the current or historical ecosystem, but which because of changes in species composition and function exist outside their historical range of variability. Because these hybrid ecosystems can still contain original keystone species and most of the original ecosystem functions, restoration in these cases can contribute to maintaining ecosystem integrity.[96]

If changes are so drastic that the original keystone species are lost and many of the original ecosystem functions are lost or altered, however, then an ecosystem may become a 'novel ecosystem' where restoration to a historical baseline might be impossible. In that situation restoration could be aimed at restoring ecosystem functions.[97] The concept of novel ecosystems raises concerns among restoration scientists and practitioners. Concern exists that this concept might create the impression that novel ecosystems are 'the new normal', and that restoration towards a historical trajectory has become difficult, if not impossible, for many ecosystems. However, many ecosystems, even when degraded, can be restored to some extent.[98] Even when novel ecosystems involving new species assemblages

---

92 See, for example, the EU Habitats Directive; see Chapter 7.
93 Keenleyside et al., *Ecological Restoration for Protected Areas* 13.
94 See, for instance, R. Hobbs, E. Higgs and J. Harris, 'Novel ecosystems: implications for conservation and restoration' (2009) 11 *Trends in Ecology and Evolution* 599–605.
95 Keenleyside et al., *Ecological Restoration for Protected Areas* 13.
96 S. Allison, *Ecological Restoration and Environmental Change: Renewing Damaged Ecosystems* (Routledge 2012) 100.
97 Ibid., 100–102.
98 See, for instance, C. Murcia, J. Aronson, G. Kattan, D. Moreno-Mateos, K. Dixon and D. Simberloff, 'A critique of the "novel ecosystem" concept' (2014) 10 *Trends in Ecology and Evolution* 548–553.

could be considered necessary, the use of historical information may still be significant as a source of context and constraint in shaping goals of restoration projects. Historical knowledge will play a key role in restoration, for example in improving understanding of range shifts, species interactions and adaptive capacity, regardless of the extent to which it is used as the basis of goal-setting.[99]

The call for more flexibility in law and concepts such as novel ecosystems, as a means to cope with climate change, could easily be abused as a means of weakening the legal obligations to conserve and restore ecosystems.[100] If novel ecosystems are regarded as the new normal, this could probably reduce the willingness of states or other stakeholders to undertake restoration efforts towards a historical trajectory of ecosystems or it could lead them to lower the standards for restoration. If the loss of original keystone species and habitats is seen as inevitable, this could easily lead to a weakening of the protection system of these habitats and species. It might lead to declassification of protected areas, as the keystone species and habitats for which the area has been designated were lost and cannot be restored. However, species could also be gone because of the lack of sufficient management measures, or because of fragmentation and other negative impacts on biodiversity. As Craig rightly points out, increasing regulatory flexibility always opens the door to potential abuse.[101] Conservation and restoration towards a historical trajectory should remain the predominant strategy in law.

Rather than adapting the existing legislation on conservation and restoration to climate change, to make the law 'more flexible', it seems much more important to actually implement the existing legislation properly and take conservation and restoration measures that make ecosystems more resilient. In recent analysis of the EU Birds and Habitats Directives, researchers observe that 'climate change adaptation will require the interpretation and implementation of the Nature Directives to be further developed, but that their fundamental construction is as sound today as it was when they were adopted'.[102] Arie Trouwborst, in his analysis on the EU Habitats Directive, concludes that 'the law is climate proof. It just needs to be applied'.[103]

---

99 Keenleyside et al., *Ecological Restoration for Protected Areas* 13.
100 See also, for instance, D. Simberloff, C. Murcia and J. Aronson, '"Novel ecosystems" are a Trojan horse for conservation. They provide a license to trash nature if they provide ecosystem services' (January 2015) *Ensia*, <http://ensia.com/voices/novel-ecosystems-are-a trojan horse for conservation/>.
101 R. Craig, '"Stationary is dead" – Long live transformation: five principles for climate change adaptation law' (2010) 34 *Harvard Environmental Law Review* 9–73; see also Trouwborst, 'International nature conservation law'; C. Bastmeijer and K. Willems, '"Robuust, verbonden en ... beschermd". Past een klimaatbestendig natuurbeleid met aandacht voor "wilde natuur"-beleving in het juridische Natura 2000-jasje?', in C. Backes et al. (eds), *Natuur(lijk) met recht beschermd: naar een effectieve en hanteerbare natuurbescherming* (Boom Juridische uitgevers 2010) 85–115.
102 A. Dodd, A. Hardiman, K. Jennings and G. Williams, 'Protected areas and climate change: reflections from a practitioner's perspective' (2010) 6 *Utrecht Law Review* 141–150, 148; see also A. Dodd, 'EU nature directives: rights, responsibilities and results – are we striking the right balance?' (2008) 20 *Environmental Law & Management* 237–245, 244.
103 Trouwborst, 'The Habitats Directive and climate change' 324.

Conclusion

Climate change will pose additional threats to biodiversity. Restoring degraded ecosystems can help nature adapt to the impacts of climate change. Restoration might become more challenging, as it could become more difficult to restore an ecosystem to a historical reference level. Even when this is the case, historical information still remains important as a context and constraint in determining goals for restoration projects.

Additionally, restoring ecosystems can help in mitigation and adaptation to climate change. The role of restoration has been increasingly recognised, in treaty law, such as the Rio Conventions and the Ramsar Convention. In addition, voluntary commitments have been made to restore degraded lands, such as the Bonn Challenge to increase carbon sinks.

However, several concerns remain in using restoration as a strategy for mitigation and adaptation. Most of the commitments are voluntary, so it remains to be seen whether the commitments made will actually be met. In general, the international focus on restoration has been on restoration of forests for carbon sequestration and the importance of restoring other ecosystems has been largely overlooked. Because negative effects might arise for biodiversity and other ecosystem services by restoring ecosystems primarily for their services as carbon sinks, it is therefore crucial that safeguards are included in restoration planning to make sure that mutual benefits arise, and that restoration for carbon sinks is not done at the expense of biodiversity.

12 Future directions for law on ecological restoration

Multinationals gauge the success of their operation each economic quarter. Government agencies evaluate the effect of their work plan on a fiscal year. Through collective efforts such as the Millennium Development Goals and Sustainable Development Goals, the United Nations measure achievement on the basis of decade-long time frames. Most landscape restoration projects cannot achieve objectives on such limited timelines. As one landscape restoration project illustrates, there is no need for ecological ventures to confine themselves to human timelines.

In a corner of Wellington, New Zealand, a few kilometres from the central business district of the nation's government ministries, a group of ecological visionaries challenge the perception of human time. In a valley that was mined for gold, stripped of vegetation, replanted with exotic trees, and formerly used for Wellington's reservoir, the leadership of Zealandia (Karori Wildlife Sanctuary) have taken the first steps to create a 235-hectare 'mainland island' to reintroduce birds and vegetation, destroyed by predators introduced over the centuries to New Zealand, by introducing a predator-proof fence and reintroducing species that have nearly vanished.[1] While the Anthropocene epoch has been a long time in the making, the efforts of a dedicated group of conservationists in a few decades has resulted in the first 500-year restoration plan. The staff and dedicated volunteers of Zealandia are committed to restoring the reserve to the ecological conditions that reflect abundant native flora and fauna.

Contemplating the loss of sea ice, toxic smog and basic needs of a growing human population including food production, a 500-year restoration plan for 235 hectares may seem like a fool's errand given the current scale of existing environmental crises. Yet, the Zealandia effort is pivotal because it raises significant questions about individual, community and national ethics and a new definition of what constitutes 'connective justice'. Recognising that every state has individual and often unique restoration challenges, this conclusion does not contain any neatly packaged one-size-fits-all policy proposals, but instead offers some

---

1 Zealandia, Forest Restoration, <www.visitzealandia.com/what-is-zealandia/conserva tion-restoration/forest-restoration/>.

observations on the current strengths and weaknesses of ecological restoration as a pragmatic legal regime, followed by some thoughts on three legal questions for further debate and discussion by ecological restoration practitioners and policy-makers:

1   What is the relationship between ecological restoration and justice?
2   If necessary for improving restoration outcomes, how might the duty to restore from an obligation of conduct evolve to an obligation of result?
3   Is ecological restoration changing how we conceive of international environmental law?

Strengths and weaknesses of the existing ecological restoration regime

This book has reviewed a wide scope of restoration laws and policies including international laws, international frameworks, regional laws, regional cases, national laws and private practices. What becomes apparent is that ecological restoration when viewed from a legal lens is in a state of flux. Some of this state of rapid flux reflects some of the strengths of the restoration regime. For example, the concept of restoration as a national obligation supporting a broader collective interest is clearly gaining momentum. Since the start of the twenty-first century, ecological restoration has become a mainstream practice supporting sustainability initiatives and conservation efforts. Dozens of states have begun investing in large-scale restoration planning that includes changes to existing laws to facilitate restoration activities. Some of these investments, particularly in reforestation of denuded areas, are beginning to make a visible difference. As momentum builds behind restoration projects, States are increasingly joining multi-stakeholder partnerships that involve a variety of actors.

A review of the broad scope of laws on restoration also reflects some of the weaknesses in the existing regime. The term 'restoration' is used frequently but with little uniformity. Almost no laws provide a specific vision for what outcomes actors are expected to achieve. This lack of definition may be problematic in terms of guiding actors. For example, is achieving a restoration effort that produces a novel ecosystem adequate for achieving the purpose of a law? While it may be implicit in some of the laws requiring restoration that restoration efforts will take decades to achieve, this is not reflected in actual text. In fact, some of the texts involving restoration policies such as the Aichi Targets set restoration goals on a short time frame of a little over a decade. As with many environmental laws, the political constraints of law-making often do not reflect accurately emerging ecological knowledge. Ecologists understand that while restoration projects are often more successful than they have been in the past in restoring key functions and structures, ecologists are still learning more about restoration from each project. Existing restoration law is largely non-responsive to the ongoing challenges of doing restoration work in the field. Cooperation between lawyers and restoration ecologists is a fundamental precondition for future development and implementation of international restoration law.

Legal questions for further debate

Ecological restoration and justice: do communities restore for 'connective justice'?

While some aspects of ecological restoration projects, for example replanting mangroves, might serve a human functional purpose such as reviving damaged ecosystem services, there are many aspects of restoration projects that seem to have an appeal beyond utilitarian function. Why do restoration practitioners such as those working at Zealandia invest so many resources in bringing back native species and helping them to flourish and become abundant again? Why do they expend so much effort in creating habitats for rare and often difficult to reinstate species when there are plenty of other more robust species that would happily fill the existing ecological niches? Why do so much work when there is no legal requirement? One answer to these questions might be that restoration practitioners simply enjoy an intrinsically complex challenge. While this might offer one motivation for some restoration practitioners, it does not offer a full explanation of the level of dedication by communities. An alternative response to these questions about effort might be that practitioners are motivated by justice concerns.

Specifically, individuals and communities who fund and carry out restoration work may conceive of their work as a form of corrective justice. The concept of corrective justice is grounded on a pre-existing moral relationship between an individual and society-at-large that is broken or strained by some action taken by individuals. Underpinning much of modern liability theory and criminal penalties, the theory of corrective justice recognises that where an injury has occurred restitution may be necessary not just to make the harmed party whole but also to restore the relationship between a particular individual and society at large. For example, when a youth defaces a wall with graffiti, it would not be uncommon for a legal system to require that youth to both remove the graffiti and do further improvements in the community such as cleaning up graffiti that some other youth might have left. The idea in such a sentence would be to offer an offender the chance to deliver restitution.

Arguably, the same corrective justice motivations prevail within the restoration community even though one important distinction must be made. Unlike the youth expressing himself through graffiti who is the causal agent for a particular harm, there is a lack of causation between restoration practitioners and their restoration work. The individuals seeking to restore habitats and species are not the same individuals who caused the original wrongs. They may, however, be the progeny of the original injurers and possibly indirect beneficiaries of an injury to the extent that an ancestor financially benefited from environmentally destructive activities to such a degree that they were able to deliver long-term financial benefits to future generations.

Correcting the environmental wrongs of the past through contemporary ecological restoration activities offers a moral response by contemporary populations to yesterday's injuries. In the context of justice theory, ecological restoration may be considered an unusual form of corrective justice but it plays the same critical role of building connections. In this case restoration efforts can rebuild connections between people and land as well as connections between the descendants of those who benefited from damaging the environment and others. One might argue that for some, restoration is a form of atonement or a collective reparation. The current global push for ecological restoration is simply a reflection of 'restorative justice' principles where this generation treats the land and waters as equal parties.

The idea of 'corrective justice' has the potential to be a powerful concept underpinning restoration, but for the reasons described above it can also backfire as a motivator. While some individuals may replant trees as a means of healing past wrongs and restoring their relationship with the land, other individuals can rightfully argue that they have no direct responsibility for the current state of the environment. With no responsibility, there is no inherent obligation to undertake long-term and expensive restoration efforts.

Perhaps it is time for a new concept to explain why the present generation must repair the damage of former generations. Terms such as 'environmental justice' and 'natural justice' already have a pre-existing meaning in modern legal circles. In Chapter 3 we discussed how ecological restoration had the potential to broaden how we think about these established concepts. This chapter instead proposes that a better way forward is the further development of a concept of 'connective justice' building on the previous discussion of 'corrective justice' that provides an explanation for why individuals engage in more than simply minimalistic restoration efforts.[2] Connective justice does not presuppose a necessary linkage between an act of restitution and an injury; it is instead predicated on the idea that humans cannot flourish without a healthy environment and that it is the environment that is the critical medium connecting humans across communities and across generations. We undertake certain actions such as ecological restoration projects in order to understand our connectivity with past generations and to improve ecological integrity for future generations. Connective justice can liberate ecosystems from the stranglehold of development by ensuring that the past and future are taking into account.

International duty to restore: how might laws evolve from obligations of conduct to obligations of result?

Second, while this book asserts that the legal duty to restore that is codified across numerous international and regional conventions, the precise content of this legal duty is not always clear. At best, a review of the existing international conventions, regional laws and national laws suggests that states have agreed within their own

---

2 'Connective justice' is a new term for environmental law. The term has been used by Egyptian scholars to explain an Egyptian term 'Ma'at' which refers both to 'justice' and also to 'order, cosmos, and truth'. Scholars understand 'Ma'at' to be the basis for ethical moral behaviour that connects 'the divine order of the world' with the human order; J.G. Manning, 'The representation of justice in Ancient Egypt' (2012) 24(1) *Yale Journal of Law and the Humanities* 111–118, 114.

territories to take steps to reverse ecological degradation. The visions of restoration articulated by states generally do not involve rewilding the landscape or constructing networks of novel ecosystems. Rather, states generally speak of restoring ecosystems to revitalise ecosystem services and biodiversity values, or support vulnerable communities.

The content of a duty to restore is still evolving as demonstrated by the growing number of resolutions and decisions from the Conference of the Parties of several international conventions. To the extent the topic of ecological restoration is clearly on the international agenda and is increasingly a priority for regions such as the EU and countries both in the Global North and Global South, there may need to be a more focused discussion on the nature of what constitutes an international duty to restore.

As suggested in earlier chapters of this book, the obligation to restore has historically been regarded as an obligation of conduct with states complying with such an obligation through a variety of programmes and efforts that do not necessarily share common objectives. Unfortunately, not all efforts titled as 'restoration' offer equally valuable results from an ecological perspective. Simply fulfilling a due diligence obligation of conduct has not been sufficient to reverse environmental degradation. Some of the differences of quality in state-led restoration work can be justified by increasing gaps in financial and human resource capacity. While a lack of capacity might explain some differences in restoration work, differences across international restoration efforts can also be attributed to the lack of standardisation given to defining best practices for ecological restoration.

With the introduction of the Aichi Targets, states seemed to be asserting a nascent interest in the possibility of developing long-term obligations of result. Even though the existing targets do not create legally binding obligations, they do reflect a substantial shift in implementation strategies. Within the target framework, states are now publicly holding themselves accountable to making progress towards the general targets and expecting the same level of accountability from other states committed to conserving and restoring biodiversity. In 2016, the Subsidiary Body on Implementation for the Biodiversity Convention adopted a recommendation on national reporting for the Sixth National Report that requires states to reflect on the implementation of their national targets, the effectiveness of measures to implement the targets, and the assessment of the nation's contributions to achieving the Aichi Biodiversity Targets. These new recommendations for reporting clearly suggest that setting targets will not be sufficient for the international community and that states must be prepared to offer evidence of how states are achieving the Aichi Targets.[3] In theory, the text of the Aichi Targets with its quantitative targets may be realised through national targets that could become one starting place for future restoration-related obligations of result.

---

3 Subsidiary Body on Implementation, Biodiversity Convention, UNEP/CBD/SBI/REC/1/10 (2016), <www.cbd.int/doc/recommendations/SBI-01-sbi-01-rec-10-en.pdf> (states can measure their progress as 'on track to exceed target', 'on track to achieve target', 'progress towards target but at an insufficient rate', 'no significant change', or 'moving away from target').

While it is beyond the scope of this book to analyse the suitability of setting collective and obligatory quantifiable restoration targets, it appears desirable for states to develop some common language and operative framework when approaching the design, implementation and monitoring of long-term restoration projects. While a wetland in the Inner Niger Delta may be biologically and hydrologically different from a wetland in the Amazonian river basin, proponents of ecological restoration projects across a diversity of geographical areas share some basic practical objectives. Negotiating shared standards to guide ecological restoration work should assist states in moving beyond simply fulfilling an obligation of conduct to establishing restoration programmes capable of achieving long-term ecologically meaningful results.

Depending on the available political will, standards for ecological restoration can be negotiated internationally, regionally, nationally or sub-nationally. They can be negotiated exclusively by government actors or they can be negotiated in collaboration with private landowners and other essential stakeholders. Depending on the political context, negotiated standards could be advisory guidelines of best practices to guide restoration practice or might become the foundation for future land and natural resource statutes or regulations. Whether an outcome is binding or non-binding does not define the value in negotiating shared standards. There is substantial normative value in creating structured conversations about what a state or province expects to be able to achieve by developing and implementing a variety of ecological restoration projects.

Recognising the need to be able to provide some guidance about a high-quality ecological restoration project versus a low-quality ecological restoration project, the Society for Ecological Restoration Australasia Chapter (SERA), based on cooperation with 12 non-profit restoration groups, has drafted and published a set of national restoration standards designed 'to reaffirm the place of restoration in a changing and uncertain world, positively acknowledge other forms of environmental repair, and enable maximum inclusivity of practitioners without lowering standards for restoration'.[4] As explained by SERA in its public posting of the Standards:

> The Standards identifies (sic) the principles underpinning restoration philosophies and methods, and outlines the steps required to plan, implement, monitor and evaluate a restoration project to increase the likelihood of its success. The Standards are relevant to – and can be interpreted for – a wide spectrum of projects ranging from minimally resourced community projects to

---

4 T. McDonald, *SER Australia's National Restoration Standards* (Vol. 29, Issue 5, December 2015), <http://ser.org/resources/sernews/volume-29-issue-5-december-2015-Restoration-Standards>; the Standards are a collaborative product of the Australian Association of Bush Regenerators, Australian Institute of Landscape Architects, Australian Network for Plant Conservation, Australian Seed Bank Partnership, Bush Heritage Australia, Gondwana Link, Greening Australia, Indigenous Flora and Fauna Association, The Nature Conservancy, Trees For Life, Trust for Nature and WetlandCare Australia.

large-scale, well-funded industry or government projects. SERA and its Partners have produced these Standards for adoption by community, industry, regulators/government and land managers (including private landholders and managers of public lands at all levels of government) to raise the standard of restoration and rehabilitation practice across all sectors. The document provides a blueprint of principles and standards that will aid voluntary as well as regulatory organisations in their efforts to encourage, measure and audit ecologically appropriate environmental repair in all land and water ecosystems of Australia.[5]

The SERA standard-setting project is potentially significant beyond its application within Australia because it makes a specific distinction between an 'ecological restoration' project and a 'rehabilitation' project. To be an ecological restoration project, the project must be designed to achieve an 'appropriate local indigenous reference ecosystem' and must have a goal of full recovery of the reference ecosystem's attributes.[6] The SERA drafters are realistic about the parameters of what can be achieved through a given restoration effort. An 'indigenous reference ecosystem' can be 'an actual site' or 'a conceptual model synthesised from numerous reference sites, field indicators and historical or predictive records'.[7] Any project focused on partial recovery of certain ecosystem attributes can have environmental benefits but should be regarded as a 'rehabilitation' project. Over time, if states were to adopt this same distinction as a pragmatic legal distinction, this could have implications for whether states are in compliance with some of their existing international obligations of conduct. Various regional and nationally-financed projects that are currently being implemented as 'restoration' projects might actually only qualify as 'rehabilitation' projects if standards such as the National Standards for the Practice of Ecological Restoration in Australia were to apply.

In addition to the important distinction between 'ecological restoration' and 'rehabilitation' which may become increasingly relevant for measuring restoration efforts in light of sometimes substantial shifts in climate, the drafters of the SERA standards offer a series of 'first order principles' and 'second order principles' that are intended to assist decision-makers in ascertaining whether a restoration should be considered successful.[8] These principles cover both core principles such as relying on restoration as 'a substitute for sustainable managing and protecting ecosystems in the first instance' and a variety of principles designed to foster best administrative and ecological practices.

In addition to these principles, SERA has introduced a novel and innovative approach to measuring the extent to which a project achieves a 'recovery outcome'. Identifying six attributes for recovery (absence of threats, physical

---

5 Society for Ecological Restoration Australasia, *National Standards for the Practice of Ecological Restoration in Australia, Executive Summary* (2016).

6 Society for Ecological Restoration Australasia, *National Standards for the Practice of Ecological Restoration in Australia, Key Principles of Ecological Restoration Practice* (2016), Principle 1 and Principle 4.

7 Ibid., Principle 1.

8 Ibid., Appendix 4 for the 19 SERA Principles.

conditions, species composition, community structure, ecosystem function and external exchanges[9]) and assuming the existence of an appropriate local indigenous reference ecosystem, the SERA team has provided a ranking of progress along a 'recovery trajectory' based on a scale of one to five stars.[10] The proposed SERA system of recovery rating could become the basis for a harmonised global evaluation standard to determine the ecological and governance efficacy of various ecological restoration efforts. Where possible, most 'ecological restoration' projects should strive to achieve 'five stars', meaning that:

- A characteristic assemblage of biota is established to a point where structural and trophic complexity is likely to develop without further intervention.[11]
- Appropriate ecosystem exchanges[12] are enabled and commencing, and high levels of resilience are likely with return of appropriate disturbance regimes.[13]
- Long-term management arrangements are in place.[14]

In the long term, the efforts of SERA to assist in defining practice standards may also eventually redefine the context of what qualifies as 'restoration' for legal purposes. From the current situation where states are still trying to make practical sense of their individual restoration obligations in the context of numerous international commitments to 'conserve, protect and restore', states may soon be engaging in more careful technical assessments of restoration outcomes. While there is no way of predicting what norms will eventually become legal obligations, it is a fair wager that the rules of ecological restoration will continue to evolve.

Ecological restoration and international environmental law: will the concept of restoration change the practice of environmental law?

As described in the first chapter of this book, law is essential for conceptualising good governance. The rise of ecological restoration in the last two decades may be transforming the practice of international environmental law from a state-driven practice focused on transboundary impacts to a multi-stakeholder practice of

9 'External exchanges' refers to 'the 2-way flows that occur between elements in the landscape or aquatic environment including flows of energy, water, fire, genetic material, animals and seeds. Exchanges are facilitated by habitat linkages' (Society for Ecological Restoration Australasia, *National Standards for the Practice of Ecological Restoration in Australia* (2016) Glossary).
10 Society for Ecological Restoration Australasia, *National Standards for the Practice of Ecological Restoration in Australia, Key Principles of Ecological Restoration Practice* (2016) Principle 4.
11 Ibid.
12 Here 'ecosystem exchanges' is synonymous with 'external exchanges', Ibid., Appendix 4 for the 19 SERA Principles.
13 Society for Ecological Restoration Australasia, *National Standards for the Practice of Ecological Restoration in Australia, Key Principles of Ecological Restoration Practice* (2016) Principle 4.
14 Ibid.

seeking interdisciplinary global solutions capable of local implementation. As the recent work of the parties to the Biodiversity Convention and the Ramsar Convention demonstrates, restoration as a legal concept is more than simply aspirational. It is a collective project requiring the commitment of already scarce domestic government resources and the commitment of both individual citizens and corporate citizens. Restoration is both a subject for long-term sustainability planning and a conservation objective to be implemented.

With the ongoing mainstreaming of ecological restoration into national policies and strategies, how existing general international environmental principles are conceived by jurists and policy-makers may need to be rethought. For example, should ecological restoration be the basic remedy when applying the polluter pays principle? Given the long-time frames for restoration outcomes, is intergenerational equity the primary driver for existing ecological restoration efforts? If so, does that lead to the prioritisation of some restoration efforts (e.g. restoration of ecosystem services) over other restoration efforts (e.g. restoration of habitat for endangered species without instrumental value for humans)? Does the application of the precautionary principle require vastly greater social investments in existing restoration efforts to avoid the possibility of the emergence of novel ecosystems and a possible loss of ecosystem services? If novel ecosystems emerge due to a lack of precaution and there is a loss of ecological function (e.g. 'jellyfish sea' due to overfishing and eutrophication), does state responsibility follow for the loss of historic ecological function given the explicit duty to restore embodied in international environmental law?

As a concept, restoration challenges basic environmental ideas of compliance. Under a variety of existing environmental laws, parties must stop certain environmentally destructive behaviour, install certain equipment, clean up contaminated areas, conserve certain areas or pay fines. Compliance with these laws can be measured. Has a party released chemicals into the environment or installed particulate filters? Has a party implemented a clean-up plan and is there a reduction of metals in the soil? Has a party set aside land to be protected from development? Where law requires parties to engage in ecological restoration, it is harder to measure implementation success. Unlike other environmental laws, like pollution laws where compliance depends on measurements that can be taken today, laws that require restoration rely on achieving a future outcome based on past conditions that are no longer present. This requires the law to function dynamically, with law-making, law-interpreting and law-enforcing institutions forecasting future restoration outcomes that may or may not be technically achievable and then measuring incremental progress towards the anticipated outcomes. Because ecological restoration requires an ongoing process cutting across generations and often across jurisdictional lines, the practice of ecological restoration raises questions about the capacity of existing legal institutions to effectively mediate between humans and the environment. If ecological restoration is to be realised at the landscape level with long-term time horizons as urged by ecological restoration practitioners, then existing environmental law may need to broaden its reach both spatially and temporally.

Ecological restoration and law: three world visions

How does the evolving practice of ecological restoration shape law? In the age of the Anthropocene, humans are no longer bystanders to the shaping of Earth processes, but active agents. The existing trial and error practice of restoration offers potentially valuable lessons for law, including both the multi-generational time frames necessary for effective implementation of environmental law and the potential of environmental law as not just an administrative code to punish polluters but also a force to revitalise humans' relationship with wild or cultural spaces. The New Zealand nature reserve Zealandia offers an interesting long-term case study for the evolution of ecological restoration as a practice. While the decision-makers of today will not be here in 500 years as the Zealandia community strives to achieve its ecological restoration outcomes, it is this generation that has the critical opportunity to negotiate and implement rules to create a positive legal trajectory for progress towards long-term ecological restoration outcomes. Whether today's ecological restoration efforts will revive environmental values such as biodiversity that collectively matter to this generation remains to be seen many decades from now. Whether today's restoration efforts conserve values that will be core to the well-being of future generations is a great unknown.

What is certain is that the socio-ecological challenges of today reflected in the degradation of ecosystems will be compounded over time if business continues as usual. Environmental laws that conceive of a different future based in part on a known past where ecosystems were capable of self-functioning may prove to be both an essential and positive social disruptor. We close this book with three future world visions.

In the first world, the environment has intrinsic value. In this world, rivers, old growth forest, whales, gorillas and plankton – in short, biodiversity – matter simply because they exist. Unfortunately for rivers and gorillas, humans also exist; many but not all of these humans carry an intrinsic bias that their needs are more important than the needs of rivers and gorillas.

In the second world, the environment has functional value and exists to meet human needs. Rivers and forests matter because they satisfy human needs, ranging from aesthetic and spiritual needs to energy needs. In this world, functions can be commodified, packaged and traded. These two worlds are increasingly difficult to reconcile with lived human experience where the value of nature is not understood in such polarised perspectives.

In a third world, humans live in an acknowledged close relationship with the environment and are continually mediating this relationship as a way of creating purpose and meaning in human lives. What this means in practice is that if communities are honest about the cumulative human impact on the environment across nations, communities will collectively act through institutions, laws and other social mechanisms to restore broken or fraying environmental connections. As this book goes to publication, this vision of a third world seems a little distant. Media reports are heavy with stories about violence, industry, celebrity and commodities. It is rare to read a celebration of the work of a handful of concerned citizens planting trees, plucking weeds or releasing hatchlings into the wild. Yet the possibility of placing this generation on a trajectory to achieve this third-world vision where humans live in relation to nature is not beyond imagination. While laws alone will not achieve new relations between humans and the environment, law, as the product of good faith negotiations between States and other interested stakeholders, can provide a pathway where forests are replanted, reefs are regenerated and human well-being is restored.

Bibliography

Books and journals

AgInnovations, Permitting Restoration Case Study: Apanolio Creek Fish Passage Project, <www.aginnovations.org/uploads/result/1431289151-93b99a9dd9a0442df/Apanolio_Creek.pdf>.

Agyeman, J., B. Doppelt and K. Lynn, 'The climate-justice link: communicating risk with low-income and minority audiences' in S. Moser and L. Dilling (eds), *Communicating a Climate for Change: Communicating Climate Change and Facilitating Social Change* (Cambridge University Press 2007) 119–138.

Akhtar-Khavari, A., 'Beyond compliance and the sea disposal of dredged and excavated materials' (1998) 1 *Maritime Studies* 23–33.

Akhtar-Khavari, A., 'Accessing ecological justice in the Anthropocene epoch!' in P. Keyzer, V. Popovski and C. Sampford (eds), *Access to International Justice* (Routledge 2015) 199–224.

Akhtar-Khavari, A., *Global Governance of the Environment: Environmental Principles and Change in International Law and Politics* (Edward Elgar 2010).

Alexander, S. et al., 'Opportunities and challenges for ecological restoration within REDD+' (2011) 6 *Restoration Ecology* 683–689.

Allison, S., *Ecological Restoration and Environmental Change: Renewing Damaged Ecosystems* (Routledge 2012).

Aronson, J. et al., 'Restoring natural capital: definitions and rationale' in J. Aronson, S. Milton and J. Blignaut (eds), *Restoring Natural Capital: Science, Business, and Practice* (Island Press 2007) 3–9.

Australia 21 Limited, Discussion Paper on Ecosystem Services for the Department of Agriculture, Fisheries and Forestry (Final Report), 5 July 2012, <www.australia21.org.au/publication-archive/discussion-paper-on-ecosystem-services-for-the-department-of-agriculture-final-report-for-the-department-of-agriculturefisheries-and-forestry/#.VPRAKk2zXcc>.

Balaguer, L., A. Escudero, J. Martin-Dugue, I. Mola and J. Aronson, 'The historical reference in restoration ecology: re-defining a cornerstone concept' (2014) 176 *Biological Conservation* 12–20.

Balick, M. and L. Mendelsohn, 'Assessing the economic value of traditional medians from tropical rainforest' (1992) 6 *Conservation Biology* 128–130.

Baron, J., L. Gunderson, G. Allen, E. Fleishman, D. McKenzie, L. Meyerson, J. Oropeza and N. Stephenson, 'Options for National Parks and Reserves for adapting to climate change' (2009) 44 *Environmental Management* 1033–1042.

Barrett, N., 'The promise and peril of ecological restoration: why ritual can make a difference' (2011) 32(2) *American Journal of Theology and Philosophy* 139–155.

Bastmeijer, C. and K. Willems, '"Robuust, verbonden en … beschermd". Past een klimaatbestendig natuurbeleid met aandacht voor "wilde natuur"-beleving in het juridische Natura 2000-jasje?' in C. Backes et al. (eds), *Natuur(lijk) met recht beschermd: naar een effectieve en hanteerbare natuurbescherming* (Boom Juridische uitgevers 2010) 85–115.

Bellard, C., C. Bertelsmeier, P. Leadley, W. Thuiller and F. Courchamp, 'Impacts of climate change on the future of biodiversity' (2012) 15 *Ecology Letters* 365–377.

BenDor, T. et al., 'Estimating the size and impact of the ecological restoration economy' (2015) 10(6) *PLoS One* 1–15, <http://journals.plos.org/plosone/article?id=10.1371/journal.pone.0128339>.

Bennett, A.F., *Linkages in the Landscape: The Role of Corridors and Connectivity in Wildlife Conservation* (IUCN 1998, 2003).

Bennett, G. and K.J. Mulongoy, *Review of Experience with Ecological Networks, Corridors and Buffer Zones* (CBD Technical Series No. 23, 2003).

Bertzy, B. et al., *Protected Planet Report 2012: Tracking Progress towards Global Targets for Protected Areas* (IUCN and UNEP-WCMC 2012).

Beyerlin, U., 'Different types of norms in international environmental law policies, principles, and rules' in D. Bodansky, J. Brunee and E. Hey (eds), *Oxford Handbook of International Environmental Law* (Oxford University Press 2008) 425–448.

Biermann, F. et al., 'Down Earth: contextualizing the Anthropocene epoch' (2015) *Global Environmental Change*, <http://dx.doi.org/10/1016/j.gloenvcha/2015.11.004>.

Blanco, E. and J. Razzaque, 'Ecosystem services and human well-being in a globalized world: assessing the role of law' (2009) 31 *Human Rights Quarterly* 692–720.

Blanco, E. and J. Razzaque, *Globalisation and Natural Resources Law: Challenges, Key Issues and Perspectives* (Edward Elgar 2011).

Blewett, T. and G. Cottam, 'History of the University of Wisconsin Arboretum Prairies' (1984) 72 *Transactions of the Wisconsin Academy of Sciences Arts and Letters* 130–144.

Bond, B., 'Trends in the state of nature and their implications of human well-being' (2005) 8 *Ecological Letters* 1218–1234.

Born, C.-H., A. Cliquet, H. Schoukens, D. Misonne and G. Van Hoorick (eds), *The Habitats Directive in its EU Environmental Law Context: European Nature's Best Hope?* (Routledge 2015).

Bosselmann, K., *The Principle of Sustainability: Transforming Law and Governance* (Ashgate 2014).

Bradshaw, A., 'Introduction and philosophy' in M. Perrow (ed) *Handbook of Ecological Restoration* (Cambridge University Press 2002) 3–9.

Bradshaw, A., 'The reconstruction of ecosystems' (1983) 20 *Journal of Applied Ecology* 1–17.

Brown, C. and R. Day, 'The future of stock enhancements: lessons for hatchery practice from conservation biology' (2002) 3 *Fish and Fisheries* 82–83.

Brunnée, J., 'COPing with consent: law-making under multilateral environmental agreements' (2002) 15 *Leiden Journal of International Law* 1–52.

Bullock, J. et al., 'Restoration of ecosystem services and biodiversity: conflicts and opportunities' (2011) 26(10) *Trends in Ecology and Evolution* 541–549.

Burkle, L. et al., 'Plant-pollinator interactions over 120 years: loss of species, co-occurrence and function' (2013) 339(6127) *Science* 1611–1625.

Butler, R., 'A new leaf in the rainforest: longtime villain vows reform' (10 March 2014) *Yale Environment* 360.

Caddy, J. and J. Sejio, 'This is more difficult than we thought! The responsibility of scientists, managers and stakeholders to mitigate the unsustainability of marine fisheries' (2005) 360(1453) *Philosophical Transactions B Royal Society London Biological Sciences* 59–75.

Cairns, J. Jr., 'Rationale for restoration' in M. Perrow and A. Davy (eds), *Handbook of Ecological Restoration* (Vol. 1, Cambridge University Press 2002) 10–23.

Caldwell, L., *International Environmental Policy: Emergence and Dimensions* (Duke University Press 1984).

Calmon, M. et al., 'Emerging threats and opportunities for large-scale ecological restoration in the Atlantic Forest of Brazil' (2011) 19(2) *Restoration Ecology* 154–158.

Cao, S. et al., 'Impact of China's Grain for Green Project on the landscape of vulnerable arid and semi-arid agricultural regions: a case study in northern Shaanxi Province' (2009) 46 *Journal of Applied Ecology* 536–543.

Ceballos, G., P.R. Ehrlich, A.D. Barnosky, A. García, R.M. Pringle and T.M. Palmer, 'Accelerated modern human-induced species losses: entering the sixth mass extinction' (2015) 1(5) *Science Advances* 1–5, DOI: 10.1126/sciadv.1400253.

Chan, K., T. Satterfield and J. Goldstein, 'Rethinking ecosystem services to better address and navigate cultural values' (2012) 75 *Ecological Economics* 8–18 .

Chapin, F. et al., 'Consequences of changing biodiversity' (2000) 405 *Nature* 234–242.

Chaves, R., G. Durigan, P. Brancalion and J. Aronson, 'On the need of legal frameworks for assessing restoration projects success: new perspectives from São Paulo state (Brazil)' (2015) 23(6) *Restoration Ecology* 754–759.

Chopra, K. et al. (eds), *Ecosystems and Human Well-Being: Policy Responses: Findings of the Responses Working Group of the Millennium Ecosystem Assessment* (Island Press 2005).

Churchill, R. and G. Ulfstein, 'Autonomous institutional arrangements in multilateral environmental agreements: a little-noticed phenomenon in international law' (2000) *American Journal of International Law* 636–642.

Clewell, A., J. Rieger and J. Munro, *Guidelines for Developing and Managing Ecological Restoration Projects* (2nd edn, Society for Ecological Restoration International 2005), <www.ser.org/about>.

Clewell, A. and J. Aronson, *Ecological Restoration: Principles, Values and Structure of an Emerging Profession* (Island Press 2008).

Cliquet, A., C. Backes, J. Harris and P. Howsam, 'Adaptation to climate change: legal challenges for protected areas' (2009) 1 *Utrecht Law Review* 158–175.

Cliquet, A., 'International and European law on protected areas and climate change: need for adaptation or implementation?' (2014) 54 *Environmental Management* 721–722.

Cliquet, A., K. Decleer and H. Schoukens, 'Restoring nature in the EU: the only way is up?' in C.-H. Born, A. Cliquet, H. Schoukens, D. Misonne and G. Van Hoorick (eds), *The Habitats Directive in its EU Environmental Law Context: European Nature's Best Hope?* (Routledge 2015) 267–272.

Cliquet, A. and H. Schoukens, 'Terrestrial protected areas' in J. Razzaque and E. Morgera (eds), *Biodiversity and Nature Protection Law: Encyclopedia of Environmental Law* (Edward Elgar Publishing, in press).

Cole, L. and S. Foster, *From the Ground Up: Environmental Racism and the Rise of the Environmental Justice Movement* (NYU Press 2001).

Colette, A., *Climate Change and World Heritage Report on predicting and managing the impacts of climate change on World Heritage and Strategy to assist States Parties to implement appropriate management responses* (World Heritage Report No. 22, UNESCO World Heritage Centre 2007).

Corlett, R., 'New approaches to novel ecosystems' (2014) 29 *Trends in Ecological Evolution* 137–138.

Costanza, R. et al., 'Changes in the global value of ecosystem services' (2014) 26 *Global Environmental Change* 152–158.

Costanza, R. and C. Folke, 'Valuing ecosystem services with efficiency, fairness and sustainability as goals' in G. Daily (ed) *Nature's Services: Societal Dependence on Natural Ecosystems* (Island Press 1997) 49–68.

Craig, R., '"Stationary is dead" – Long live transformation: five principles for climate change adaptation law' (2010) 34 *Harvard Environmental Law Review* 9–73.

Cross, M.S., J.A. Hilty, G.M. Tabor, J.J. Lawler, L.J. Graumlich and J. Berger, 'From connect-the-dots to dynamic networks: maintaining and enhancing connectivity as a strategy to address climate change impacts on wildlife' in J.F. Brodie, E.S. Post and D.F. Doak (eds), *Wildlife Conservation in a Changing Climate* (University of Chicago Press 2013) 307–329.

Crutzen, P., 'Geology of mankind' (2002) 415(3) *Nature* 6867.

Curran, M., S. Hellweg and J. Beck, 'Is there any empirical support for biodiversity offset policy?' (2014) 24(4) *Ecological Applications* 617–632.

Dahl, T.E., *Wetlands Losses in the United States 1780's to 1980's* (US Department of the Interior, Fish and Wildlife Service, Washington, DC 1990).

Dahl, T.E., *Status and Trends of Wetlands in the Conterminous United States 2004 to 2009*, Report to Congress, Department of Interior, US Fish and Wildlife Service (2011).

Daily, G., 'Introduction: what are ecosystem services' in G. Daily (ed) *Nature's Services: Societal Dependence on Natural Ecosystems* (Island Press 1997) 1–10.

Daily, G., *Nature's Services: Societal Dependence on Natural Ecosystems* (4th edn, Island Press 1997).

Dean, A.R., *New Orleans and the Wetlands of Southern Louisiana* (The Bridge, National Academy of Engineering 2006).

Decleer, K., 'Vallei van de Zuidleie: Leiemeersen (Oostkamp)' in K. Decleer (ed) *Ecological Restoration in Flanders (Belgium)* (published on the occasion of the 6th European Conference on Ecological Restoration, Ghent, Belgium, 8–12 September, INBO 2008) 48–49.

Decleer, K., 'The new European biodiversity strategy: a challenge to the restoration community' (2012) 30(2) *Ecological Restoration* 93–94.

Deguignet, M., D. Juffe-Bignoli, J. Harrison, B. Macsharry, N. Burgess and N. Kingston, *2014 United Nations List of Protected Areas* (UNEP-WCMC 2014).

Deland, C. and Z. Yuan, *China's Grain for Green Program: A Review of the Largest Ecological Restoration and Rural Development Program in the World* (Springer 2014).

de Sadeleer, N., *Environmental Principles: From Political Slogans to Legal Rules* (Oxford University Press 2002).

De Smedt, P. and M. van Rijswick, 'Nature conservation and water management' in C.-H. Born, A. Cliquet, H. Schoukens, D. Misonne and G. Van Hoorick (eds), *The Habitats Directive in its EU Environmental Law Context: European Nature's Best Hope?* (Routledge 2015) 417–433.

Di Minin, E. and T. Toivonen, 'Global protected area expansion: creating more than paper parks' (2015) 65(7) *BioScience* 637–638.

Dobson, A., *Fairness and Futurity: Essays on Environmental Sustainability and Social Justice* (Oxford University Press 1999).

Dodd, A., A. Hardiman, K. Jennings and G. Williams, 'Protected areas and climate change: reflections from a practitioner's perspective' (2010) 6 *Utrecht Law Review* 141–150.

Dodd, A., 'EU nature directives: rights, responsibilities and results – are we striking the right balance?' (2008) 20 *Environmental Law & Management* 237–245.

Dohren, P. and D. Haase, 'Ecosystem disservices research: a review of the state of the art with a focus on cities' (2015) 52 *Ecological Indicators* 490–497.

Donald, P.F., F.J. Sanderson, I.J. Burfield, S.M. Bierman, R.D. Gregory and Z. Waliczky, 'International conservation policy delivers benefits for birds in Europe' (2007) 317 (5839) *Science* 810–813.

Douglas, A., *The Symbiotic Habit* (Princeton University Press 2010).

Doussan, I. and H. Schoukens, 'Biodiversity and agriculture: greening the CAP beyond the status quo?' in C.-H. Born, A. Cliquet, H.Schoukens, D.Misonne and G. Van Hoorick (eds), *The Habitats Directive in its EU Environmental Law Context: European Nature's Best Hope?* (Routledge 2015) 437–451.

Dube, O., 'Challenges of wildland fire management in Botswana; towards a community inclusive fire management approach' (2013) 1 *Weather Climate Extremes* 26–41.

Dudley, N. (ed) *IUCN, Guidelines for Applying Protected Area Management Categories* (IUCN 2008).

Dudley, N. et al. (eds), *Natural Solutions: Protected Areas Helping People Cope with Climate Change* (IUCN WCPA, TNC, UNDP, WCS, The World Bank and WWF 2010).

Dudley, N., *Authenticity in Nature* (Earthscan 2011).

Dworkin, R., *Taking Rights Seriously* (Harvard University Press 1977).

Egan, D., 'Historic initiatives in ecological restoration' (1990) 8(2) *Restoration and Management Notes* 83–90.

Egerer, H., K. Kuras and G. Luciani, 'The protection of biodiversity and ecological connectivity in the Carpathian Convention' in M. Alberton (ed) *Towards the Protection of Biodiversity and Ecological Connectivity in Multi-Layered Systems* (Nomos 2013) 81–99.

Ehrenfield, J., 'Defining the limits of restoration: the need for realistic goals' (2000) 8(1) *Restoration Ecology* 2–9.

Ellis, E. et al., 'All is not loss: plant biodiversity in the Anthropocene' (2012) 7 *PLos One*, e30535.

Ervin, J., K.J. Mulongoy, K. Lawrence, E. Game, D. Sheppard, P. Bridgewater, G. Bennett, S.B. Gidda and P. Bos, *Making Protected Areas Relevant: A Guide to Integrating Protected Areas into Wider Landscapes, Seascapes and Sectoral Plans and Strategies* (CBD Technical Series No. 44, 2010).

European Commission, *Communication from the Commission to the Council and the European Parliament on a European Community biodiversity strategy* (COM(1998) 42 final, European Commission 1998).

European Commission, *Managing Natura 2000 sites: The provisions of Article 6 of the 'Habitats' Directive 92/43/EEC* (Office for Official Publications of the European Communities 2000).

European Commission, *Guidance document on Article 6(4) of the 'Habitats Directive' 92/43/EEC* (European Commission 2007/2012).

European Commission, *Guidance document on the strict protection of animal species of Community interest under the Habitats Directive 92/43/EEC* (European Commission 2007).

European Commission, *Report from the Commission to the Council and the European Parliament Composite – Report on the Conservation Status of Habitat Types and Species as required under Article 17 of the Habitats Directive* (COM(2009) 358 final, European Commission 2009).

European Commission, *Report from the Commission to the Council, the European Parliament, the European Economic and Social Committee and the Committee of the Regions* under Article 14(2) of Directive 2004/35/CE on the environmental liability with regard to the prevention and remedying of environmental damage (COM(2010) 581 final, European Commission 2010).

European Commission, *Communication from the Commission to the European Parliament, the Council, the Economic and Social Committee and the Committee of the Regions, Our life insurance, our natural capital: an EU biodiversity strategy to 2020* (COM(2011) 244 final, European Commission 2011).

European Commission, *Report from the Commission to the European Parliament and the Council on the Implementation of the Water Framework Directive (2000/60/EC). River Basin Management Plans* (COM(2012) 670 final, European Commission 2012).

European Commission, *Communication from the Commission to the European Parliament, the Council, the European Economic and Social Committee and the Committee of the Regions, Green Infrastructure (GI) – Enhancing Europe's Natural Capital* (COM(2013) 249 final, European Commission 2013).

European Commission, *Report from the Commission to the Council and the European Parliament, The first phase of implementation of the Marine Strategy Framework Directive (2008/56/EC). The European Commission's assessment and guidance* (COM(2014) 97 final, European Commission 2014).

European Environment Agency, *Assessing biodiversity in Europe – the 2010 report* (No. 5/2010, European Environment Agency 2010).

European Environment Agency, *Landscape fragmentation in Europe* (Joint EEA-FOEN report, No. 2/2011, European Environment Agency 2011).

European Environment Agency, *European waters – current status and future challenges. Synthesis* (EEA Report No. 9/2012, European Environment Agency 2012).

European Environment Agency, *Protected areas in Europe – an overview* (European Environment Agency 2012).

European Environment Agency, *The European environment – state and outlook 2015: synthesis report* (European Environment Agency 2015).

European Environment Agency, *State of nature in the EU. Results from reporting under the nature directives 2007–2012* (EEA Technical report No. 2/2015, European Environment Agency 2015).

European Environment Agency, *Exploring nature-based solutions. The role of green infrastructure in mitigating the impact of weather- and climate change-related natural hazards* (EEA Technical report No. 12, European Environment Agency 2015).

European Environment Agency, *Abundance and distribution of selected species (SEBI 001)* (European Environment Agency 2015), <www.eea.europa.eu/data- and-maps/indicators/abundance-and-distribution-of-selected-species/abundance-and-distribution-of-selected-2>.

European Environment Agency, *State of Europe's seas* (EEA Report No. 2/2015, European Environment Agency 2015).

Feng, X. et al., 'How ecological restoration alters ecosystem services: an analysis of carbon sequestration in China's Loess Plateau' (2013) 3(2846) *Scientific Reports* 1, <www.nature.com/srep/2013/131003/srep02846/full/srep02846.html>.

Finley, C., *All the Fish in the Sea; Maximum Sustainable Yield and the Failure of Fisheries Management* (University of Chicago Press 2011).

Fischman, R. and J. Hyman, 'The legal challenges of protecting animal migrations as phenomena of abundance' (2010) 28 *Vanderbilt Environmental Law Journal* 173–239.

Fisher, B., K. Turner and P. Morling, 'Defining and classifying ecosystem services for decision making' (2009) 68 *Ecological Economics* 643–653.

Fogleman, V., 'The threshold for liability for ecological damage in the EU' in C.-H. Born, A. Cliquet, H. Schoukens, D. Misonne and G. Van Hoorick (eds), *The Habitats Directive in its EU Environmental Law Context: European Nature's Best Hope?* (Routledge 2015) 181–214.

Foley, J.A., 'Global consequences of land use' (2005) 309 *Science* 570–574.

Gabriel, W. and P. Mace, *A Review of Biological Reference Points in the Context of the Precautionary Approach*, Proceedings, 5th NMFS NSAW, 1999. NOAA Tech. Memo. NMFS-F/SPO-40, <www.st.nmfs.noaa.gov/Assets/stock/documents/workshops/nsaw_5/gabriel_.pdf>.

Galaz, V., *Global Environmental Governance, Technology, and Politics: The Anthropocene Gap* (Edward Elgar 2014).

Gantioler, S., M. Rayment, S. Bassi, M. Kettunen, A. McConville, R. Landgrebe, H. Gerdes and P. ten Brink, *Costs and Socio-Economic Benefits associated with the Natura 2000 Network* (Final report to the European Commission, DG Environment on Contract ENV.B.2/SER/2008/0038, Institute for European Environmental Policy/GHK/Ecologic 2010).

Garcia, L.C. et al., 'Restoration challenges and opportunities for increasing landscape connectivity under the new Brazilian Forest Act' (2013) 11(2) *Brazilian Journal of Nature Conservation* 181–185.

Gardner, T. et al., 'A framework for integrating biodiversity concerns into national REDD+ programmes' (2012) 154 *Biological Conservation* 61–71.

Gillespie, A., *Protected Areas and International Environmental Law* (Martinus Nijhoff Publishers 2007).

Griffin, P., *The Ramsar Convention: A New Window for Environmental Diplomacy* (Institute for Environmental Diplomacy and Security, University of Vermont 2012) 25–92, <www.uvm.edu/ieds/sites/default/files/Ramsar_IEDSResearchSeries.pdf>.

Griffiths, C. et al., 'The use of extant non-indigenous tortoises as a restoration tool to replace extinct ecosystem engineers' (2010) 18 *Restoration Ecology* 1–7.

Griffiths, C. et al., 'Resurrecting extinct interactions with extant substitutes' (2011) 21 *Current Biology* 762–765.

Gross, M., 'Beyond expertise: ecological science and the making of socially robust restoration strategies' (2006) 14 *Journal for Nature Conservation* 172–179.

Hall, M., 'Restoring the countryside: George Perkins Marsh and the Italian land ethic' (1861–1882) (1998) 4(1) *Environment and History* 91–103.

Hall, M., *Earth Repair: A Transatlantic History of Environmental Restoration* (University of Virginia Press 2005).

Hannah, L. et al., 'Protected area needs in a changing climate' (2007) 5 *Frontiers in the Ecology and the Environment* 131–138.

Harvey, C.A., B. Dickson and C. Kormos, 'Opportunities for achieving biodiversity conservation through REDD' (2010) 3 *Conservation Letters* 53–61.

Heller, N. and E. Zavaleta, 'Biodiversity management in the face of climate change: a review of 22 years of recommendations' (2009) 142 *Biological Conservation* 14–32.

Hering, D. et al., 'The European Water Framework Directive at the age of 10: a critical review of the achievements with recommendations for the future' (2010) 408 *Science of the Total Environment* 4007–4019.

Higgs, K., *Collision Course: Endless Growth on a Finite Planet* (MIT Press 2016).

Hiss, T., 'Can the world really set aside half of the planet for wildlife?' Smithsonian.com (2014), <www.smithsonianmag.com/science-nature/can-world-really-set-aside-half-planet-wildlife-180952379/?no-ist>.

Hobbs, R. and J. Harris, 'Restoration ecology: repairing the Earth's ecosystems in the new millennium' (2001) 9(20) *Restoration Ecology* 239–246.

Hobbs, R., E. Higgs and J. Harris, 'Novel ecosystems: implications for conservation and restoration, trends' (2009) 24(11) *Ecology and Evolution* 599–605.

Hobbs, R. et al., 'Intervention ecology: applying ecological science in the twenty-first century' (2011) 61 *BioScience* 442–450.

Hobbs, R., E. Higgs and C. Hall (eds), *Novel Ecosystems: Intervening in the New Ecological World Order* (Wiley-Blackwell 2013).

Hobbs, R. et al., 'Defining novel ecosystems' in R. Hobbs, E. Higgs and C. Hall (eds), *Novel Ecosystems: intervening in the New Ecological World Order* (Wiley-Blackwell 2013) 58–60.

Hochkirch, A. et al., 'Europe needs a new vision for a Natura 2020 Network' (2013) 00 *Conservation Letters* 1–6.

Hockings, M., S. Stolton, F. Leverington, N. Dudley and J. Courrau, *Evaluating Effectiveness: A Framework for Assessing Management Effectiveness of Protected Areas* (2nd edn, IUCN 2006).

Holland, B., *Allocating the Earth* (Oxford University Press 2014).

Hunter, M., 'Benchmarks for managing ecosystems: are human activities natural?' (1996) 10 *Conservation Biology* 695–697.

Inger, R., R. Gregory, J.P. Duffy, I. Stott, P. Voříšek and K.J. Gaston, 'Common European birds are declining rapidly while less abundant species' numbers are rising' (2015) 18(1) *Ecology Letters* 28–36.

Jack, Z., *Liberty Hyde Bailey: Essential Agrarian and Environmental Writings* (Cornell University Press 2008).

Jans, J. and H. Vedder, *European Environmental Law* (Europa Law Publishing 2012).

Jordan, W., *The Sunflower Forest. Ecological Restoration and the New Communion with Nature* (University of California Press 2003).

Jørgensen, D., 'Ecological restoration in the Convention on Biological Diversity targets' (2013) 22 *Biodiversity and Conservation* 2977–2982.

Juffe-Bignoli, D., N.D. Burgess, H. Bingham, E.M.S. Belle, M.G. de Lima, M. Deguignet, B. Bertzky, A.N. Milam, J. Martinez-Lopez, E. Lewis, A. Eassom, S. Wicander, J. Geldmann, A. van Soesbergen, A.P. Arnell, B. O'Connor, S. Park, Y.N. Shi, F.S. Danks, B. MacSharry and N. Kingston, *Protected Planet Report 2014* (UNEP-WCMC 2014).

Kareiva, P., 'Domesticated nature: shaping landscapes and ecosystems for human welfare' (2007) 316 *Science* 1866–1869.

Kareiva, P., M. Marvier and R. Lalasz, 'Conservation in the Anthropocene: beyond solitude and fragility' (2012) *The Breakthrough*, <http://thebreakthrough.org/index.php/journal/past-issues/issue-2/conservation-in-the-anthropocene>.

Keenleyside, K.A., N. Dudley, S. Cairns, C.M. Hall and S. Stolton, *Ecological Restoration for Protected Areas: Principles, Guidelines and Best Practices* (IUCN 2012).

Kettunen, M., A. Terry, G. Tucker and A. Jones, *Guidance on the maintenance of landscape features of major importance for wild flora and fauna – Guidance on the implementation of Article 3 of the Birds Directive (79/409/EEC) and Article 10 of the Habitats Directive (92/43/EEC)* (Institute for European Environmental Policy (IEEP) 2007).

Kingsford, R. et al., 'A Ramsar wetland in crisis – the Coorong, Lower Lakes and Murray Mouth' (2011) 62 *Australia Marine Freshwater Research* 255–265.

Klare, M., *Resource Wars: The New Landscape of Global Conflict* (Owl Books 2001).

Kotze, L., 'Rethinking global environmental law and governance in the Anthropocene' (2014) 32(2) *Journal of Energy and Natural Resources Law* 121–156.

Krämer, L., 'Regional economic integration organizations: the European Union as an example' in D. Bodansky, J. Brunnée and E. Hey (eds), *The Oxford Handbook of International Environmental Law* (Oxford University Press 2007) 853–876.

Krämer, L., *Environmental Judgments by the Court of Justice and Their Duration* (Research Papers in Law No. 4/2008, College of Europe 2008).

Krämer, L., *EU Environmental Law* (Sweet & Maxwell 2012).

Krämer, L., 'Implementation and enforcement of the Habitats Directive' in C.-H. Born, A. Cliquet, H. Schoukens, D. Misonne and G. Van Hoorick (eds), *The Habitats Directive in its EU Environmental Law Context: European Nature's Best Hope?* (Routledge 2015) 229–244.

Lammerant, J., R. Peters, M. Snethlage, B. Delbaere, I. Dickie and G. Whiteley, *Implementation of 2020 EU Biodiversity Strategy: Priorities for the restoration of ecosystems and their services in the EU. Report to the European Commission* (ARCADIS, in cooperation with ECNC and Eftec 2013).

Lausche, B., D. Farrier, J. Verschuuren, A.G.M. La Viña, A. Trouwborst, C.-H. Born and L. Aug, 'The legal aspects of connectivity conservation: a concept paper' (IUCN 2013).

Lawler, J., 'Climate change adaptation strategies for resource management and conservation planning' (2009) 1162 *The Year in Ecology and Conservation Biology* 79–98.

Leadley, P.W., C.B. Krug, R. Alkemade, H.M. Pereira, U.R. Sumaila, M. Walpole, A. Marques, T. Newbold, L.S.L. Teh, J. van Kolck, C. Bellard, S.R. Januchowski-Hartley and P.J. Mumby, *Progress towards the Aichi Biodiversity Targets: An Assessment of Biodiversity Trends, Policy Scenarios and Key Actions* (Secretariat of the Convention on Biological Diversity, Technical Series 78, 2014).

Leopold, A., *A Sand County Almanac and Sketches Here and There* (Oxford University Press 1949).

Leopold, A., 'Conservation' in L. Leopold (ed) *Round River* (Oxford University Press 1966) 146–147.

Leopold, A.C., 'Living with the land ethic' (2004) 54(2) *BioScience* 149–154.

Lester, S.E., B.S. Halpern, K. Grorud-Colvert, J. Lubchenko, B.I. Ruttenberg, S.D. Gaines, S. Airamé and R.R. Warner, 'Biological effects within no-take marine reserves: a global synthesis' (2009) 384 *Marine Ecology Progress Series* 33–46.

Leverington, F., K. Lemos Costa, J. Courrau, H. Pavese, C. Nolte, M. Marr, L. Coad, N. Burgess, B. Bomhard and M. Hockings, *Management Effectiveness Evaluation in Protected Areas – A Global Study* (2nd edn, University of Queensland 2010).

Lewis, J., 'The Pinchot family and the battle to establish American forestry' (1999) 66(2) *Pennsylvania History* 143–165.

Long, J., A. Tecle and B. Burnette, 'Cultural foundations for ecological restoration on the White Mountain Apache Reservation' (2003) 8(1) *Conservation Ecology* 4.

Lopoukhine, N., 'Protected areas – for life's sake' in Secretariat of the Convention on Biological Diversity, *Protected Areas in Today's World: Their Values and Benefits for the Welfare of the Planet* (CBD Technical Series No. 36, 2008).

Lü, Y. et al., 'A policy-driven large scale ecological restoration: quantifying ecosystem services changes in the Loess Plateau of China' (2012) *PlosOne* 1, <www.plosone.org/article/info%3Adoi%2F10.1371%2Fjournal.pone.0031782>.

Lugo, E., 'Ecosystem services, the millennium ecosystem assessment, and the conceptual difference between benefits provided by ecosystems and benefits provided by people' (2008) 23(1) *Journal of Land Use and Environmental Law* 243–261.

Lynch, J., 'History of Zealandia' (2007) <www.visitzealandia.com/what-is-zealandia/our-history/one-mans-vision/>.

Lyytimaki, J., 'Ecosystem disservices: embrace the catchword' (2015) 12 *Ecosystem Services* 136–142.

Maathai, W., *The Green Belt Movement: Sharing the Approach and the Experience* (Lantern Books 2004).

Macaskill, C., *The National Agricultural Directory 2011* (Department Agriculture, Forestry and Fisheries, Republic of South Africa 2011).

Madgwick, F. and T. Jones, 'Europe' in M. Perrow and A. Davy (eds), *Handbook of Ecological Restoration. Volume 2: Restoration in Practice* (Cambridge University Press 2002) 32–56.

Mann, H., 'The Rio Declaration' (1992) 86 *Proceedings of the American Society of International Law* 405–411.

Manning, J.G., 'The representation of justice in Ancient Egypt' (2012) 24(1) *Yale Journal of Law and the Humanities* 111–118.

Marsh, G., *Man and Nature* (Scribner 1864).

Marview, M., 'New conservation is true conservation' (2013) 28(1) *Conservation Biology* 1–3.

Mawdsley, J., R. O'Malley and D. Ojima, 'A review of climate-change adaptation strategies for wildlife management and biodiversity conservation' (2009) 23 *Conservation Biology* 1080–1089.

McConnell, J. et al., '20th-century industrial black carbon emissions altered Arctic climate forcing' (2007) 17 *Science* 1381–1384.

McCormick, J., *The Global Environmental Movement* (2nd edn, John Wiley & Sons 1995).

McDonald, T., *SER Australia's National Restoration Standards* (Vol. 29, Issue 5, December 2015), <http://ser.org/resources/sernews/volume-29-issue-5-december-2015-Restoration-Standards>.

McGillivray, D., 'Compensatory measures under Article 6(4) of the Habitats Directive: no net loss for Natura 2000?' in C.-H. Born, A. Cliquet, H. Schoukens, D. Misonne and G. Van Hoorick (eds), *The Habitats Directive in its EU Environmental Law Context: European Nature's Best Hope?* (Routledge 2015) 101–118.

McKinney, M. and J. Lockwood, 'Biotic homogenization: a few winners replacing many losers in the next mass extinction' (1999) 14 *Trends in Ecology and Evolution* 450–453.

McRae, B. et al., 'Where to restore ecological connectivity? Detecting barriers and quantifying restoration benefits' (2012) 7(12) *PLOS One* e52604.

Meadows, D., D. Meadows, J. Randers and W. Behrens III, *The Limits to Growth* (Universe Books 1972).

Meine, C., *Aldo Leopold: His Life and Work* (University of Wisconsin Press 2010).

Menz, M.H. et al., 'Hurdles and opportunities for landscape-scale restoration' (2013) 339 *Science* 526–527.

Mertens, K., A. Cliquet and B. Vanheusden, 'Ecosystem services: what's in it for a lawyer?' (2012) 21(1) *European Environmental Law Review* 31–40.

Millennium Ecosystem Assessment, *Ecosystems and Human Well-being: Biodiversity Synthesis* (World Resources Institute 2005).

Milton, S.J., 'Economic incentives for restoring natural capital in Southern African rangelands' (2003) 1 *Frontiers in Ecology and the Environment* 247–254.

Mora, C. and P.F. Sale, 'Ongoing global biodiversity loss and the need to move beyond protected areas: a review of the technical and practical shortcomings of protected areas on land and sea' (2011) 434 *Marine Ecology Progress Series* 251–266.

Moreno-Mateos, D., V. Maris, A. Béchet and M. Curran, 'The true loss caused by biodiversity offsets' (2015) 192 *Biological Conservation* 552–559.

Morgera, J., 'European environmental law' in S. Alam, J. Bhuiyan, T. Chowdhury and E. Techera (eds), *Routledge Handbook of International Environmental Law* (Routledge 2013) 427–442.

Mooney, H., P. Ehrlich and G. Daily, 'Ecosystem services: a fragmentary history' in G. Daily (ed) *Nature's Services: Societal Dependence on Natural Ecosystems* (Island Press 1997) 11–19.

Morris, R.K.A., I. Alonso, R.G. Jefferson and K.J. Kirby, 'The creation of compensatory habitat – can it secure sustainable development?' (2006) 14 *Journal for Nature Conservation* 106–116.

Mulongoy, K.J. and S.B. Gidda, *The Value of Nature: Ecological, Economic, Cultural and Social Benefits of Protected Areas* (Secretariat of the Convention on Biological Diversity 2008).

Murcia, C., J. Aronson, G. Kattan, D. Moreno-Mateos, K. Dixon and D. Simberloff, 'A critique of the "novel ecosystem" concept' (2014) 29(1) *Trends in Ecology & Evolution* 548–553.

Nahlik, A., M. Kentula, S. Fennessy and D. Landers, 'Where is the consensus? A proposed foundation for moving ecosystem service concepts into practice' (2012) 77 *Ecological Economics* 27–35.

Nash, R., *The Rights of Nature: A History of Environmental Ethics* (University of Wisconsin Press 1989).

National Research Council, *Restoration of Aquatic Ecosystems: Science, Technology, and Public Policy* (National Academy Press 1992).

Nelleman, C. and E. Corcoran (eds), *Dead Planet, Living Planet: Biodiversity and Ecosystem Restoration for Sustainable Development – A Rapid Response Assessment* (United Nations Environment Programme 2010).

Newbold, T. et al., 'Global effects of land use on local terrestrial biodiversity' (2015) 520 *Nature* 45–50.

Normander, B. et al., *State of Biodiversity in the Nordic Countries* (Norde Press 2008).

Onida, M., 'The protection of biodiversity and ecological connectivity in the Alpine Convention' in M. Alberton (ed) *Towards the Protection of Biodiversity and Ecological Connectivity in Multi-Layered Systems* (Nomos 2013) 57–79.

Palmer, M., D. Falkand and J. Zedler, 'Ecological theory and restoration ecology' in D. Falk, M. Palmer and J. Zedler (eds), *Foundations of Restoration Ecology* (Island Press 2006) 1–10.

Palmer, M. and J.B. Ruhl, 'Aligning restoration science and the law to sustain ecological infrastructure for the future' (2015) 13(9) *Frontiers in Ecology and the Environment* 512–519.

Passamore, J., *Man's Responsibility for Nature* (Duckworth 1974).

Pearce, D. and D. Moran, *The Economic Value of Biodiversity* (Earthscan 1994).

Pearce, F., 'True nature: revising ideas on what is pristine and wild' (13 May 2013) *Yale Environment* 360.

Perring, M. et al., 'Novel urban ecosystems and ecosystem services' in R. Hobbs et al. (eds), *Novel Ecosystems: Intervening in the New Ecological World Order* (John Wiley & Sons 2013) 310–325.

Perry, J. and C. Falzon, *Climate Change Adaptation for Natural World Heritage Sites: A Practical Guide* (World Heritage Paper Series No. 37, UNESCO 2014).

Phillips, G., 'Progress towards the implementation of the European Water Framework Directive (2000–2012)' (2014) 17(4) *Aquatic Ecosystem Health and Management* 424–436.

Pinto, S. et al., 'Governing and delivering a biome-wide restoration initiative: the case of Atlantic Forest Restoration Pact in Brazil' (2014) 5 *Forests* 2212–2229.

Pittock, J. and D. Connell, 'Australia demonstrates the planet's future: water and climate in the Murray-Darling Basin' (2010) 26(4) *International Journal of Water Resources Development* 561–578.

Poláková, J., G. Tucker, K. Hart, J. Dwyer and M. Rayment, *Addressing biodiversity and habitat preservation through Measures applied under the Common Agricultural Policy* (Report Prepared for DG Agriculture and Rural Development, Contract No. 30-CE-0388497/00–44, Institute for European Environmental Policy 2011).

Postel, S. and S. Carpenter, *Nature's Services: Societal Dependence of Natural Ecosystems* (Island Press 1997).

Power, A.G., 'Ecosystem services and agriculture: tradeoffs and synergies' (2010) 365 *Philosophical Transactions of the Royal Society B* 2959–2971, DOI: 10.1098/rstb.2010.0143.

Pressey, B. and E. Ritchie, 'We have more parks than ever, so why is wildlife still vanishing?' (2014) *The Conversation*, <http://theconversation.com/we-have-more-parks-than-ever-so-why-is-wildlife-still-vanishing-34047>.

Pruitt, B., S. Miller, C. Theiling and C. Fischenich, 'The use of reference ecosystems as a basis for assessing restoration benefits', <http://citeseerx.ist.psu.edu/viewdoc/download?doi=10.1.1.310.2493&rep=rep1&type=pdf>.

Putz, F.E. and K.H. Redford, 'Dangers of carbon-based conservation' (2009) 19 *Global Environ* 400–401.

Rackham, O., *The History of the Countryside* (Phoenix 1986).

Ramsar Convention Secretariat, *Addressing Change in Wetland Ecological Character: Addressing Change in the Ecological Character of Ramsar Sites and Other Wetlands. Ramsar Handbooks for the Wise Use of Wetlands* (4th edn, vol. 19, Ramsar Convention Secretariat 2010).

Rees, R., 'Do protected areas for wildlife really work?' (2012) *The Ecologist*, <www.theecologist.org/News/news_analysis/1304082/do_protected_areas_for_wildlife_really_work.html>.

Rey Benayas, J., A. Newton, A. Diaz and J. Bullock, 'Enhancement of biodiversity and ecosystem services by ecological restoration: a meta-analysis' (2009) 325(5944) *Science* 1121–1124.

Sala, O. et al., 'Biodiversity – Global biodiversity scenarios for the year 2100' (2000) 287 *Science* 1770–1774.

Salzman, J., B. Thompson and G. Daily, 'Protecting ecosystem services: science, economics, and law' (2001) 20 *Stanford Environmental Law Journal* 309–332.

Sanderson, F., R. Pople, C. Ieronymidou, I. Burfield, R. Gregory, S. Willis, C. Howard, P. Stephens, A. Beresford and P. Donald, 'Assessing the performance of EU nature legislation in protecting target bird species in an era of climate change' (2015) *Conservation Letters* DOI: 10.1111/conl.12196.

Sandom, C., C. Donlan, J.-C. Svenning and D. Hansen, 'Rewilding' in D. Macdonald and K. Willis (eds), *Key Topics in Conservation Biology 2* (Oxford University Press 2013) 430–451.

Sands, P., *Principles of International Environmental Law* (2nd edn, Cambridge University Press 2003).

Savaresi, A., 'The protection of biodiversity and ecological connectivity in the international arena' in M. Alberton (ed) *Towards the Protection of Biodiversity and Ecological Connectivity in Multi-Layered Systems* (Nomos 2013) 11–28.

Sayer, J. et al., 'Ten principles for a landscape approach to reconciling agriculture, conservation, and other competing land uses' (2013) 110(21) *PNAS* 8349–8356.

Schlosberg, D., *Defining Environmental Justice: Theories, Movements, and Nature* (Oxford University Press 2007).

Schlosberg, D., 'Theorising environmental justice: the expanding sphere of a discourse' (2013) 2291 *Environmental Politics* 37–55.

Schoukens, H., 'Omlegging Griekse rivier: de mythe van "groene" infrastructuurprojecten' (2013) 1 *Tijdschrift voor Omgevingsrecht en Omgevingsbeleid* 67–69.

Schoukens, H., 'Going beyond the status quo: towards a duty for species restoration under EU law?' in V. Sancin and M.K. Dine (eds), *International Environmental Law: Contemporary Concerns and Challenges* (IUS Software, d.o.o., GV Zalozba 2014).

Schoukens, H., 'The ruling of the Court of Justice in Sweetman: how to avoid a death by a thousand cuts?' (2014) *ELNI* 2–12.

Schoukens, H. and A. Cliquet, 'Mitigation and compensation under EU Nature Conservation Law in the Flemish region: beyond the deadlock for development projects?' (2014) 10(2) *Utrecht Law Review* 194–215.

Schoukens, H. and H. Woldendorp, 'Site selection and designation under the Habitats and Birds Directives: a Sisyphean task?', in C.-H. Born, A. Cliquet, H. Schoukens, D. Misonne and G. Van Hoorick (eds), *The Habitats Directive in its EU Environmental Law Context: European Nature's Best Hope?* (Routledge 2015) 31–55.

Schoukens, H. and A. Cliquet, 'Biodiversity offsetting and restoration under the EU Habitats Directive: balancing between no net loss and deathbed conservation?' *Ecology & Society* (2016) 21(4):10.

Schrack, E., M. Beck, R. Brumbaugh, K. Crisley and B. Hancock, *Restoration Works: Highlights from a Decade of Partnership between The Nature Conservancy and the National Oceanic and Atmospheric Administration's Restoration Center* (The Nature Conservancy 2012).

Secretariat of the Convention on Biological Diversity, *Global Biodiversity Outlook 4* (2014).

Shrader-Frechette, K., *Environmental Justice: Creating Equality, Reclaiming Democracy* (Oxford University Press 2002).

Shue, H., *Climate Justice: Vulnerability and Protection* (Oxford University Press 2014).

Sizer, N. et al., 'Bonn Challenge 2.0: forest and landscape restoration emerges as a key climate solution' (Insights Blog, World Resources Institute 2015).

Soares-Filho, B., R. Rajao, M. Macedo, A. Carneiro, W. Costa, M. Coe, H. Rodrigues and A. Alencar, 'Cracking Brazil's forest code' (2014) 344 *Science* 363–364.

Society for Ecological Restoration International Science & Policy Working Group, *The SER International Primer on Ecological Restoration* (Society for Ecological Restoration International 2004).

Society for Ecological Restoration Australasia, *National Standards for the Practice of Ecological Restoration in Australia, Executive Summary* (2016).

Soule, M. and R. Noss, 'Rewilding and biodiversity: complementary goals for continental conservation' (1998) *Fall Wild Earth* 18–28.

Spencer, D. and J.S. Collie, 'Patterns of population variability in marine fish stocks' (1997) 6 *Fisheries Oceanography* 188–204.

Steffen, W. et al., 'Planetary boundaries: guiding human development on a changing planet' (2015) 347(6223) *Science* 1259855.

Stevens, S. (ed) *Conservation through Cultural Survival: Indigenous Peoples and Protected Areas* (Island Press 1997).

Stickler, C. et al., 'The potential ecological costs and cobenefits of REDD: a critical review and case study from the Amazon region' (2009) 15 *Global Change Biology* 2803–2824.

Strack, M. (ed) *Peatlands and Climate Change* (International Peat Society 2008).

Sze, J. and J. London, 'Environmental justice at the crossroads' (2008) 2(4) *Sociological Compass* 1331–1354.

Sze, J. et al., 'Defining and contesting environmental justice: socio-natures and the politics of scale in the Delta' in R. Holifield, M. Porter and G. Walker (eds), *Spaces of Environmental Justice* (Wiley-Blackwell 2010) 219–256.

TEEB, *The Economic of Ecosystems and Biodiversity for National and International Policy Makers – Summary: Responding to the Value of Nature* (2009).

TEEB, *The Economics of Ecosystems and Biodiversity in National and International Policy Making*, Edited by P. ten Brink (Earthscan 2011).

ten Brink, P., *The Economic Benefits of the Natura 2000 Network* (Final Synthesis Report to the European Commission, DG Environment 2011).

Thomas, C. et al., 'Extinction risk from climate change' (2004) 427 *Nature* 145–148.

Tittensor, D. et al., 'A mid-term analysis of progress toward international biodiversity targets' (2014) 346(6206) *Science* 241–244.

Tokar, B., *Toward Climate Justice: Perspectives on the Climate Crisis and Social Change* (New Compass Press 2014).

Travis, J., 'Climate change and habitat destruction: a deadly anthropogenic cocktail' (2003) 270 *Proceedings of the Royal Society of London B* 467–473.

Trouwborst, A., 'International nature conservation law and the adaptation of biodiversity to climate change: a mismatch?' (2009) 21 *Journal of Environmental Law* 419–442.

Trouwborst, A., 'Conserving European biodiversity in a changing climate: the Bern Convention, the European Union Birds and Habitats Directives and the adaptation of nature to climate change' (2011) 20 *RECIEL* 62–77.

Trouwborst, A., 'Transboundary Wildlife conservation in a changing climate: adaptation of the Bonn Convention on migratory species and its daughter instruments to climate change' (2012) 4 *Diversity* 258–300.

Trouwborst, A., 'The Habitats Directive and climate change: is the law climate proof?' in C.-H. Born, A. Cliquet, H. Schoukens, D. Misonne and G. Van Hoorick (eds), *The Habitats Directive in its EU Environmental Law Context: European Nature's Best Hope?* (Routledge 2015) 303–324.

Tucker, G., E. Underwood, A. Farmer, R. Scalera, I. Dickie, A. McConville and W. van Vliet, *Estimation of the financing needs to implement Target 2 of the EU Biodiversity Strategy. Report to the European Commission* (Institute for European Environmental Policy 2013).

United Nations Environment Programme, *Healthy Environment, Health People. Thematic Report for the Ministerial policy review session of the Second session of the United Nations Environment Assembly of the United Nations Environment Programme, Nairobi, 23–27 May 2016* (2016).

van Andel, J. and J. Aronson, 'Getting Started' in J. van Andel and J. Aronson (eds), *Restoration Ecology: The New Frontier* (2nd edn, Wiley-Blackwell 2012).

van Boven, G., *An Evaluation of the Use & Utility of Ramsar Guidance: A Report to Ramsar Scientific & Technical Review Panel and Ramsar Secretariat* (2007).

van den Broek, G.M., 'Environmental liability and nature protected areas: will the EU Environmental Liability Directive actually lead to the restoration of damaged natural resources?' (2009) 5(1) *Utrecht Law Review* 118–119.

van Dover, C.L. et al., 'Ecological restoration in the deep sea: Desiderata' (2014) 44 *Marine Policy* 98–106.

van Eijk, P. and R. Kumar, 'Bio-rights in theory and in practice' (2009) 21, <www.wetla nds.org/Portals/0/publications/Report/WI_Bio-rights%20in%20theory%20and%20pra ctice.pdf>.

Van Hoorick, G., 'Compensatory measures in European Nature Conservation Law' (2014) 10(2) *Utrecht Law Review* 161–171.

Van Hoorick, G., 'Biodiversity outside protected areas: an outlaw waiting to be saved?' in C.-H. Born, A. Cliquet, H. Schoukens, D. Misonne and G. Van Hoorick (eds), *The Habitats Directive in its EU Environmental Law Context: European Nature's Best Hope?* (Routledge 2015).

van Leeuwen, B. and P. Opdam, 'Klimaatsverandering vergt aanpassing van het natuurbeleid' (2003) 104 *De Levende Natuur* 122–124.

Veldman, J.W. et al., 'Tyranny of trees in grassy biomes' (2015) 347 *Science* 484–485.

Verschuuren, J., 'Climate change: rethinking restoration in the European Union's Birds and Habitats Directives' (2010) 28(4) *Ecological Restoration* 431–439.

Verschuuren, J., 'Connectivity: is Natura 2000 only an ecological network on paper?' in C.-H. Born, A. Cliquet, H. Schoukens, D. Misonne and G. Van Hoorick (eds), *The Habitats Directive in its EU Environmental Law Context: European Nature's Best Hope?* (Routledge 2015) 285–302.

Walker, G., 'Globalizing environmental justice: the geography and politics of frame contextualization and evolution' (2009) 9(3) *Global Social Policy* 355–382.

Walker, L. and P. Bellingham, *Island Environments in a Changing World* (Cambridge University Press 2012).

Wegener, M., P. Zedler, B. Herrick and J. Zedler, 'Curtis Prairie: 75-year old restoration research site' (2008) *Arboretum Leaflets*, Leaflet 16.

Westra, L., *Environmental Justice and the Rights of Unborn and Future Generations: Law, Environmental Harm and the Right to Health* (Taylor & Francis 2008).

Wheeler, K., 'Bird protection & climate changes: a challenge for Natura 2000?' (2006) 13 *Tilburg Foreign Law Review* 283–299.

Wiersema, A., 'The new international law-makers? Conferences of the Parties to multilateral environmental agreements' (2009) 31 *Michigan Journal of International Law* 231–287.

Willis, K. and S. Bhagwat, 'Biodiversity and climate change' (2009) 326 *Science* 806–807.

Wilson, E., *The Diversity Life* (Harvard University Press 1992).

Wilson, E., *Half-Earth: Our Planet's Fight for Life* (W.W. Norton & Company 2016).

Wirth, D., 'The Rio Declaration on Environment and Development: two steps forward and one back, or vice versa' (1995) 29 *Georgia Law Review* 599–653.

Woldendorp, H., 'Dynamische natuur in een statische rechtsorde' (2009) 3 *Milieu & Recht* 134–143.

World Bank Environment Department, *Convenient Solutions to an Inconvenient Truth: Ecosystembased Approaches to Climate Change* (2009).

World Business Council for Sustainable Development, 'New global partnership will intervene in landscapes at risk' (21 April 2015), <www.wbcsd.org/Pages/eNews/eNews Details.aspx?ID=16480&NoSearchContextKey=true>.

Yin, R. et al., 'Assessing China's ecological restoration: what's been done and what remains to be done' (2010) 45 *Environmental Management* 442–453.

Yin, R. and G. Ying, *China's Ecological Restoration Programs: Initiation, Implementation and Challenges in an Integrated Assessment of China's Ecological Restoration Programs* (Springer 2009).

Young, T., D. Petersen and J. Clary, 'The ecology of restoration: historical links, emerging issues and unexplored realms' (2005) 8 *Ecology Letters* 662–673.

Young, T., 'Restoration ecology and conservation biology' (2000) 92 *Biological Conservation* 73–83.

Zlasiewic, A., M. Williams, W. Steffen and P. Crutzen, 'The new world of the Anthropocene' (2010) 44 *Environmental Science Technology* 2228–2231.

International and European Union law (selected materials)

International Law (international conventions and other instruments)

Convention between the United States, Great Britain, Japan and Russia providing for the Preservation and Protection of the Fur Seals, 7 July 1911, 37 Stat. 1542.

Charter of the United Nations, 24 October 1945, 1 *UNTS* XVI.

International Convention for the Regulation of Whaling, 2 December 1946, 161 *UNTS* 361.

Convention on Wetlands of International Importance especially as Waterfowl Habitat, Ramsar, 2 February 1971, (1972) 11 *ILM* 963.

Convention Concerning the Protection of the World Cultural and Natural Heritage, Paris, 16 November 1972, (1972) 11 *ILM* 1358.

Convention on the Conservation of Migratory Species of Wild Animals, Bonn, 23 June 1979, (1979) 19 *ILM* 15.

Convention on the Conservation of European Wildlife and Natural Habitat, Bern, 19 September 1979.

United Nations Convention on the Law of the Sea, Montego Bay, 10 December 1982, (1982) 21 *ILM* 1261.

Agreement on the Conservation of Seals in the Wadden Sea, Bonn, 16 October 1990.

Convention on the Protection and Use of Transboundary Watercourses and International Lakes, Helsinki, 17 March 1992, (1992) 31 *ILM* 1312.

Convention on Biological Diversity, 5 June 1992, (1992) 31 *ILM* 818.

United Nations Framework Convention on Climate Change, Rio de Janeiro, 9 May 1992, (1992) 31 *ILM* 849.

United Nations Convention on Combating Desertification, Paris, 17 June 1994, (1994) 33 *ILM* 1328.

African-Eurasian Migratory Waterbirds Agreement, The Hague, 16 June 1996, 2365 *UNTS* I-42632.

Agreement for the Implementation of the Provisions of the United Nations Conventions on the Law of the Sea of 10 December 1982 Relating to the Conservation and Management of Straddling Fish Stocks and Highly Migratory Fish Stocks, 4 August 1995, 2167 *UNTS* 88.

Protocol Between the Government of Canada and the Government of the United States of America Amending the 1916 Convention Between the United Kingdom and the United States of America for the Protection of Migratory Birds in Canada and the United States, 14 December 1995.

Agreement on the Conservation of Albatrosses and Petrels, 19 June 2001, 2258 *UNTS* 257.

Framework Convention on the Protection and the Sustainable Development of the Carpathians, Kiev, 22 May 2003.

Agreement on the Conservation of Gorillas and their Habitat, Paris, 26 October 2007, 2544 *UNTS* I-45400.

Protocol on Conservation and Sustainable Use of Biological and Landscape Diversity to the Framework Convention on the Protection and Sustainable Development of the Carpathians, Bucharest, 19 June 2008.

Soft law

Declaration of Principles for the Preservation and Enhancement of the Human Environment, Report of the UN Conference on the Human Environment, Stockholm 5–16 June 1972, UN Doc. A/CONF. 48/14/Rev.1(1973), 3, reprinted in (1972) 11 *ILM* 1416.

World Charter for Nature, United Nations General Assembly, A/RES/37/7 (28 October 1982).

UN Conference on Environment and Development, Rio de Janeiro, Rio Declaration on Environment and Development, 13 June 1992, UN Doc. A/CONF. 151/26 (Vol. 1) (1992), reprinted in (1992) 31 *ILM* 874.

United Nations Agenda 21 (1992).

Johannesburg Plan of Implementation of the World Summit on Sustainable Development, UN Doc. A/CONF.199/20 (4 September 2002).

Hyogo Framework for Action, UN Doc. A/CONF.206/6 (22 January 2005).

Aichi Biodiversity Targets, UNEP/CBD/COP/DEC/X/2 (29 October 2010).

The Future We Want, A/RES/66/288 (27 July 2012).

Sendai Framework for Disaster Risk Reduction 2015–2030, UN Doc. A/CONF.224/L.2 (7 April 2015).

Transforming Our World: the 2030 Agenda for Sustainable Development, UN Doc. A/Res/70/1 (21 October 2015).

European Union law

Treaty on European Union, consolidated version *OJ C* 326, 26 October 2012.

Treaty on the functioning of the European Union, consolidated version *OJ C* 326, 26 October 2012.

Directive 79/409/EEC of 2 April 1979 on the Conservation of Wild Birds, *OJ L* 103, 25 April 1979.

Directive 92/43/EEC of 21 May 1992 on the conservation of natural habitats and of wild fauna and flora, *OJ L* 206, 22 July 1992.

Directive 2000/60/EC of the European Parliament and of the Council of 23 October 2000 establishing a framework for Community action in the field of water policy, *OJ L* 327, 22 December 2000.

Directive 2004/35/EC of the European Parliament and of the Council of 21 April 2004 on environmental liability with regard to the prevention and remedying of environmental damage, *OJ L* 143, 30 April 2004.

Directive 2007/60/EC of the European Parliament and of the Council of 23 October 2007 on the assessment and management of flood risks, *OJ L* 288, 6 November 2007.

Directive 2008/56/EC of the European Parliament and of the Council of 17 June 2008 establishing a framework for community action in the field of marine environmental policy, *OJ L* 164, 25 June 2008.

Directive 2009/147/EC of the European Parliament and of the Council of 30 November 2009 on the conservation of wild birds, *OJ L* 20, 26 January 2010.

Regulation (EU) No. 1305/2013 of the European Parliament and of the Council of 17 December 2013 on support for rural development by the European Agricultural Fund for Rural Development (EAFRD) and repealing Council Regulation (EC) No. 1698/2005, *OJ L* 347, 20 December 2013.

European Court of Justice cases

Case C-117/00 *Commission v Ireland* [2002] ECR I-5335

Case C-6/04 *Commission v United Kingdom* [2005] ECR I-9017.

Case C-209/04 *Commission v Austria* [2006] ECR I-2755.

Case C-235/04 *Commission v Spain* [2007] ECR I-0000.

Case C-418/04 *Commission v Ireland* [2007] ECR I-10947.

Case C-518/04 *Commission v Greece* [2006] ECR I-42.

Case C-183/05 *Commission v Ireland* [2007] ECR I-137.

Case C-383/09 *Commission v France* [2011] ECR I-4869.

Case C-404/09 *Commission v Spain* [2011] ECR I-11853.

Case C-43/10 *Nomarchiaki Aftodioikisi Aitoloakarnanias and Others* (2012).

Case C-301/12 *Cascina Tre Pini s.s. v Ministero dell'Ambiente e della Tutela del Territorio e del Mare and Others (Cascina Tre Pini)* (2013).

Case C-258/11 *Peter Sweetman and Others v An Bord Pleanála (Sweetman)* (2013).

Case C-521/12 *Briels and Others v Minister van Infrastructuur en Milieu* (2014).

Index

Taylor & Francis eBooks

Helping you to choose the right eBooks for your Library

Add Routledge titles to your library's digital collection today. Taylor and Francis ebooks contains over 50,000 titles in the Humanities, Social Sciences, Behavioural Sciences, Built Environment and Law.

Choose from a range of subject packages or create your own!

Benefits for you

» Free MARC records
» COUNTER-compliant usage statistics
» Flexible purchase and pricing options
» All titles DRM-free.

REQUEST YOUR FREE INSTITUTIONAL TRIAL TODAY

Free Trials Available
We offer free trials to qualifying academic, corporate and government customers.

Benefits for your user

» Off-site, anytime access via Athens or referring URL
» Print or copy pages or chapters
» Full content search
» Bookmark, highlight and annotate text
» Access to thousands of pages of quality research at the click of a button.

eCollections – Choose from over 30 subject eCollections, including:

Archaeology	Language Learning
Architecture	Law
Asian Studies	Literature
Business & Management	Media & Communication
Classical Studies	Middle East Studies
Construction	Music
Creative & Media Arts	Philosophy
Criminology & Criminal Justice	Planning
Economics	Politics
Education	Psychology & Mental Health
Energy	Religion
Engineering	Security
English Language & Linguistics	Social Work
Environment & Sustainability	Sociology
Geography	Sport
Health Studies	Theatre & Performance
History	Tourism, Hospitality & Events

For more information, pricing enquiries or to order a free trial, please contact your local sales team: www.tandfebooks.com/page/sales

 **Routledge**
Taylor & Francis Group

The home of Routledge books

www.tandfebooks.com